PRINCIPLES

OF

ECOTOXICOLOGY

PRINCIPLES OF ECOTOXICOLOGY

SECOND EDITION

C. H. WALKER

S. P. HOPKIN

R. M. SIBLY

D. B. PEAKALL

Taylor & Francis
Taylor & Francis Group

Boca Raton   London   New York   Singapore

Published in 2001 by
Taylor & Francis Group
6000 Broken Sound Parkway NW, Suite 300
Boca Raton, FL 33487-2742

Published in Great Britain by
Taylor & Francis Group
2 Park Square
Milton Park, Abingdon
Oxon OX14 4RN

International Standard Book Number-0-7484-0940-8 (Softcover)

Library of Congress Cataloging-in-Publication Data

Principles of Ecotoxicology/C.H. Walker...[et al.].—2nd ed.
    p. cm.
    Includes bibliographical references and index (p.).
    ISBN 0-7484-0940-8
    1. Pollution—Environmental aspects.   I. Walker, C.H. (Colin Harold), 1936-

QH945.A1 P745   2001
577.27—dc21

Taylor & Francis Group
is the Academic Division of T&F Informa plc.

**Visit the Taylor & Francis Web site at
http://www.taylorandfrancis.com**

Contents

Preface

When writing the first edition of *Principles of Ecotoxicology*, we were conscious of the need for such a book as well as of the difficulty of writing one for a new discipline which represents a synthesis of several older ones. Since publication, there has been both encouraging support and constructive criticism pointing to certain topics which were lightly treated in the original. The purposes of this second edition are to fill gaps in the original which have become apparent with the benefit of constructive criticism and of hindsight and to update some aspects of this rapidly evolving discipline.

It is important to emphasize that our purpose is to explain principles. Other texts emphasize practice. Detailed descriptions of ecotoxicity testing procedures or of analytical procedures lie outside the scope of the present text and would deflect it from its primary purpose. Such descriptions of practice are used economically and are given as examples to illustrate principles.

In the new edition, there has been some expansion of Chapter 3 to explain the importance of the properties of chemicals in determining their environmental fate. The chapter entitled 'Toxicity testing' has been expanded to give more examples, to say more about the problems of testing the toxicity of mixtures and to address the currently very topical issue of alternative testing methods. Chapter 8, 'Physiological effects of pollutants', has been expanded to include more on neurotoxicological effects, behavioural effects and effects on plants. Chapter 9 has also been expanded to deal at greater length with additive effects of mixtures. Chapters 10 and 15 have been updated to describe recent developments in the field of biomarkers. Chapter 12 has been expanded to include recent evidence for the decline of certain species of birds on farmland, and Chapter 14 has been enlarged to deal with structural changes in communities in response to the action of pollutants. Summaries have now been added at the end of all chapters and the bibliography has been updated.

C. H. Walker
S. P. Hopkin
R. M. Sibly
D. B. Peakall

Preface to the first edition

The origins of this book lie in the MSc course 'Ecotoxicology of Natural Populations' which was first taught at Reading in 1991. In recent years ecotoxicology has emerged as a distinct subject of interdisciplinary character. The structure of the course reflects this, and it is taught by people of widely differing backgrounds ranging from chemistry and biochemistry through to population genetics and ecology. Putting the different disciplines together in an integrated way was something of a challenge.

Experience of teaching the course persuaded the authors of the need for a textbook which would deal with the basic principles of such a wide-ranging subject. The intention has been to approach ecotoxicology in a broad interdisciplinary way, cutting across traditional subject boundaries. However, the nature of the text is bound to reflect the experience and interests of the authors, which will now be briefly reviewed.

Steve Hopkin is a Zoologist who has worked on electron microscopy and X-ray analysis for his PhD, and later investigated the effects of metals on soil ecology at the University of Bristol. Since coming to Reading his teaching and research has focused on the role of essential and non-essential metals in the biology of soil invertebrates.

David Peakall originally graduated as a chemist, and commenced his research as a physical chemist. Over a period he moved into biochemistry and finally into environmental toxicology. The last move was in keeping with his long-standing interest and active involvement in ornithology. During the last 15 years of his scientific career, he was chief of the Wildlife Toxicology division of the Canadian Wildlife Service, where he had a major involvement in studies of the Great Lakes.

Richard Sibly applied a degree in mathematics first in animal behaviour and then more widely in population biology. He has particular interests in life history evolution and trade-offs, and in how these may be affected by environmental pollutants.

Colin Walker originally qualified as an agricultural chemist, and was responsible for chemical and biochemical studies on environmental pollutants at Monks Wood Experimental Station during the mid-1960s when the major concern was about the effects of organochlorine insecticides. He subsequently moved to the University of Reading where he has developed teaching and research into molecular basis of toxicity, with particular reference to ecotoxicology.

Acknowledgements

Many people have contributed to this book in all sorts of ways. While we cannot acknowledge them all, we would particularly like to mention our MSc students, who have contributed much in discussion and feedback, and Amanda Callaghan, Peter Dyte, Glen Fox, Andy Hart, Graham Holloway, Alan McCaffery, Mark Macnair, Ian Newton, Demetris Savva, Ken Simkiss, Nick Sotherton, and George Warner. Last, but not least, Gill Bogue and Val Walker who have given invaluable secretarial support.

The publishers have made every effort to contact authors/copyright holders of works reprinted in *Principles of Ecotoxicology*. This has not been possible in every case, and we would welcome correspondence from those individuals/companies that we have been unable to trace.

Introduction

The term 'ecotoxicology' was introduced by Truhaut in 1969 and was derived from the words 'ecology' and 'toxicology'. The introduction of this term reflected a growing concern about the effects of environmental chemicals upon species other than man. It identified an area of study concerned with the harmful effects of chemicals (toxicology) within the context of ecology. Up to this time, the subject of environmental toxicology had been principally concerned with the harmful effects of environmental chemicals upon man, e.g. the effects of smoke upon urban communities. However, environmental toxicology, in its widest sense, encompasses the effects of chemicals upon ecosystems as well as upon man. Thus, ecotoxicology is a discipline within the wider field of environmental toxicology. In the present text, it is defined as 'the study of harmful effects of chemicals upon ecosystems', to include effects upon individuals as well as consequent effects at the levels of population and above.

Despite the definition given above, much early work answering to the description of ecotoxicology had little 'ecology' or 'toxicology' about it. It was concerned with the detection and determination of chemicals in samples of animals and plants. Seldom could the analytical results be related to effects upon individual organisms, let alone effects upon populations or communities. Analytical techniques such as gas chromatography, thin-layer chromatography and atomic absorption facilitated the detection of very low concentrations of chemicals in biota; establishing the biological significance of these residues was a more difficult matter! One of the main themes of the present text is the problem of progressing from the measurement of concentrations of environmental chemicals to establishing their effects at the levels of the individual, the population and the community.

New disciplines frequently present problems of terminology, and ecotoxicology is no exception to this trend. Several important terms in ecotoxicology are used inconsistently in the literature. Their use in the present text will now be explained. Both 'pollutants' and 'environmental contaminants' are regarded as chemicals which exist at levels judged to be above those that would normally occur in any particular component of the environment. This immediately raises the question: what is to be considered normal? With most man-made organic chemicals, such as pesticides, the situation is simple – any detectable level is abnormal as the compounds did not exist in the environment until released by man. On the other hand, chemicals such as heavy metals, sulphur dioxide, nitrogen oxides, polycyclic aromatic hydrocarbons (PAHs) and methyl mercury are naturally occurring and were present in the environment before the appearance of man. In the nature of things, there is a variation in the concentration of these chemicals from place to place and from time to time. This makes it difficult to judge their normal ranges.

The distinction sometimes made between 'pollutants' and 'contaminants' raises further difficulties. The term 'pollutant' is taken to indicate that the chemical it describes is causing actual environmental harm, whereas the term 'contaminant' implies that the chemical is not harmful. The difficulties with this distinction are threefold. First, there is the general toxicological principle that toxicity is related to dose (Chapter 5). Thus, a compound may answer to the description of pollutant in one situation but not another – a problem mentioned earlier. Second, there is no general agreement about what constitutes environmental harm or 'damage'. Some scientists would regard deleterious biochemical changes in an individual organism as harmful – others would reserve the term for declines in populations. Third, the effects of measured levels of chemicals in living organisms – or in their environment – are seldom known, yet the term 'pollutant' is frequently applied to them. Judgement of this issue is made more difficult by the possibility that there may be potentiation of toxicity when organisms are exposed to mixtures of environmental chemicals. To minimize these problems of terminology, the term 'pollutant' will be applied to environmental chemicals which exceed normal background levels and have the potential to cause harm. It would be attractive to reserve the term for particular chemicals in situations where they have been shown to cause harm, but because of the measurement problems referred to above this usage would be too restrictive. 'Harm' will be taken to include biochemical or physiological changes which adversely affect individual organisms' birth, growth or mortality rates. Such changes would necessarily produce population declines were it not that other processes (e.g. density dependence) may compensate (Chapter 12).

Whether or not a contaminant is a pollutant therefore depends on its level in the environment, the organism being considered and on whether or not the organism is harmed. Thus, a compound may answer to the description of 'pollutant' for one organism but not for another. Because of the problems in demonstrating harmful effects in the field, the terms 'pollutant' and 'contaminant' will, to a large extent, be used synonymously because it can

seldom be said that contaminants have no potential to cause environmental harm in any situation. The term 'environmental chemical' will be used to describe any chemical that occurs in the environment without making any judgement as to whether it should be regarded as a pollutant or a contaminant.

Another word that has been used inconsistently in the literature is the term 'biomarker'. Here, biomarkers are defined as *biological responses to environmental chemicals* at the individual level or below demonstrating departure from normal status. Biomarker responses may be at the molecular, cellular or 'whole-organism' level. Some workers would regard population responses (changes in number or gene frequency) as biomarkers. However, as the latter tend to be much longer term than the former, it may be unwise to use the same term for both. In the present text, the term biomarker will be restricted to biological responses at the level of the whole organism or below. An important thing to emphasize about biomarkers is that they represent measurements of effects, which can be related to the presence of particular levels of environmental chemical; they provide a means of interpreting environmental levels of pollutants in biological terms.

Finally, the organic pollutants to be considered here are examples of 'xenobiotics' ('foreign compounds'). They play no part in the normal biochemistry of living organisms. The concept of 'xenobiotics' will be discussed further in Chapter 5.

An exciting feature of ecotoxicology is that it represents a 'molecules to ecosystems' approach which relates to the 'genes-to-physiologies' approach originally identified by Clarke (1975) and extensively developed in North America in the 1980s (see, for example, Feder *et al.*, 1987). Moreover, it analyses 'experimental' manipulations on the largest of scales (although the 'experiments' were not designed as such). Thus, heavy metal pollution, acid rain and application of pesticides have affected whole ecosystems, sometimes with dramatic consequences for the populations within them. In ecotoxicology, the ecosystem response is studied at all levels. Initially (see figure below), the molecular structures of pollutants, their properties and environmental fate are considered (Part 1 of book).

Ecophysiologists generally analyse the impact of pollutants on an organism's growth, birth and death rates; indeed, as explained above, pollutants can adversely affect these 'vital rates'. This makes it desirable to understand how adverse effects on vital rates have implications for populations (Chapters 12 and 13). Thus, the relationship between the vital rates and 'population growth rate' is described in detail in Chapter 12. Consequently it is, in principle, possible to evaluate pollutants quantitatively in terms of their population effects. This emphasis on vital rates as crucial intervening variables, linking physiological effects to population effects, is a particular feature of this book. The approach is continued in Chapter 13 to consider whether and how fast resistant genes increase in populations. The rate at which re-

sistant genes increase is measured by the 'population growth rate' of the 'population' of resistant genes. The 'population growth rate' of resistant genes is a measure of their Darwinian fitness. Although this is not the conventional population–genetic measure of fitness, it is particularly useful in ecotoxicology because it (alone) shows explicitly how the fitness of resistant genes depends on the effects those genes have on their carriers. To summarize, the approach taken in this book allows linkage to be made between the different levels of organization shown in the figure below, from molecules to physiologies to populations, right through to ecosystems. This is the underlying basis for the biomarker strategy which seeks to measure sequences of responses to pollutants from the molecular level to the level of ecosystems (Chapters 10 and 15). The use of biomarkers in biomonitoring is described in Chapter 11. These three chapters are placed at the end of their respective sections of the book. They represent the practical realization of theoretical aspects described in earlier chapters.

The text is divided into three parts, as follows.

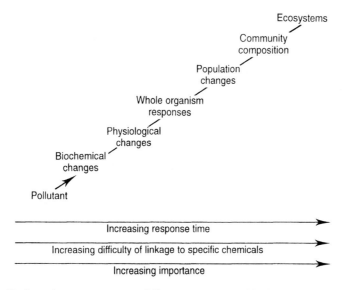

Schematic relationship of linkages between responses at different organizational levels.

Part 1 describes major classes of organic and inorganic pollutants, their entry into the environment and their movement, storage and transformation within the environment. Thus, it bears a certain resemblance to toxicokinetics in 'classic' toxicology, which is concerned with the uptake, distribution, metabolism and excretion of xenobiotics by living organisms (Chapter 5). The difference is one of complexity. Ecotoxicology deals with movements of pollutants in air, water, soils and sediments and through food chains, with chemical transformation and biotransformation.

Part 2 deals with the effects of pollutants upon living organisms, thus resembling toxicodynamics in classic toxicology. The difference is again one of complexity. Whereas toxicodynamics focuses upon interactions between xenobiotics and their sites of action, ecotoxicology is concerned with a wide range of effects upon individual organisms at differing organizational levels (molecular, cellular and whole animal). Toxicity data obtained in the laboratory are used for the purposes of risk assessment. Effects of pollutants are studied in the laboratory, an approach that can lead to the development of biomarker assays (Chapter 10). The use of biomarker assays in biomonitoring is discussed in Chapter 11, which also considers some effects at the population level, thereby looking ahead to the final part of the text.

Part 3 addresses questions which are of the greatest interest to ecologists. What effects do pollutants have at the level of population, community and whole ecosystem? This takes the discussion into the disciplines of population biology and population genetics. Whereas classic toxicology is concerned with chemical toxicity to individuals, ecotoxicologists are particularly interested in effects at the level of population community and whole ecosystem. Effects at the population level may be changes in numbers of individuals (Chapter 12), changes in gene frequency (as in resistance) (Chapter 13) or changes in ecosystem function (e.g. soil nitrification) (Chapter 14). They may be due to sublethal effects (e.g. on physiology or behaviour) rather than lethal toxicity. Sometimes they may be indirect (e.g. the decline in a predator because of direct chemical toxicity may lead to an increase in numbers of its prey). It is often very difficult to establish effects of pollutants on natural populations. However, the development of appropriate biomarker assays can help to resolve this problem.

Part 3 illustrates the truly interdisciplinary character of ecotoxicology. The study of the harmful effects of chemicals upon ecosystems draws on the knowledge and skills of ecologists, physiologists, biochemists, toxicologists, chemists, meteorologists, soil scientists and others. It is nevertheless a discipline with its own distinct character. Apart from the important applied aspects which address current public concerns, it has firm roots in basic science. Chemical warfare is nearly as old as life itself, and the evolution of detoxication mechanisms by animals to avoid the toxic effects of xenobiotics produced by plants is paralleled by the recent development of resistance by pests to pesticides made by humans.

PART 1

Pollutants and their fate in ecosystems

CHAPTER 1

Major classes of pollutant

Many different chemicals are regarded as pollutants, ranging from simple inorganic ions to complex organic molecules. In the present chapter, representatives will be identified of all the major classes of pollutant, and their properties and occurrence will be briefly reviewed. These pollutants will be used as examples throughout the text. Their fate in the living environments will be the subject of the remainder of Part 1. Their effects upon individuals and ecosystems will be considered in Parts 2 and 3 respectively.

1.1 *Inorganic ions*

1.1.1 METALS

A metal is defined by chemists as being an element which has a characteristic lustrous appearance, is a good conductor of electricity and generally enters chemical reactions as positive ions or cations.

Although metals are usually considered as pollutants, it is important to recognize that they are natural substances. With the exception of radioisotopes produced in man-made nuclear reactions (bombs and reactors), all metals have been present on the Earth since its formation. There are a few examples of localized metal pollution resulting from natural weathering of ore bodies (e.g. Hågvar and Abrahamsen, 1990). However, in most cases, metals become pollutants where human activity, mainly through mining and smelting, releases them from the rocks in which they were deposited during volcanic activity or subsequent erosion and relocates them into situations where they can cause environmental damage.

The extent to which human activity contributes to global cycles of metals can be described by the anthropogenic enrichment factor (AEF) (table 1.1). From this table it is clear that human activity is responsible for the majority of the global movement of cadmium, lead, zinc and mercury but is relatively unimportant in the cycling of manganese. The AEF for lead is due mostly to the widespread use and subsequent release of lead-based additives to petrol. For most radioactive isotopes, the AEF is 100%.

Elements considered to be metals are iden-

TABLE 1.1 *Anthropogenic enrichment factors (AEF) for total global annual emissions of cadmium, lead, zinc, manganese and mercury in the 1980s (all values 10^6 kg $year^{-1}$)**

Metal	Anthropogenic sources (A) (industry etc.)	Natural sources (volcanoes etc.)	Total (T)	AEF (A/T) × 100
Cadmium (Cd)	8	1	9	89%
Lead (Pb)	300	10	310	97%
Zinc (Zn)	130	50	180	72%
Manganese (Mn)	40	300	340	12%
Mercury (Hg)	100	50	150	66%

*From various sources.

tified within the **periodic system** for the classification of all elements shown in figure 1.1. Groups of elements sharing similar chemical properties are contained within individual vertical columns. The first two columns contain elements that readily lose one or two outer electrons to yield monovalent cations (column 1) or divalent cations (column 2). Among these are many of the most widespread metals, found in surface waters and in soils in their stable ionic forms, e.g. Na^+, K^+, Mg^{2+} and Ca^{2+}. The following 10 columns contain what are termed transition elements, and these are also regarded as metals, although their chemistry is more complex than that of the alkali and alkali earth elements which constitute the first two groups. Moving from left to right through the three main series of transition elements, the nuclei become larger and the outer electrons show less tendency to escape (i.e. to form cations) than is the case with elements listed in columns 1 and 2. Consequently, there is a tendency to share electrons with other elements, leading to the formation of covalent bonds and complex ions (e.g. by copper, iron, cobalt or nickel). Some of the larger atoms tend to retain electrons and

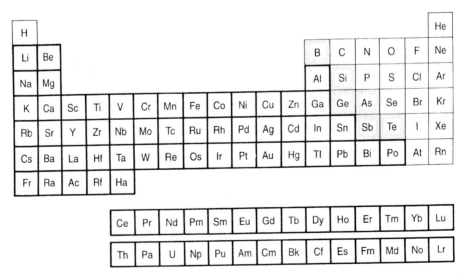

FIGURE 1.1 *Periodic table of the elements. Those considered to be metals are surrounded by bold lines. Metalloids (with properties of metals and non-metals) are shaded. Reproduced from Hopkin (1989) with permission from Elsevier Applied Science.*

remain in the elemental state (e.g. silver and gold, the so-called 'noble metals'). Other characteristics of iron, copper and certain other transition elements are variable valency and participation in electron transfer reactions. Electron transfer reactions involving oxygen can lead to the production of toxic oxyradicals, a toxicity mechanism now known to be of considerable importance in both animals and plants: it is now recognized that some oxyradicals, such as superoxide anion (O_2^{-}) and the **hydroxyl radical** ($OH^{.}$), can cause serious cellular damage. In the remaining vertical groups, as one moves from left to right, there is a reducing tendency to form cations. There is a progression from metals to **metalloids**, the latter showing characteristics of both metals and non-metals, until the non-metals are reached (C, N, O, P, S, Cl, Br, etc.). The final vertical column contains the very stable inert gases, which have hardly any chemical reactivity. The two horizontal boxes below the main periodic classification contain the generally rare elements of the lanthanide and actinide series, which are metallic in character.

The tendency to form covalent bonds shown by metalloids, and also by metals located close to them in the periodic classification, has two important toxicological consequences. First, these elements are able to bind covalently to organic groups, thereby forming lipophilic compounds and ions. Some of these compounds are highly toxic, e.g. tetra-alkyl lead, tributyl tin oxide, methyl mercury salts and methylated forms of arsenic. Because of their lipophilicity, their distribution within animals and plants and their toxic action usually differs from simple ionic forms of the same elements. **Organometallic compounds** are discussed later in section 1.3. Second, these elements can have toxic effects by binding to non-metallic constituents of cellular macromolecules, e.g. the binding of copper, mercury, lead and arsenic to sulphydryl groups of proteins.

The term **heavy metals** has been used extensively in the past to describe metals which are environmental pollutants. For a metal to be considered 'heavy', it must have a density relative to water of greater than five. However, the term 'heavy metals' has been replaced in recent years by a classification scheme that considers their chemistry rather than relative density (Nieboer and Richardson, 1980; table 1.2). This approach is more logical because there are some metals that are not 'heavy' which can be important environmental pollutants. Aluminium, for example, which is a metal, has a relative density of only 1.5. However, it is an extremely important pollutant in acidified lakes, where it becomes soluble and is toxic to fauna. The gills of fish are particularly susceptible to aluminium poisoning. Aluminium has also been implicated in Alzheimer's disease in humans and may be deposited in the brain.

Metals are *non-biodegradable*. Unlike some organic pesticides, metals cannot be broken down into less harmful components. Detoxification by organisms consists of 'hiding' active

TABLE 1.2 *Separation of some essential and non-essential metal ions of importance as pollutants into class A (oxygen seeking), class B (sulphur or nitrogen seeking) and borderline elements based on the classification scheme of Nieboer and Richardson (1980)**

Class A	Borderline	Class B
Calcium	Zinc	Cadmium
Magnesium	Lead	Copper
Manganese	Iron	Mercury
Potassium	Chromium	Silver
Strontium	Cobalt	
Sodium	Nickel	
	Arsenic	
	Vanadium	

*This distinction is important in determining rates of transport across cell membranes and sites of intracellular storage in metal-binding proteins and metal-containing granules (e.g. section 8.2).

metal ions within a protein such as metallo-thionein (binding covalently to sulphur), or depositing them in an insoluble form in intracellular granules for long-term storage or excretion in the faeces (see Chapter 8).

Essential elements all have a 'window of essentiality' within which dietary concentrations in animals, or soil concentrations in plants, have to be maintained if the organism is to grow and reproduce normally (figure 1.2). In addition to carbon, hydrogen, oxygen and nitrogen, all animals need the seven major mineral elements calcium, phosphorus, potassium, magnesium, sodium, chlorine and sulphur for ionic balance and as integral parts of amino acids, nucleic acids and structural compounds. Thirteen other so-called 'trace elements' are definitely required, namely iron, iodine, copper, manganese, zinc, cobalt, molybdenum, selenium, chromium, nickel, vanadium, silicon and arsenic. Zinc, for example, is an essential component of at least 150 enzymes, copper is essential for the normal function of cytochrome oxidase and iron is part of haemoglobin, the oxygen-carrying pigment in red blood cells. Boron is required exclusively by plants. A few other elements, such as lithium, aluminium, fluorine and tin, may be essential at ultratrace levels. The window of

essentiality for some elements is very narrow. Selenium, for example, was considered for a long time to be only a dangerous toxin until its role in the enzyme glutathione peroxidase was discovered. The dose determines the poison.

Non-essential metals such as mercury or cadmium, in addition to being toxic above certain levels, may also affect organisms by inducing deficiencies of essential elements through competition at active sites in biologically important molecules (table 1.3) (see Chapter 7). Such **antagonism** also occurs between essential elements. A concentration of only 5 μg Mo g^{-1} in the diet of cattle is sufficient to reduce copper intake by 75%, which often leads to symptoms of copper deficiency.

1.1.2 ANIONS

There are some inorganic pollutants which are not particularly toxic, but which cause environmental problems because they are used in such large quantities. These include anions such as nitrates and phosphates.

Nitrate fertilizers are used extensively in agriculture. During the growth period of crops, most of the fertilizer applied is absorbed by

 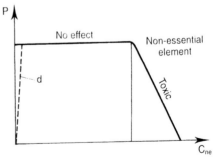

FIGURE 1.2 *Relationships between performance (P) (growth, fecundity, survival) and concentrations of an essential (C₂) or non-essential (Cₙₑ) element of the diet of animals. Possible deficiency effects at ultratrace levels (d) of an apparently non-essential element may be discovered as the sensitivities of analytical techniques are improved. Reproduced from Hopkin (1989) with permission from Elsevier Applied Science.*

TABLE 1.3 *Level of activity of carbonic anhydrase, expressed relative to that of zinc, of different metals substituted in the protein**

Metal	Normal activity (hydration of CO_2) (%)
Zinc	100
Cobalt	56
Nickel	5
Cadmium	4
Manganese	4
Copper	1
Mercury	0.05

*From Coleman (1967).

plant roots. However, when growth ceases, nitrate released during the decomposition of dead plant material passes down through the soil and may enrich adjacent water courses. The increase in available nitrogen may cause blooms in algal populations. This effect is called **eutrophication** and eventually leads to oxygen starvation as microorganisms break down the dead algal tissues.

The safe limit for nitrates in drinking water in the UK has been set at 50 parts per million (p.p.m.). A human health problem may arise if young babies ingest bottled milk made up with nitrate-contaminated water. During their first few months of life, human infants have an anaerobic stomach. The nitrates are converted to nitrites in this oxygen-poor environment. The nitrites bind to haemoglobin, reduce its capacity to carry oxygen and the infant may develop 'blue baby syndrome' or **methaemoglobinaemia**. The problem does not arise with breast-fed babies (definitely a case of 'breast is best'!). In regions of intensive agriculture, the 100 p.p.m. level is exceeded in water extracted from rivers or in bore holes where nitrates have leached down to aquifers. The problem can be solved by removing the nitrate chemically at the water treatment works or by diluting the contaminated water with water from a relatively nitrate-free source. The long-term solution is of course to reduce nitrate usage, and this is being done in so-called 'exclusion zones' around sources of water for human consumption.

Similar problems of eutrophication can also arise with phosphates used as fertilizers. However, there is an additional source: washing powders. These have been made less resistant to breakdown in recent years because of co-operation between soap manufacturers and water-treatment companies. In the 1950s and 1960s, it was common to see a huge build-up of foam below weirs and waterfalls downstream of the outfalls of sewage treatment works.

1.2 *Organic pollutants*

The great majority of compounds that contain carbon are described as 'organic', the few exceptions being simple molecules such as CO_2 and CO. Carbon has the ability to enter into the formation of a bewildering diversity of complex organic compounds, many of which provide the basic fabric of living organisms. The reason for this is the tendency of carbon atoms to form stable bonds with one another, thereby creating rings and extended chains. Carbon can also form stable bonds with hydrogen, oxygen and nitrogen atoms.

Molecules built of carbon alone (e.g. graphite and diamond) or of carbon and hydrogen (hydrocarbons) have very little polarity and consequently low water solubility. Polar molecules have electrical charge associated with them; non-polar molecules have little or none. Molecules with a strong charge are described as highly polar; molecules of low charge have low polarity. Polar compounds tend to be water soluble because the charges on them are attracted to opposite charges on water molecules. For example, a positive charge on an organic molecule will be attached to a negative charge on a water molecule. Carbon compounds tend to be more polar and more chemically reactive when they contain functional groups such as OH, HC=O and NO_2. In these examples, the oxygen atom attracts electrons away from neighbouring carbon atoms, thereby creating a charge imbalance on the molecule. Molecules of high polarity tend to enter into chemical and biochemical reactions more readily than do molecules of low polarity.

The behaviour of organic compounds is dependent upon their molecular structure – molecular size, molecular shape and the presence of functional groups being important determinants of metabolic fate and toxicity. Thus, it is important to know the formulae of pollutants in order to understand or predict what happens to them in the living environment. The principles operating here are illustrated by examples given in Chapters 5 and 7. Readers with a limited knowledge of chemistry are referred to the text of Manahan (1994), which contains two useful concise chapters on basic principles.

The pollutants that will be described here are predominantly man-made ('anthropogenic') compounds which have appeared in the natural environment only during the last century. This is only a very short time in evolutionary terms, and there has been only limited opportunity for the evolution of protective mechanisms against their toxic effects (e.g. detoxication by enzymes) beyond pre-existing mechanisms acting against 'natural' xenobiotics. In this respect, they differ from inorganic pollutants, and from those naturally occurring xenobiotics which have substantial toxicity (e.g. nicotine, pyrethrins and rotenone are compounds produced by plants which are highly toxic to certain species of insect). Aromatic hydrocarbons represent a special case. They have been generated by the combustion of organic matter since the appearance of higher plants on Earth (e.g. as a result of forest fires started by volcanic lava). Like heavy metals that

are mined, their environmental levels increase substantially as a consequence of human activity (as with the combustion of coal or petrol to produce aromatic hydrocarbons).

1.2.1 HYDROCARBONS

These are compounds composed of the elements carbon and hydrogen only. Some hydrocarbons of low molecular weight (e.g. methane, ethane and ethylene) exist as gases at normal temperature and pressure. However, the great majority of hydrocarbons are liquids or solids. They are of low polarity (i.e. electrical charge, see above) and, consequently, have low water solubility, but they have high solubility in oils and in most organic solvents. (They are not very soluble in polar organic solvents such as methanol or ethanol.)

Hydrocarbons are divisible into two classes: (i) alkanes, alkenes and alkynes and (ii) **aromatic** hydrocarbons (figure 1.3). The distinguishing feature of aromatic hydrocarbons is the presence of one or more benzene rings in their structure. Benzene rings are six-membered carbon structures which are 'unsaturated' in the sense that not all available carbon valences are taken by linkage to hydrogen. In fact, benzene rings have delocalized electrons which can move freely over the entire ring system and do not remain in the immediate vicinity of any one atom. Other hydrocarbons do not have this feature. They vary greatly in molecular size and may be fully saturated (e.g. hexane and octane) or unsaturated. Unsaturated hydrocarbons contain carbon–carbon (C–C) **double bonds** (e.g. ethylene) or carbon–carbon **triple bonds** (e.g. acetylene). Saturated hydrocarbons are referred to as alkanes, unsaturated hydrocarbons with a carbon–carbon double bond are alk**enes** and unsaturated hydrocarbons with a carbon–carbon triple bond are alk**ynes**. They may exist as single chains, branched chains or rings (figure

Propane   Cyclohexane

Benzene   3,4-Benzo(a)pyrene

FIGURE 1.3 *Hydrocarbons. Composed of only hydrogen and carbon, these compounds have low polarity and thus low water solubility but high solubility in oils and organic solvents. Propane and cyclohexane are examples of alkanes; benzene and benzo(a)pyrene of aromatic compounds. The latter contain six-membered carbon rings (benzene rings) with delocalized electrons. This is indicated by representing the benzene rings as a hexagon (the six-membered carbon frame) and a circle (the cloud of delocalized electrons) situated within it. Aromatic hydrocarbons undergo certain characteristic biotransformations influenced by the delocalized electrons (section 5.1.5).*

1.3). The properties of these two groups of hydrocarbons will now be considered separately.

The properties of non-aromatic hydrocarbons depend upon molecular weight and degree of unsaturation. Alkanes are essentially stable and unreactive and have the general formula C_nH_{2n+2}. The first four members of the series exist as gases ($<n = 4$). Where $n = 5$–17, they are liquids at normal temperature and pressure. Where $n = 18$ or more, they are solids. Alkenes and alkynes are more chemically reactive because they contain carbon–carbon double or triple bonds. As with alkanes, the lower members of the series are gases, the higher members liquids or solids.

Aromatic hydrocarbons exist as liquids or solids – none of them has a boiling point below 80°C at normal atmospheric pressure. They are

more reactive than alkanes, being susceptible to chemical and biochemical transformation. There are many polycyclic aromatic hydrocarbons (PAHs) which are planar (i.e. flat) molecules consisting of three or more six-membered (benzene) rings directly linked together.

The major sources of hydrocarbons are deposits of petroleum and natural gas in the upper strata of the earth's crust. These fossil fuels originate from the remains of plants and animals of earlier geological times (notably the carboniferous period). Although non-aromatic hydrocarbons predominate in these deposits, crude oils also contain significant amounts of PAHs. PAHs are also formed as a consequence of the incomplete combustion of organic materials. Thus, they are generated when coal, oil or petrol are burned, when trees or houses burn and when people smoke cigarettes. Major sources of hydrocarbon pollution are spillage of crude oils (e.g. tanker disasters) and the combustion of fossil fuels (notably the use of brown coal in parts of Eastern Europe).

1.2.2 POLYCHLORINATED BIPHENYLS (PCBs)

These are commercial mixtures of related compounds (congeners) which are useful for their physical properties. They are stable, unreactive viscous liquids of low volatility which have been used as hydraulic fluids, coolant–insulation fluids in transformers and plasticizers in paints. There are altogether 209 possible PCB congeners, and some 120 of these are present in commercial products such as Aroclor 1254, Aroclor 1260 and Clophen A60. The last two digits in the numbers refer to the percentage chlorine in the PCB mixture. In either case, the larger the number the greater the proportion of higher chlorinated PCBs in the mixture.

PCB mixtures have very low solubility in water but high solubility in oils and organic solvents of low polarity. The water solubilities of Aroclor 1254 and Aroclor 1260 are only 21 µg l[-1] and 2.7 µg l[-1] respectively.

The individual congeners of PCB vary in their stereochemistry, depending on the positions of substitution of chlorine atoms. Where there is no substitution in the ortho positions, the two benzene rings tend to remain in the same plane (co-planar PCBs); 3,3',4,4'-tetrachloro-biphenyl is an example of a co-planar PCB (figure 1.4). By contrast, substitution of two, three or four ortho positions with chlorine leads to the movement of the rings out of plane because of the interaction of adjacent

chlorines in different rings (chlorine atoms are bulky). The molecular conformation is not then a co-planar one, but is a more 'globular' structure.

PCBs were once used for a number of purposes – as dielectric fluids, heat transformer fluids, lubricants, vacuum pump fluids, as plasticizers (e.g. in paints) and for making carbonless copy paper. In many countries, the use of PCBs is now banned or severely restricted.

Major sources of pollution are (or have been) manufacturing wastes and the careless disposal or dumping of the liquids referred to above (Waid, 1985–7).

FIGURE 1.4 *Organohalogen compounds are organic compounds containing **halogen** atoms (the halogen elements are fluorine, chlorine, bromine and iodine). All the examples given here are of organochlorine compounds, although it should be noted that organofluorine compounds (e.g. chlorfluorocarbons) and organobromine compounds (e.g. polybrominated biphenyls) are also environmental pollutants. The compounds shown here are stable solids of low polarity and water solubility. They are not found in nature, and in many cases they are only slowly metabolized and consequently are persistent in living organisms (Chapter 5).*

1.2.3 POLYCHLORINATED DIBENZODIOXINS (PCDDs)

The best known member of this group of compounds is 2,3,7,8-tetrachlorodibenzodioxin (2,3,7,8-TCDD) (figure 1.4), usually referred to simply as 'dioxin'. This is a compound of extremely high toxicity to mammals (LD_{50} 10–200 µg kg l^{-1} in rats and mice). In structure, these are 'flat' molecules, formed by the linking of two benzene rings by two oxygen bridges with varying substitutions of chlorine on the available ring positions. There are 75 possible congeners of PCDD. PCDDs are chemically stable compounds with very low water solubilities (less than 1 µg l^{-1} at 20°C) and limited solubility in most organic solvents, even though they have a lipophilic character.

PCDDs are not produced commercially, but are unwanted by-products generated during the synthesis of other compounds. They are also formed during the combustion of PCBs (fires or chemical waste disposal) and by the interaction of chlorophenols during disposal of industrial wastes. In general, they are formed when chlorophenols interact.

PCDD residues have been detected very widely in the environment (especially in the aquatic environment), albeit at low concentrations, e.g. in fish and fish-eating birds.

1.2.4 POLYCHLORINATED DIBENZOFURANS (PCDFs)

These compounds are similar to PCDDs both in structure (figure 1.4) and in origin. Once again, there are many congeners, and the compounds arise as unwanted by-products – they are not synthesized intentionally. They have not, however, received as much attention as PCDDs and do not appear to have raised such serious environmental problems.

1.2.5 POLYBROMINATED BIPHENYLS (PBBs)

Mixtures of polybrominated biphenyls have been marketed as fire retardants (e.g. 'Firemaster'). These mixtures bear a general resemblance to PCB mixtures and are lipophilic,

Box 1.1 *PCDDs.*

The problem of environment pollution by PCCDs is best illustrated by reference to three well-publicized examples involving 2,3,7,8-TCDD.

1   2,3,7,8-TCDD ('dioxin') can occur as a contaminant of commercial preparations of herbicides such as 2,4-D and 2,4,5-T. There have been a number of investigations into pollution caused by these preparations, most notoriously the spraying by the US airforce of the Vietnamese jungle with the defoliant herbicide 'Agent Orange' during 1960–69 (see section 1.2.10).

2   The release of PCDDs by combustion furnaces used to dispose of PCB wastes, e.g. at the 'Rechem' plant in Scotland in the 1980s. PCDD residues were detected in soil and cattle in the surrounding area. The release of these compounds indicated incorrect operation of the furnace.

3   The release of PCDDs into the air from the Seveso Chemical Works in northern Italy in 1976. The PCDDs were formed in a chemical cloud which contained trichlorophenols. Some people who had been exposed developed the skin condition chloracne. There were, however, no fatalities or serious toxic effects attributable to PCDD.

stable and unreactive. As with PCBs, some congeners are very persistent in living organisms and have long biological half-lives. In one incident in the USA, a PBB mixture was accidentally fed to cattle, leading to the appearance of substantial residues in meat products and humans in Wisconsin and neighbouring states (Carter, 1976).

1.2.6 ORGANOCHLORINE INSECTICIDES

This is a relatively large group of insecticides with considerable diversity of structure, properties and uses (Brooks, 1974). Three major types will be mentioned here – DDT and related compounds, the chlorinated cyclodiene insecticides (e.g. aldrin and dieldrin) and hexachlorocyclohexanes (HCHs) such as lindane (figure 1.4).

Organochlorine insecticides are stable solids of limited vapour pressure, very low water solubility and high lipophilicity. Some of them are highly persistent in their original form or as stable metabolites. All the examples given here are nerve poisons (see Chapter 7).

Commercial DDT contains 70–80% of the insecticidal isomer p,p'-DDT. Related insecticides include rhothane (DDD) and methoxychlor. The insecticidal properties of DDT were discovered by Paul Müller of the firm Ciba-Geigy in 1939. DDT was used, mainly for vector control, during the Second World War, but came to be very widely used thereafter for the control of agricultural pests, vectors of disease (e.g. malarial mosquitoes), ectoparasites of farm animals and insects in domestic and industrial premises. Because of its low solubility in water (< 1 mg l^{-1}), DDT has been formulated as an emulsifiable concentrate for application as a spray. (**Emulsifiable concentrates** are solutions of pesticides in organic liquids; when added to water, they form a creamy emulsion which can be sprayed on crops.)

DDT has an acute oral LD_{50} of 113-450 mg kg^{-1} and is considered to be only moderately toxic to vertebrates (Chapter 6). However , it has been shown to cause eggshell thinning in certain sensitive species of birds at very low doses through the action of its stable metabolite p,p'-DDE (see Chapters 7, 12 and 15).

In addition to p,p'-DDT, some 20% of the commercial insecticide is represented as o,p'-DDT, which is more readily biodegradable than p,p'-DDT and has very low toxicity to insects and vertebrates . On the other hand, it has been shown to have **oestrogenic activity**, for example in rats, and has sometimes, together with other compounds of the DDT group, been implicated in cases of alleged endocrine disruption in the natural environment (see Chapter 7).

Kelthane (dicofol) is another pesticide related in structure to DDT, which has been marketed as an acaricide. It has only weak insecticidal activity, is of limited persistence and there is some evidence that it may act as an 'endocrine disruptor' (see Chapter 7).

The chlorinated cyclodiene insecticides were introduced after DDT (very widely during the 1950s), and some of them gave rise to serious environmental problems because they (or their stable metabolites) have both high toxicity to vertebrates and marked biological persistence. Aldrin, dieldrin and heptachlor are examples of cyclodiene insecticides showing this undesirable combination of properties (acute oral LD_{50} to rats 40–60 mg kg^{-1}). Chlordane is a similar insecticide, but is of lower vertebrate toxicity than the foregoing examples. Endrin and, to a lesser extent, endosulphan are **cyclodiene insecticides** of very high vertebrate toxicity, but only limited biological persistence. In general, the cyclodienes resemble DDT in being stable lipophilic solids of very low water solubility, but differ from it in their mode of action (see Chapter 7). Endosulphan is an exception to this rule, having appreciable water solubility.

The cyclodienes were introduced into Western countries during the 1950s and were used in diverse formulations for many different purposes. Because of their insolubility in water, emulsifiable concentrates and wettable powders were the formulations normally used for spraying. Sprays were used for the control of certain crop pests and for certain vectors of disease (e.g. tsetse fly). They were also used in dips and sprays to control ectoparasites of livestock and were very widely used as seed dressings for cereals and other crops. The use of aldrin, dieldrin and heptachlor for the latter purpose had very serious ecological consequences which will be discussed in some detail in Chapters 12 and 15.

By the 1990s, few uses of either DDT and its relatives or of the cyclodiene insecticides remained. The use of these compounds for most purposes had been banned on the grounds of perceived human health risks or hazards to the environment. However, some of these compounds continue to be used on a limited scale in some countries. For example, some use is still made of DDT in certain developing countries to control vectors such as the malarial mosquito, which is seen to represent a greater hazard to humans than the side-effects of the chemical. Although these compounds are little used today, their ecotoxicology is discussed in some detail in parts of the ensuing text for two different reasons. On the one hand, the very marked persistence of compounds such as dieldrin and p,p'-DDE has ensured that significant residues are still present in once heavily contaminated soils and/or sediments and will only slowly disappear over the decades to come. These residues are still slowly released into aquatic and terrestrial food chains and can reach significant concentrations in animals at higher trophic levels. The second reason is that they have been studied in considerable depth and detail – probably more so than any other type of organic pollutant. A great deal has been learned about the ecological hazards associated with them, and they serve as useful examples and models for persistent lipophilic pollutants more generally in the present text.

Hexachlorocyclohexane (HCH) has been marketed as a crude mixture of isomers ('BHC') but more extensively as a refined product containing mainly the γ isomer, known as γ-HCH, γ-BHC or lindane (figure 1.4). γ-HCH has similar properties to other organochlorine insecticides, but is somewhat more polar and water soluble (7 mg l^{-1}). Emulsifiable concentrates of HCH have been used for controlling agricultural pests or parasites of farm animals. It has also been used as an insecticidal seed dressing (e.g. on cereal seed). HCH is only moderately toxic to rats (LD_{50} 60–250 mg kg^{-1}).

1.2.7 ORGANOPHOSPHOROUS INSECTICIDES (OPs)

During the Second World War, interest developed in organophosphorous compounds which act as nerve poisons (neurotoxins) because of their ability to inhibit the enzyme acetylcholinesterase (AChE) (Chapter 7). These compounds were produced for two main uses – as insecticides and as chemical warfare agents (nerve gases) (Ballantyne and Marrs, 1992). They are organic esters of phosphorus acids (figure 1.5). Today, a large number of organophosphorous compounds are marketed as insecticides, and nearly all of them correspond to the basic formula shown in figure 1.5.

Most organophosphorous insecticides (OPs) are liquids of lipophilic character and some volatility; a few are solids. They are, in general, less stable than organochlorine insecticides and are more readily broken down by chemical or biochemical agencies (Eto, 1974; Fest and Schmidt, 1982). Thus, they tend to be relatively short-lived when free in the environment, and the environmental hazards that they present are

Organophosphorus insecticides: general formula

Carbamate insecticides: general formula

FIGURE 1.5 *Organophosphorus and carbamate insecticides. These compounds are toxic to insects because they inhibit the enzyme acetylcholinesterase (Chapter 7). They vary in their polarity and water solubility. They are generally more reactive and less stable and persistent than the organochlorine insecticides. The leaving group of the organophosphorus compounds breaks away from the rest of the molecule when hydrolysis occurs (Chapter 5).*

largely, but not exclusively, associated with short-term (acute) toxicity. They are more polar and water soluble than the main types of organochlorine insecticides. Their water solubility is highly variable, with some compounds (e.g. dimethoate) having appreciable solubility. The active forms of some OPs have sufficient water solubility to be effective systemic insecticides, reaching high enough concentrations in the phloem of plants to poison sap-feeding insects (cf. organochlorine compounds and pyrethroids).

The formulation of organophosphorous compounds is important in determining the environmental hazards that they present. Many are formulated as emulsifiable concentrates for spraying. Others are incorporated into seed dressings or into granular formulations. Granular formulations are required for the most toxic OPs (e.g. disyston and phorate) because they are safer to handle than emulsifiable concentrates or certain other types of formulation. The insecticide is 'locked up' within the granule, and is only slowly released into the environment.

In many countries, OPs are still applied to crops as sprays, granules, seed dressings and root dips. They are used to control ectoparasites of farm and domestic animals (commonly in sheep dipping), and sometimes also for controlling internal parasites (e.g. ox warble fly). Other uses include control of certain vertebrate pests (e.g. the bird *Quelea* in parts of Africa), locusts, stored product pests, especially beetles, insect vectors of disease, such as mosquitoes, and parasites of salmon at 'fish farms'.

1.2.8 CARBAMATE INSECTICIDES

These are derivatives of carbamic acid which have been developed more recently than organochlorine compounds (OCs) and OPs (figure 1.5) (Kuhr and Dorough, 1977). Like OPs, however, they act as inhibitors of acetylcholinesterase. Carbamates are frequently solids, sometimes liquids. They vary greatly in water solubility. Like OPs, they are readily degradable by chemical and biochemical agencies and do not usually raise problems of persistence. The main hazards that they present relate to short-term toxicity. Some of them (e.g. aldicarb and carbofuran) act as systemic insecticides. A few (e.g. methiocarb) are used as molluscicides for controlling slugs and snails. It is important to distinguish between the insecticidal carbamates and herbicidal carbamates (e.g. propham, chlorpropham) which have only low toxicity to animals.

Carbamate insecticides are formulated in a similar way to OPs, with the most toxic ones (e.g. aldicarb and carbofuran) being available only as granules They are used, principally, to control insect pests on agricultural and horticultural crops, although they also have some use for control of nematodes (i.e. as nematicides) and molluscs (i.e. as molluscicides).

1.2.9 PYRETHROID INSECTICIDES

Naturally occurring pyrethrin insecticides which are found in the flowering heads of *Chrysanthemum* spp. provided the model for the development of synthetic pyrethroids. Synthetic pyrethroids are, in general, more stable chemically and biochemically than are natural pyrethrins (Leahy, 1985). Pyrethroids are solids of very low water solubility which act as neurotoxins in a way similar to DDT (see Chapter 7). They are esters formed between an organic acid (usually chrysanthemic acid) and an organic base (see figure 1.6). Although pyrethroids are more stable than pyrethrins,

they are readily biodegradable and do not have long biological half-lives. They can, however, bind to particles in soils and sediments and show some persistence in these locations. With their low water solubilities, they do not show significant systemic properties and are not used as systemic insecticides. The hazards that they present relate mainly to short-term toxicity. However, it should be emphasized that they are highly selective between insects on the one hand and mammals and birds on the other. The main environmental concerns relate to their toxicity to fish and non-target invertebrates.

Pyrethroids are formulated mainly as emulsifiable concentrates for spraying. They are used to control a wide range of insect pests of agricultural and horticultural crops throughout the world and are coming to be used extensively to control insect vectors of disease (e.g. tsetse fly in parts of Africa).

1.2.10 PHENOXY HERBICIDES (PLANT GROWTH REGULATOR HERBICIDES)

These constitute the single most important group of herbicides. Familiar examples are 2,4-D, MCPA, CMPP, 2,4-DB, and 2,4,5-T (for general formula, see figure 1.7). They act by

R = CH₃ or Cl

Phenoxyacetic acids (general formula)

FIGURE 1.7 *Phenoxyalkanoic acid herbicides. The example given is a general formula for phenoxyacetic acids, which include 2,4-D and MCPA. There are also phenoxypropionic acids (e.g. CMPP) and phenoxybutyric acids (e.g. 2,4-DB). All of them have plant growth-regulating properties and have some resemblance to the natural growth regulator indole acetic acid.*

Permethrin

FIGURE 1.6 *Pyrethroid insecticides. These have low polarity and limited water solubility. They are related in structure to the natural pyrethrins, which are also toxic to insects.*

disturbing growth processes in a manner akin
to that of the natural plant growth regulate 'indole acetic acid'. They are derivatives of
phenoxyalkane carboxylic acids. When formulated as alkali salts, they are highly water soluble; when formulated as simple esters, they are
lipophilic and of low water solubility.

Most phenoxy herbicides are readily biodegradable and so are not strongly persistent in
living organisms or in soil. They are selective,
i.e. selectively toxic between monocotyledonous
and dicotyledonous plants. Their principal use
is to control 'dicot' weeds in 'monocot' crops
(e.g. cereals and grass). The environmental hazards are of two kinds. First, there is the problem of unwanted phytotoxicity as a consequence
of spray or vapour drift. Second, formulations
of certain herbicides of this type have sometimes been contaminated with the highly toxic
compound TCDD, e.g. 'Agent Orange', a formulation containing 2,4-D and 2,4,5-T which
was used as a defoliant in Vietnam (see section
1.2.3).

Water-soluble salts (e.g. Na^+, K^+) are formulated as aqueous solutions (aqueous concentrates) whereas lipophilic esters are formulated
as emulsifiable concentrates. The latter have
sometimes caused unwanted phytotoxicity ow-
ing to the volatility of some of the esters (vapour drift).

1.2.11 ANTICOAGULANT RODENTICIDES

For many years, the compound warfarin (figure 1.8) has been used as a rodenticide. It is a
lipophilic molecule of low water solubility which
acts as an antagonist to vitamin K (Chapter 7).
More recently, as wild rodents have developed
resistance to warfarin, a number of second-generation anticoagulant rodenticides (sometimes
called super warfarins) have been marketed,
which are structurally related to warfarin. These

Warfarin

FIGURE 1.8 *Rodenticides. Warfarin and related compounds, such as diphenacoum and brodifacoum, are
anticoagulant rodenticides. They are complex molecules
bearing some structural resemblance to vitamin K. Their
toxic action is due to competition with vitamin K in the
liver (vitamin K antagonism).*

include diphenacoum, bromadiolone, brodifacoum and flocoumafen and resemble warfarin in their general properties but are more
toxic to mammals and birds and are markedly
persistent in the livers of vertebrates. Thus, they
may be transferred from rodents to the vertebrate predators and scavengers that feed upon
them. Owls, for example, have been found to
contain residues of them in the UK. Rodenticides are usually incorporated into bait, which
is then placed in buildings or out of doors, where
it will be taken by wild rodents.

1.2.12 DETERGENTS

Detergents are organic compounds which have
both polar and non-polar characteristics. They
tend to exist at phase boundaries, where they
are associated with both polar and non-polar
media. Some examples are shown in figure 1.9.
Detergents are of three types: (i) anionic, (ii)
cationic and (iii) non-ionic. The first two types
have permanent negative or positive charges,
attached to non-polar (hydrophobic) C–C
chains. Non-ionic detergents have no such permanent charge; rather, they have a number of
atoms which are weakly electropositive and
electronegative. This is due, in the examples
shown, to the electron-attracting power of oxygen atoms.

Anionic

Sodium tetrapropylene
benzene sulphonate (hard)

Dobane J.N. sulphonate
(soft)

$CH_3(CH_2)_n$ —O—S—O⁻Na⁺   Sodium alkyl sulphonate (soft)

Cationic

H—C—$(CH_2)_{14}$—C—N^+ Br⁻

Cetyl pyridinium bromide

Nonionic

Polyglycol ethers of alkylated HO—$(C_2H_4O)_n$—◯—R   R = alkyl group
phenols e.g. lissapol N. stergene

FIGURE 1.9 *Detergents. Detergents are molecules that have both polar and non-polar elements. They may have permanent negative charge (anionic detergents), permanent positive charge (cationic detergents) or a collection of small positive and negative charges over their structure (non-ionic detergents).*

Detergents are very widely used in both domestic and industrial premises. The major entry point into water is via sewage works into surface waters. They are also used in pesticide formulations and for dispersing oil spills at sea.

The degradation of alkylphenol polyethoxylates (non-ionic detergents) can lead to the formation of alkylphenols (particularly nonylphenols), which act as endocrine disruptors (see Chapter 10).

1.2.13 CHLOROPHENOLS

A number of different polychlorinated phenols (PCPs) occur as environmental pollutants (figure 1.10). A major source is pulp mill effluent, from which they arise because of the chemical action of chlorine (used as a bleaching agent) upon phenolic substances present in wood pulp (Södergren, 1991). Pentachlorophenol is used as a wood preservative and this is an important source of pollution.

OH

2,4,5-Trichlorophenol

FIGURE 1.10 *Chlorinated phenols. These have acidic properties, releasing H⁺ ions when they dissolve in water. They can interact to form dioxin (see section 1.2.3).*

Chlorinated phenols have acidic properties and are water soluble, chemically reactive and of limited persistence. The tendency of some PCPs to interact and form PCDDs was discussed in section 1.2.3.

1.3 *Organometallic compounds*

Some metal ions are so insoluble that they are relatively non-toxic to animals if ingested. Liquid metal mercury, for example, can be swallowed in small amounts by humans with little long-term effect. Indeed, until the last century, drinking liquid mercury was recommended as a cure for constipation! The low toxicity of tin is demonstrated by its use as a lining in food containers.

Nevertheless, the toxicity of several metals is greatly enhanced if they become bound either deliberately or accidentally to an organic ligand. This changes their chemical behaviour in the environment and within organisms. Metals are modified in this way to increase their toxicity for use as pesticides. Organomercury compounds were used widely as antifungal seed dressings in the UK until as recently as 1993. Organolead compounds have been used extensively for control of caterpillars on fruit crops. Organotin compounds, particularly tributyl tin,

are extremely toxic. Their main use is for preserving timbers from the activities of aquatic boring animals and as a component of antifouling paints which are applied to the outer surface of boats and fish cages to inhibit settlement by marine organisms. When these substances leach into the environment, they can affect non-target organisms. Tributyl tin, for example, has devastated populations of the dog whelk *Nucella lapillus* near sites of boating activity in many countries (see sections 13.6.4 and 15.3).

A tragic example of the effects of organomercury compounds occurred in Minimata Bay, Japan, in the 1950s. Metallic mercury released from a paper factory on the shores of the bay was methylated in the sediments by bacteria to form methyl mercury (see Kudo *et al.*, 1980). Mercury in its methylated form is much more bioavailable than liquid mercury and it passed rapidly along the food chain until it reached high concentrations in fish. The local people relied heavily on locally caught fish and were thus vulnerable to poisoning. About 100 people died and many suffered severe disabilities from mercury poisoning. Such incidents are most severe when a local population is highly dependent on a single food source but are rare in more developed regions where food is obtained from much wider sources. Similar problems are likely to occur in the near future in the Amazon Basin. Huge quantities of mercury are being dumped into the river as a by-product of gold refining, and there is evidence that this is becoming methylated and is passing into food chains (Pfeiffer *et al.*, 1989).

1.4 *Radioactive isotopes*

1.4.1 INTRODUCTION

Since the development of nuclear energy and atomic weapons, there has been an ongoing debate as to the safety of low levels of radioactivity in the environment. We are all exposed to background radiation from cosmic rays and the natural decay of radioactive isotopes. Some consider that this exposure is beneficial as it promotes natural DNA repair mechanisms (a type of 'immunization'). Others consider that there is no safe level of radiation. The contribution of different sources to overall natural background radiation depends to a large extent on local geology. One of the most important sources is radon gas, which may reach levels that give cause for concern in poorly ventilated houses, especially if they are sited on igneous rocks (Mose *et al.*, 1992).

Three factors determine whether or not radioactive isotopes are harmful to organisms. First, the *nature* and *intensity* of the radioactive decay in terms of the mass and energy of the particles produced. Second, the **half-life** of the isotope. Third, the *biochemistry* of the radioactive element. Regarding biochemistry, radioactive isotopes of essential elements will follow the same pathways as their stable forms and accumulate in particular organs. Furthermore, some radioactive non-essential elements may be biochemical analogues of essential elements and follow similar routes in living tissues.

1.4.2 THE NATURE AND INTENSITY OF THE RADIOACTIVE DECAY PRODUCTS

When an atom of a radioactive substance decays, it can produce one of four types of particle: alpha (α), beta (β), gamma (γ) or neutrons. It can subsequently decay one or more times until the atom is stable.

The intensity of a radioactive substance is measured in the SI unit of becquerels (Bq) and represents the number of atoms that disintegrate per second. Formerly, radioactivity was measured in curies (Ci) and was equal to the number of disintegrations per second of 1 g of radium. 1 Ci $= 3.7 \times 10^{10}$ Bq and 1 Bq $= 2.7 \times 10^{-11}$ Ci.

The alpha particle consists of two protons and two neutrons and is a positively charged helium nucleus. Alpha particles are relatively massive compared with other radioactive emissions. Although they travel only a few centimetres in air, and in biological tissue only a few millimetres at most, their large mass makes them very damaging if they collide with cells, especially if inhaled into the lungs of vertebrates.

The beta particle is an electron and is negatively charged. Beta particles have more penetrating ability than alpha particles, with a range of a few metres in air. Sources of beta radiation can be shielded by a thin layer of Perspex. However, their small mass means that they do much less damage than alpha particles to tissues.

Gamma rays are quanta of electromagnetic radiation. They are highly penetrating and can pass through several centimetres of lead. As a rule of thumb, the damage they cause is similar to that of beta particles.

Neutrons have no charge and are liberated only when certain elements are bombarded with alpha or gamma rays. They react with other elements only by direct collision. Production of neutrons is the basis for nuclear fission in a reactor. They have a range of several centimetres of lead.

In biological terms, if we were to measure the radioactivity of a substance only in becquerels, we would get little information about the effects it might have on tissues. This is because the becquerel takes no account of the nature of the nuclear disintegration, merely its frequency. Two other SI units are needed, the gray and the sievert.

The **gray** (Gy) is equal to the amount of radiation causing 1 kg of tissue to absorb 1 joule of energy. However, different kinds of radiation do different amounts of damage to living tissue for the same amount of energy. This is sometimes difficult to understand, and a simply analogy is helpful. If a heavyweight boxer were to gently tap your chin with his fist 100 times and impart X joules of energy each time, you would find this irritating but no long-term damage to your jaw would result. However, if the boxer was to impart 100 times X joules of energy in a single punch, then a broken jaw and several hours of unconsciousness would result. The boxer has imparted the same amount of energy to your chin, but the rate at which it is applied affects the amount of damage that results. Thus, another SI unit, the sievert, is needed. The **sievert** (Sv) takes account of the different ways in which the same amount of energy can be imparted to tissues. The 'safe' annual exposure of the general public is usually given as 5 mSv. A dose of 20 Sv (a dose received by several hundred workers involved in the Chernobyl clean up; Edwards, 1994) is equal to 20 Gy of beta or gamma emissions or only 1 Gy of alpha particles. Thus, alpha particles have about 20 times the effect of beta or gamma radiation for the same number of grays.

1.4.3 HALF-LIVES

Another important consideration is the half-life of radioactive isotopes. This is the time taken for half of the atoms of a radioactive isotope to decay. The decay curve follows an exponential decline. For example, ^{32}P is a radioactive isotope of phosphorus used extensively in experiments on molecular biology and plant physiology. It has a half-life of 14 days. If we were to start on day 0 with 1 g of ^{32}P, on day 14 we would have 0.5 g of ^{32}P and 0.5 g of ^{32}S (stable sulphur). On day 28, we would have 0.25 g of ^{32}P and 0.75 g of ^{32}S and so on.

After the passage of 10 half-lives, the radioactivity of an isotope is no longer significantly different from the background level and is considered to be 'safe'. Thus, ^{32}P can be discarded with normal refuse providing that it is held in shielded conditions for 140 days before disposal.

The method of disposal of radioactive waste is dictated by the half-life of the longest-lived isotope. Thus, if short-lived isotopes can be separated from long-lived ones before disposal, the volume of waste that has to be held in long-term storage can be greatly reduced. This is much easier in a laboratory or hospital where different isotopes can be separately maintained. Low-level waste is usually placed in concrete and buried in surface trenches. However, medium and high-level radioactive waste from nuclear reactors contains a complex 'cocktail' of highly radioactive isotopes, some with very long half-lives indeed. The only option in this case is secure long-term storage underground. In the UK, the long-term aim is to bury the waste deep underground, although, at the time of writing, the exact location of such a store(s) is still the subject of much political debate. Some examples of the wide range in half-lives of different isotopes are given in table 1.4.

1.4.4 BIOCHEMISTRY

Some radioactive isotopes are especially dangerous because they follow the same biochemical pathways in the body as stable elements. For example, more than 80% of the total iodine in the human body is contained in

TABLE 1.4 *Half-lives of some radioactive isotopes*

Isotope	Half-life
Plutonium-241	13 years
Plutonium-239	24 000 years
Iodine-131	8 days
Iodine-129	160×10^6 years
Strontium-90	28 years
Caesium-137	30 years

the thyroid gland, where it forms an essential component of the growth hormone thyroxin. If radioactive iodine is ingested, it becomes concentrated in the thyroid gland and thyroid cancer may result. There is evidence that thyroid cancer has shown a dramatic increase in people living in the vicinity of Chernobyl (Kazakov *et al.*, 1992). The radioactive isotope strontium-90 (table 1.4) follows the same pathways as calcium in humans. Most people in the world contain trace amounts of this isotope in their bones following its global dispersion after atmospheric atom bomb tests in the 1950s and 1960s. Caesium-137 follows potassium pathways and has been a particular problem in areas subjected to Chernobyl fallout, such as north-west England (Crout *et al.*, 1991; Simkiss, 1993) and Scandinavia (Åhman and Åhman, 1994). Where nutrients are strongly recycled by the vegetation, it may take many decades before levels of Chernobyl contamination decline to background levels.

1.5 *Gaseous pollutants*

The most important gaseous pollutants are ozone (O_3) and oxides of carbon, nitrogen and sulphur.

On a global level, the main concern with ozone is the reduction in its concentration in the upper atmosphere. The well-publicized ozone 'hole' which occurs over the Antarctic (but now detected at high latitudes in the northern hemisphere) is caused by the degradative effects of chlorofluorocarbons (CFCs) on ozone molecules (see Chapter 3). CFCs are released from aerosol containers, from the coolants in domestic refrigerators when they are broken up and from foam packaging. The ozone layer absorbs ultraviolet light, so that one hazard associated with its thinning is an increase in the rates of skin cancer. Environmentally, it has been

suggested that the increased radiation could decrease photosynthesis of phytoplankton in the Antarctic.

At a local level, ozone is produced in photochemical smogs, in which oxides of nitrogen from car exhausts (NO and NO_2, sometimes known as 'NO_x' or 'NOX' gases) and fumes from other fossil fuel consumption react with moisture under the action of sunlight (Bower *et al.*, 1994). High concentrations of ozone in the air irritate respiratory epithelia of animals and can directly affect the growth of some plants (Mehlhorn *et al.*, 1991). Tobacco is particularly sensitive to ozone-induced damage (Heggestad, 1991). Ozone has been considered a major factor in forest dieback in Germany (Postel, 1984).

Levels of carbon dioxide rarely reach toxic concentrations, except in very confined places. The effects of CO_2 on global warming will be discussed in Chapter 14.

Sulphur dioxide is produced mainly from volcanoes and fossil fuel burning. The SO_2 dissolves in water droplets, forming sulphurous and sulphuric acids which fall as 'acid rain'. Rain with a low pH may directly damage the leaves and roots of plants. Furthermore, nutrients may be washed out of acidified soils as the hydrogen ions displace essential elements from soil particles. Plants growing on such soils may become deficient in one or more trace elements. A deficiency of magnesium, for example, is thought to be one of the causes of forest dieback in Germany. Acid rain may also increase the mobility of metal pollutants in soils if the pH drops sufficiently low (see figure 4.3). The effects of acid rain are much greater on soils and lakes with a low buffering capacity. One of the reasons why Scandinavia has been so badly affected by acid rain is the poor buffering capacity of the soils developed from the granite bedrock which underlies much of the region. The effects of acid rain on ecosystems are discussed in section 14.3.1.

1.6 *Summary*

In this chapter, major classes of pollutants have been identified that have been described under five different categories: (i) inorganic ions, (ii) organic pollutants, (iii) organometallic compounds, (iv) radioactive isotopes and (v) gases. Their structures, main properties and occurrence are briefly described. Pollutants described here will be used as examples throughout the subsequent text. It is important to relate the properties of individual compounds to their environmental fate (Chapters 2–5), their toxicity to individual organisms (Chapters 6–11) and their effects upon populations, communities and ecosystems (Chapters 12–15). The influence of properties such as polarity, vapour pressure, partition coefficient and molecular stability on the movement and distribution of environmental chemicals will be described in section 3.2. Chemical properties have been used in the development of models that predict the environmental fate (Chapters 3 and 5) and the toxicity (Chapter 6) of chemicals.

1.7 *Further reading*

ÅHMAN, B. and ÅHMAN, G. (1994) A very interesting study of the consequences of Chernobyl fall-out in reindeer and the local human population which rely on them.

BUNCE, N. (1991) *Environmental Chemistry*. An account of the pollutants discussed here; very little, however, on pesticides and radionuclides.

CROSBY, D. G. (1998) *Environmental Toxicology and Chemistry*. A useful background on many organic pollutants.

EDWARDS, T. (1994) An excellent article which captures the impact of Chernobyl in the usual National Geographic style.

GUTHRIE, F. E. and PERRY, J. J. (eds) (1980) *Introduction to Environmental Toxicology*. A multiauthor work giving an in depth account of most of the important organic and inorganic pollutants.

HASSALL, K. A. (1990) *The Biochemistry and Uses of Pesticides*, 2nd edn. A very readable account of the chemical properties of the major types of pesticide.

JUKES, T. (1985) An interesting discussion of the consequences of the narrow window of essentiality for selenium.

MANAHAN, S. E. (1994) *Environmental Chemistry*, 6th edn. A detailed and comprehensive text covering both inorganic and organic chemicals. Two useful chapters on basic principles for those with limited background in chemistry.

MERIAN, E. (ed.) (1991) *Metals and their Compounds in the Environment*. Almost everything you could possibly want to know about metals in this comprehensive text (1438 pages!).

NIEBOER, E. and RICHARDSON, D. H. S. (1980) A classic influential paper on the chemistry of metal ions which had a major impact on studies of metal toxicity.

PAASIVIRTA, J. (1991) *Chemical Ecotoxicology*. A brief overview of most of the important organic and inorganic pollutants.

CHAPTER 2

Routes by which pollutants enter ecosystems

Pollutants may enter ecosystems as the consequence of human activity in the following ways:

1. 'unintended' release in the course of human activities (e.g. in nuclear accidents, mining operations, shipwrecks and fires);
2. disposal of wastes (e.g. sewage, industrial effluents);
3. deliberate application of biocides (e.g. pest control and vector control).

Some of the chemicals so released can also reach unusually high levels locally as a result of natural processes such as weathering of rocks (many metals and inorganic anions) and volcanic activity with associated forest fires (SO_2, CO_2 and aromatic hydrocarbons). As noted in the Introduction, there are problems of defining what actually constitutes pollution. Some authorities prefer to restrict the terms 'pollution' and 'pollutants' to the consequences of human activity. However, it is sometimes difficult or impossible to determine the relative contribution of hu-

man processes and natural ones to residues that are present in the general environment.

2.1 *Entry into surface waters*

The discharge of sewage into surface waters represents a major source of pollutants globally (table 2.1). Domestic wastes are discharged mainly into sewage systems. Industrial wastes are discharged *either* into the sewage system *or* directly into surface waters.

The quality of the sewage that is discharged into surface waters depends on (i) the quality of the raw sewage received by the sewage works and (ii) the treatment of the sewage that takes place within the works (Benn and McAuliffe, 1975). Urine, faeces, paper, soap and synthetic detergents are important constituents of domestic waste. Industrial wastes are many and varied and their quality depends on the nature of the operations that are followed in particular

TABLE 2.1 *Major routes of entry to surface waters*

Route	Major pollutants	Comments
Sewage outfalls	A very wide range of organic and inorganic pollutants from commercial and domestic sources; detergents generally present	Highly variable; dependent not only on what sewage works receive but also on the treatments that sewage is given
Outfalls from commercial premises	Dependent on the commercial activity; wide range of pollutants from chemical industry; heavy metals from mining operations; pulp mills an important source of pollutants in some areas	Concentration of pollutants in effluents must stay below statutory limits
Outfalls of nuclear power stations	Radionuclides	Subject to regular monitoring and close control in most countries
Run-off from land	A variety of pollutants dumped on land surface; pesticides	Generally uncontrolled and difficult to measure
From the air	(i) Precipitation with rain or snow	Sometimes pollutants transported over great distances
	(ii) Direct application of biocides	Control of pests, parasites, vectors of disease and aquatic weeds
	(iii) Accidental contamination by sprays or dusts	Aerial spraying a potential problem
Dumping at sea	Raw sewage; radiochemicals and toxic wastes in sealed containers dumped in deep ocean	Concern sometimes expressed about release from containers in the longer term when they degenerate
Release from oil rigs and terminals	Hydrocarbons	Sometimes accidental, sometimes as a result of war (e.g. Gulf War in Kuwait)
Shipwrecks	Hydrocarbons and some other organic pollutants	Wrecks of oil tankers particularly a problem

establishments. A variety of treatments may be carried out at sewage works to improve the quality of sewage before it is discharged into surface waters (figure 2.1). In primary treatment, sewage is passed through a sedimentation tank, where it has a retention period of several hours. At this stage, 'primary sludge' will settle out. Subsequent to this, during secondary treatment, biological oxidation and flocculation of most of the remaining organic matter takes place. Typically, this is carried out by either the 'activated sludge' process or by 'biological filtration'. Conversion of ammonia to nitrites and nitrates by microorganisms is a feature of the process. Also, detergents are removed by biological oxidation. Much of the organic matter entering sewage works is converted into sludge – and this is disposed of by spreading on land as a fertilizer or by dumping on land or at sea. The sewage effluent resulting

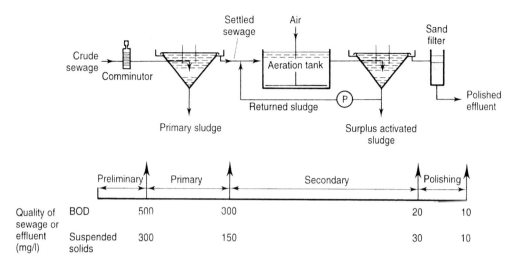

FIGURE 2.1 *Conventional treatment of sewage by the activated sludge process. The top of the diagram illustrates typical stages in sewage treatment. The lower figure indicates the quality of sewage at different stages of treatment. Reproduced after Benn and McAuliffe (1975), p. 80, with permission from Macmillan.*

from secondary treatments may be subjected to further treatments to remove constituents such as phosphate, nitrate, silicates and borates, depending on the quality of the final effluent that is required.

Important properties of sewage are levels of suspended solids, chemical oxygen demand (COD) and biochemical oxygen demand (BOD). COD is a measurement of the amount of oxygen required to achieve a complete chemical oxidation of 1 l of a sewage sample. BOD is a measurement of the amount of dissolved oxygen used by microorganisms to oxidize the organic matter in 1 l of a sewage sample. Sewage that is discharged into surface waters should have values for COD and BOD that fall below agreed limits. If the values are too high, this means that the organic content of the sewage is too high for discharge into receiving waters. Such a discharge could cause substantial reduction in the oxygen level of the water with serious consequences for aquatic organisms. In practice, the quality of sewage effluent varies enormously from country to country and from place to place. Because of the high cost of sewage treatment, no more is done than the situa-

tion requires, even in developed countries. Full advantage is taken of the capacity of receiving waters to achieve degradation of sewage components. Sewage is a rich source of organic and inorganic pollutants, prominent among them detergents which are extensively used in both domestic and industrial premises. These detergents have given rise to serious pollution problems. For further discussion of this question, see Benn and McAuliffe (1975).

The type of pollutants in industrial effluents depends largely on the industrial processes that are being followed. Heavy metals are associated with mining and smelting operations, chlorophenols and fungicides with pulp mills, insecticides with moth-proofing factories, a variety of organic chemicals with the chemical industry and radionuclides with atomic power stations. In developed countries, there are close controls over the permitted levels of release of chemicals in industrial effluents. Offshore industrial activities, such as oil extraction and manganese nodule extraction, lead to the direct discharge of pollutants to the sea.

Apart from direct discharge, pollutants are sometimes dumped into surface waters at considerable distances from the premises where they are produced. Such dumping is largely restricted to the sea. Sometimes sludge from sewage works is taken well out to sea and dumped there. Also, radioactive wastes and chemical weapons have been dumped at sea in sealed containers and questions have been asked about release in the longer term because these containers will eventually disintegrate. It is usual to dump dangerous wastes where the sea is deep to minimize the risk of contamination of surface layers of the ocean.

A further problem is the release of oil from tankers, most dramatically in the case of shipwrecks, when large quantities are discharged in a relatively short time in one area. However, it is important to put such incidents as these – and the disastrous release of oil during the Gulf War – into a wider perspective. The total input of petroleum hydrocarbons to the marine environment has been estimated at 3.2 million tons per year. Although oil tanker disasters can cause great damage, the input from them ranks below those from normal tanker operations and discharges from industrial and municipal waste.

Biocides are sometimes deliberately applied to surface waters in order to control invertebrates or plants. Herbicides have been used to control aquatic weeds in lakes and water courses. Insecticides are applied to control fish parasites at fish farms in both freshwater and marine locations and to control pests in watercress beds. Tributyl tin fungicides have been incorporated into antifouling paints for use on boats, and this has led to marine pollution.

Most of the examples given so far have been of deliberate actions which are, at least in theory, carefully controlled. There are, however, many cases of 'accidental' pollution which are not under direct human control. Pollutants present in air may enter surface waters as a consequence of precipitation of dust or droplets or with rain or snow – or simply as a result of partition from air into water. Pollutants present on the land

surface, for example metals or pesticides, may be washed into rivers, streams or oceans when there is heavy rainfall. They may be in the free or particulate state or attached to soil or mineral particles. There is a particular problem with aerial spraying of pesticides because of the risk of spray drift into surface waters. Some pesticides are extremely toxic to aquatic organisms and it is very important that spray droplets should not directly contaminate waters.

Release of pollutants into moving surface water is followed by dilution and degradation. Consequently, biological effects are most likely to be seen at or near the point of release. Where pollutants enter rivers, there may be a biological gradient downstream from the outfall. Sensitive organisms may be absent near the outfall, but may reappear downstream. In fast-flowing rivers, the dilution effect is marked and pollutants are unlikely to reach high concentrations at a reasonable distance below the outfall.

By virtue of their size, and the action of currents, oceans can effectively dilute incoming pollutants. Of greater concern are lakes and small inland seas. Here, pollutants are brought in by rivers and other routes. Because they have no effective outlet, pollutants will tend to build up in them as water evaporates, sometimes with detrimental consequences. Much depends upon the rates of degradation or precipitation that will remove pollutants from the water. The pollution of the Great Lakes of North America provides a good example and will be discussed in Chapter 15.

2.2 Contamination of land

As with pollution of surface waters, contamination of land may or may not be deliberate. Deliberate contamination may involve either the disposal of wastes or the control of animals, plants or microorganisms with biocides. Accidental contamination may be the result of short-term or long-term aerial transport, flooding by rivers or seas, or collision of tankers/lorries carrying toxic chemicals (table 2.2).

The dumping of wastes at landfill sites is a widespread practice. Indeed, many old sites could be regarded as ecological 'time bombs'. On the one hand, there is the question of disposal of domestic and general industrial wastes. On the other hand, there are toxic wastes which require more careful handling. Of particular concern are radioactive wastes from nuclear power stations. With the latter, there are stringent regulations for safe disposal to minimize contamination of the land surface and neighbouring surface waters. One practice is to embed the disposed radioactive material in concrete.

The use of sewage sludge as fertilizer on agricultural land is another source of pollution. Heavy metals, nitrates, phosphates and deter-

gents are all added to soil in this way. Land is also contaminated by aerially transported materials. Smoke and dust from chimneys can fall on neighbouring land, carrying with them a variety of organic and inorganic pollutants. Gases such as sulphur dioxide, nitrogen oxides and hydrogen fluoride released from chimneys cause damage to vegetation in the neighbourhood of industrial premises. Thus, pollution of the land surface may occur in the immediate vicinity of domestic and industrial premises which cause air pollution. Additionally, as with surface waters, pollutants reach the land after travelling considerable distances. They are carried down by rain or snow, in solution or suspension or with associated dust particles.

Land is sometimes also contaminated by pollutants when there is flooding by rivers or seas. Considerable areas of the land surface are treated with biocides to control vertebrate and invertebrate pests, plant diseases, weeds and

TABLE 2.2 *Major routes of contamination of land*

Route	Major pollutants	Comments
Waste dumping including rubbish dumps/landfall sites/industrial dumps	A very wide range of different pollutants	Some industrial dumps are high in particular pollutants, e.g. oil, metal ore deposits, PCBs, etc.
Application of pesticides to agricultural land and forests	Insecticides, rodenticides, herbicides and fungicides as sprays, dusts, seed dressings, etc.	In most countries, there are strict regulations controlling the application of pesticides
Control of insect vectors of disease	Insecticides	Major pollution over large areas as a consequence of control measures against malarial mosquito and tsetse fly
Application of sewage to agricultural land	Heavy metals, nitrates, detergents	
Flooding by rivers or seas	A variety of pollutants including those associated with sewage	
Precipitation from air as dust or droplets or in rain or snow	Pollutants associated with soot and dust, acid rain, pesticides	Transport may be over short distances (spray drift, soot and dust from chimneys) or long distance (brought down especially by rain and snow)

vectors of disease. This is important in areas of intensive agriculture, where a variety of different pesticides are used during the course of a farming year. Pesticides are applied as different formulations – as sprays, granules, dusts and seed dressings. The manner of application and the nature of the formulation influence the way the pesticides are distributed in the crop and in the soil. There is a potential problem of spray drift to areas outside the target sites, during the course of pesticide application. This is particularly so with aerial spraying, and is very dependent upon the strength and direction of winds at the time of the operation. There is sometimes also a problem of pesticides moving through soil to contaminate groundwater, especially where there are cracks in the soil profile, allowing rapid percolation of water. A field experiment conducted by the Ministry of Agriculture, Fisheries and Food (MAFF) at Rosemaund, England, demonstrated that a variety of pesticides can find their way into drains and water courses after heavy rain.

Apart from agricultural applications, pesticides are applied over large areas for other purposes. Thus, insecticides are used extensively in Africa to control tsetse fly and locust swarms. Aerial spraying of insecticides has taken place over forests in Canada to control insect pests (see Chapter 15), and over nesting colonies of *Quelea* (a bird pest) in parts of Africa.

2.3 *Discharge into the atmosphere*

Pollutants enter the atmosphere in the gaseous state, as droplets or particles, or in association with the latter. When in the gaseous state, they may be transported over considerable distances, with the movements of air masses. Particles and droplets, on the other hand, are more likely to move over only relatively short distances before falling to the ground. However, they too may sometimes undergo long-distance transport when they are of small diameter.

Chimneys of both industrial and domestic premises are important sources of atmospheric pollution (table 2.3). Carbon dioxide (CO_2), sulphur dioxide (SO_2), oxides of nitrogen (NO_x), hydrogen fluoride and chlorofluorocarbons (CFCs) are examples of gases released in this way. The combustion of fuels releases CO_2, SO_2, NO_x and a variety of organic compounds (e.g. PAHs) which are products of incomplete combustion. The level of pollution depends on the quality of the fuel and the extent to which flue gases are cleaned up. Some forms of coal (e.g. brown coal from Silesia) are high in sulphur and can cause very serious pollution with SO_2. Many of the organic compounds released from chimneys are present in smoke particles. The subsequent movement of pollutants is dependent upon atmospheric conditions and on the height and the form of the chimney releasing them. Under clear and warm conditions, pollutants will be quickly diluted because of the mixing of air. As the earth's surface is warmed by sunlight, hot air will rise from the vicinity of the chimney, producing convection currents and carrying pollutants with it. Cold, clean air will flow in to replace it. If there is a side wind, airborne pollutants will be carried away from the initial point of release, with further dilution. In the evening, this process may be reversed, as the air cools. Then, if there is no wind, a layer of mist or fog may form, trapping a well of cold air beneath it (figure 2.2). The following morning the sun will be unable to penetrate the layer of mist–fog, thus preventing a warming of the air and consequent dispersal of air pollutants. The air pollutants will become trapped in the vicinity of the chimney from which they were emitted. Thus, in general, the dispersal of air pollutants is favoured by warm, dry conditions with a steady side wind. The dispersal of pollutant will be more effective from high chimneys than from low chimneys. In general, the

TABLE 2.3 *Major points of entry into the atmosphere*

Route	Major pollutants	Comments
Domestic chimneys	Many organic compounds, including hydrocarbons associated with smoke particles or as vapours. SO_2, CO_2, NO_2 and other gases	Level of pollution dependent upon quality of fuel burned and clean-up of flue gases
Chimneys of industrial premises, power stations, etc	As for domestic chimneys but also many other pollutants depending on the practices followed on the site.* Radiochemicals from nuclear power stations	With hazardous substances, procedures for cleaning up effluent gases are very important
Internal combustion engines and jet engines	CO_2, (NO_x) hydrocarbons and other organic pollutants; lead compounds (mainly inorganic but some organic lead) where leaded fuel is used	Level of pollution very dependent upon design of engine and exhaust system; the growing use of lead-free petrol is restricting lead pollution
Pesticide applications	Insecticides, fungicides and herbicides	Volatile pesticides enter the air in the vapour state; also, droplets of pesticide spray and pesticide dust formulations reach the atmosphere
Escape from refrigerators	Chlorofluoromethanes	
Aerosols	Chlorofluoromethanes, e.g. CF_2Cl_2, $CFCl_3$	Many countries now have strict controls over the use of these compounds as propellants

*For further details, see text.

higher the point of release, the greater the height that pollutants will reach in the atmosphere and the greater the distance that they are likely to travel (Chapter 3).

The internal-combustion engine represents another important source of air pollution. Apart from vehicles on roads, the engines of aeroplanes and ships cause pollution of air and sea. Jet engines are a significant source of air pollution. During the operation of internal-combustion engines, chemical reactions take place which generate substances not originally present in the petrol and air mixture delivered by the carburettor. Carbon monoxide and nitrogen oxides are released, together with a variety of organic molecules which are the products of incomplete combustion. The last include PAHs, aldehydes and ketones, in addition to the normal constituents of petrol. Petrol is also a source of organolead and inorganic lead compounds which arise from tetra-alkyl lead, used in some petrols as an 'antiknock' (to control semi-explosive burning during the operation of the engine). Nowadays, there is a strong movement towards the greater use of lead-free petrols to reduce the emissions of lead in its various forms from car exhausts.

The control of emissions from internal-combustion engines is a complex subject that will only be considered in outline here. In an 'un-

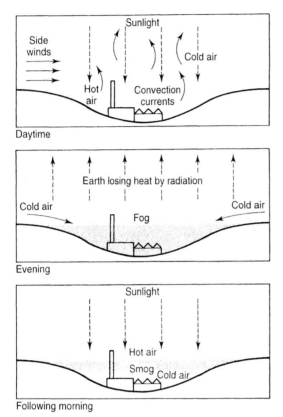

Daytime

Evening

Following morning

FIGURE 2.2 *Inversion effects in air pollution. Reproduced after Benn and McAuliffe (1975), p. 80, with permission from Macmillan.*

controlled' vehicle, effectively all of the carbon monoxide, nitrogen oxides and inorganic lead compounds and about 65% of hydrocarbons are released from the exhaust system. Evaporation from the fuel tank and carburettor accounts for a significant loss of hydrocarbons and of volatile lead tetra-alkyl. Finally, there can be a substantial loss of hydrocarbons due to leakage around the pistons and into the crankcase (crankcase 'blowby').

With modern engines, there have been considerable improvements in design which reduce all these sources of pollution. Exhaust emissions are substantially reduced by the incorporation of catalytic converters and filters into the exhaust system. These remove nitrogen oxides,

carbon monoxide and hydrocarbons. A problem with catalytic converters is that they are rapidly poisoned by tetra-alkyl lead. Recognition of this has hastened the phasing out of leaded petrols. Further improvements in exhaust emission have come through the optimization of engine performance (critical factors are air–fuel ratios, ignition timing and cylinder design). Crankcase 'blowby' has been reduced by recycling 'blowby' gases via the carburettor system. Finally, evaporative loses from the carburettor and fuel tank have been reduced by improvements in design.

Speaking generally, considerable improvements in the control of the release of pollutants by internal-combustion engines have been effected in recent years. However, with the rising use of motor vehicles, this remains one of the major sources of air pollution. The extent to which improvements have been made varies greatly from country to country. The standards that currently operate in Western Europe, North America and Australia are not usually found in other parts of the world. Recently, there has been considerable concern about the release of particulates from diesel engines (Tolba, 1992).

Air pollution also arises because of the use of pesticides. The application of pesticides as sprays or as dusts is not a very efficient process. A substantial proportion of the pesticide that is applied does not reach the crop or the soil surface. Aerosol droplets, dust particles with adhering pesticides and pesticides in the gaseous state pass into the air. This is a particularly difficult problem when pesticides are applied aerially. Climatic factors influence the extent to which pesticides contaminate the atmosphere. Strong side winds tend to move them away from original areas of application, with the risk that neighbouring areas lying downwind will be contaminated. Volatilization is most rapid where air temperatures are highest. Thus, pesticides show a greater tendency to volatilize into the air under tropical conditions than they do under temperate conditions. This point needs to be borne in mind when attempting to extrapolate from field studies performed in the temperate zone to make predictions about pesticide fate under tropical conditions.

Another factor of importance is droplet size. Very small spray droplets produced during low volume spraying fall more slowly to the ground than large droplets because their sedimentation velocity is slower and they are liable to travel for relatively long distances before reaching the ground. In general, environmental factors such as wind speed, temperature and humidity need to be taken into account when planning spray operations in order to maximize the amount of pesticide reaching its target and to minimize air pollution.

Radiochemical pollution of the air due to the explosion of atomic devices on or above the land surface was a problem for many years after the last war. By international agreement, however, this practice has now been discontinued, but concern remains over accidental release from establishments such as power stations and atomic research stations which handle nuclear materials. The seriousness of the problem was clearly illustrated by the Chernobyl accident of 1986 in the Ukraine, when a nuclear reactor caught fire with consequent widespread air pollution with radionuclides (see Chapter 1). Half of the reactor contents were dispersed.

An important group of air pollutants are low-molecular-weight halogenated hydrocarbons, such as chlorofluorohydrocarbons (CFCs), which are used as propellants and in refrigerators, and chlorinated compounds (e.g. CH_2Cl_2), which are used for dry-cleaning. These volatile substances can escape into the air during normal usage and after waste disposal. A major problem with CFCs is that they can reach zones of the upper atmosphere where they can cause damage to the ozone layer (ozonosphere) (Chapter 3).

Many of the pollutants found in air exist in the same forms that were originally released from the land or water surface. There are, however, some pollutants that are generated by chemical reactions within the atmosphere. This happens, for example, in the case of the photochemical 'smog' which has caused problems in Los Angeles and in other cities where there are large numbers of vehicles and high solar radiation. Under these conditions, if there is no wind, nitrogen dioxide and organic compounds released from car exhausts together with oxygen are involved in a complex series of reactions. The products include ozone and organic compounds such as peroxyacetyl nitrate (an eye irritant).

2.4 *Quantification of release of pollutants*

Legislation which has the purpose of controlling pollution focuses on the *amounts* of pollutants that may enter the environment and the *rates* at which they may be released. For major pollutants, international agreements are necessary to control the input. In the case of carbon dioxide, the United Nations conference on Environment and Development at Rio de Janeiro in June 1992 recommended reducing carbon dioxide emissions to 1990 levels by the year 2000. Even this modest goal was not supported by many countries, including the USA, the largest producer of CO_2. A more recent conference in Kyoto, Japan, in 1997 recommended an 8% decrease from 1990 levels of six greenhouse gases before 2012. Although definite targets have been produced by the European Union, the agreement has not been ratified by the USA. With the cautious approach of the USA and opposition of the oil-producing countries, the outcome is uncertain. More has been achieved in agreements about CFCs, which industrialized countries agreed to phase out by the year

2000 (a 10-year time lag was allowed to developing countries). Subsequently, this date advanced to 1996. In Europe, the European Economic Community (EEC) has been active in negotiating reductions of SO_2 and NO_x to limit the effects of 'acid rain' (table 2.4).

In ecotoxicity testing of new industrial chemicals, the testing protocols are influenced by the amount of chemical produced per annum because the amount produced per annum gives some indication of the possible scale of any consequent pollution problem. Under the current regulations of the European Commission, if production is at a low level then only a minimal base set of tests is usually required. However, when production exceeds certain thresholds, additional testing is required (for further details, see Walker *et al.*, 1991a).

Knowledge of the rate and pattern of release of pollutants is necessary when modelling the environmental fate of pollutants (see Chapter 3).

tal unregulated release (e.g. nuclear accidents, shipwrecks and fires). Some pollutants are released by natural processes in addition to the activities of man, thus making the source of the problem difficult to determine. Examples include the release of heavy metals due to the weathering of rock and the production of SO_2, CO_2 and aromatic hydrocarbons as a consequence of volcanic activity and associated forest fires.

The statutory regulation of pollution depends on definition of the amounts of pollutants that may be released into the environment and on the rates at which this may occur by specified routes. Thus, permitted rates of release may be defined for chemicals in sewage or factory effluents or in car exhausts. Permitted rates of application are defined for pesticides used in agriculture. The release of pollutants such as CO_2 and CFCs into the atmosphere is a matter of international concern, and there have been attempts to define the permitted rates of release by individual countries in the longer term.

2.5 *Summary*

In this chapter, the major routes by which pollutants enter surface waters, the land surface or the atmosphere have been identified. A distinction is drawn between deliberate and regulated release (e.g. application of biocides to control pests or vectors of disease) and acciden-

2.6 *Further reading*

BENN, F. R. and MCAULIFFE, C. A. (eds) (1975) *Chemistry and Pollution*. Useful chapters on sewage treatment and major sources of air pollution.

BUTLER, J. D. (1979) *Air Pollution Chemistry*. An authoritative work on air pollution.

TABLE 2.4 *Amounts of gaseous pollutants released globally per year (tons)**

	Anthropogenic sources	**Natural sources†**
CO_2	6 000 000 000	100 000 000 000
SO_x	100 000 000	50 000 000
NO_x	68 000 000	20 000 000
CFCs	1 100 000	0

*Data from Tolba (1992) and UNEP (1993).
†There is considerable uncertainty in the natural sources data.

CLARK, R. B. (1992) *Marine Pollution*, 3rd edn. A standard text giving a straightforward account of marine pollution.

MANAHAN, S. E. (1994) *Environmental Chemistry*, 6th edn. A great deal of information on the release of pollutants in this comprehensive text.

SALOMONS, W., BAYNE, B. L., DUURSMA, E. K. and FÜRSTNER, V. (eds) (1988) *Pollution of the North Sea: an Assessment*. Specialized chapters on sources of pollution in the North Sea.

CHAPTER 3

Long-range movements and global transport of pollutants

Pollutants are capable of movement over considerable distances. They can be carried over boundaries between different countries, thereby raising political as well as environmental problems because the export of environmental chemicals tends not to be appreciated by countries that receive them. Scandinavian countries, for example, have objected to the deposition of aerially transported sulphur dioxide (SO_2) originating from the British Isles. For the most part, transport over large distances is the consequence of the mass movements of air or water. However, movement may also be by diffusion, which can be rapid in air but less rapid in water. Movement by diffusion may be very localized or may occur over large distances, especially in air.

The present chapter will be restricted to the long-range transport of pollutants in surface waters and air and to the global distribution of pollutants through the different compartments of the environment: air, water, land surface and

'biota'. These are processes governed very largely by abiotic physical processes such as mass movements of air and water and by diffusion. Other, generally more local, movements which are dependent on biotic factors will be described elsewhere. Thus, movements in food chains, in migrating animals and birds and in soil will be considered in Chapters 4 and 5. This chapter will conclude with a discussion of models to describe or predict the environmental distribution of chemicals.

3.1 *Factors determining movement and distribution of pollutants*

For convenience, the environment can be divided into four distinct, yet interconnected, 'compartments' or phases. These are the air (atmosphere), surface waters (hydrosphere), the land surface (principally the soil or lithosphere) and living organisms (biosphere). As noted above, movements dependent upon biotic factors, for example along food chains, will not be considered in the present simple analysis. Biota are represented by the exposed surfaces of animals and plants, e.g. the integuments of insects and the cuticles of plants. The movement of chemicals within water and air and their movement across the interphases between different compartments are determined by physical processes; movement depends on properties of the chemicals themselves and properties of the environmental compartments. The principal factors involved will now be reviewed before describing their role in determining the environmental fate of chemicals and their incorporation into descriptive and predictive models.

3.1.1 POLARITY AND WATER SOLUBILITY

Water is an example of a polar liquid. The oxygen atom strongly attracts electrons away from the two hydrogen atoms (see section 1.1.1), with the consequence that oxygen develops a partial negative charge whereas the hydrogens develop partial positive charges. The positive and negative charges are separated and the molecule is said to be polar. By contrast, there is hardly any charge separation in non-polar compounds, such as non-aromatic hydrocarbons (figure 1.3).

Opposite charges tend to attract one another, and the positive charges on hydrogen atoms of water molecules (figure 3.1) are attracted to the

negative charges upon the oxygen atoms of their neighbours, so forming weak hydrogen bonds. Thus, water molecules tend to form aggregates in which each molecule is surrounded by four others. Cations or anions have affinity for those parts of water molecules which bear the opposite charge (thus cations are attracted by the partial negative charge on oxygen). This leads to disruption of the water aggregates and the ions go into solution. In general, many inorganic salts and polar organic compounds (e.g. simple alcohols and amines) have appreciable water solubility, whereas non-polar organic liquids and solids have virtually none. Solubility depends on the strength of charge on the solute. Among inorganic salts, those formed by alkali and alkali earth metals, the first two series of the periodic classification, readily release their ions, whereas those of metals on the right-hand side of the table (e.g. lead, mercury and tin) have a greater tendency to form covalent bonds rather than ionic ones and are, accordingly, of lower water solubility. Among organic compounds, the presence of polarizing atoms such as oxygen and nitrogen in the molecular structure tends to increase charge separation and consequently water solubility.

Some examples of the water solubility of pollutants are given in figure 3.2.

An important consequence of polarity in biochemistry is the so-called '**hydrophobic** effect' (Tanford, 1980). The tendency of water molecules to form aggregates actively excludes non-polar (hydrophobic) substances such as lipids and hydrocarbons. Hence, these attractive forces between water molecules contribute to the formation of phospholipid bilayers and, more generally, phase boundaries between lipids and water in living cells. An important aspect of this water–lipid interface is the movement of lipophilic pollutants into and through membranes and the toxicological consequences thereof. This will be discussed later in Chapter 5.

$$\delta^+H \qquad H^{\delta+}$$
$$O_{\delta^-}$$

FIGURE 3.1 *Water molecule.*

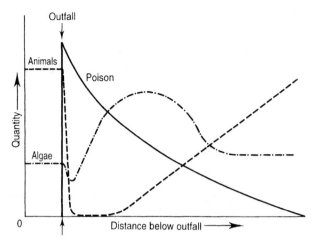

FIGURE 3.2 *Diagram of the effects of poisonous effluent on a river. Quantity (vertical axis) refers to the concentration of chemical or number of individuals per unit volume in water. Reproduced after Hynes (1960) with permission.*

3.1.2 PARTITION COEFFICIENTS

Non-polar liquids such as octanol, hexane and olive oil are immiscible with water. If a non-polar liquid is mixed with water, two phases will separate out, with the less dense of the two liquids on top. Solutes partition between the two phases and, when equilibrium is reached, the ratios of the concentrations in the two phases is given by the partition coefficient. Thus, in the case of octanol and water, the relationship is as follows:

$$K_{OW} = \frac{\text{Conc. in octanol}}{\text{Conc. in water}}$$

K_{ow} is the octanol–water **partition coefficient**, which has a high value for substances of low polarity and so provides an index of hydrophobicity. The value should be constant for any particular solute at equilibrium distribution between two defined immiscible liquids at a particular temperature. K_{ow} values are used in the prediction of the environmental distribution (section 3.4) and bioconcentration of environmental chemicals (section 5.1.8). Some examples of K_{ow} values of environmental chemicals are given in table 3.1.

3.1.3 VAPOUR PRESSURE

The tendency for a liquid or solid to volatilize is expressed by its vapour pressure. Vapour pressure is defined as the pressure exerted by the vapour of a substance on its own solid or liquid surface at equilibrium. It may be expressed in units of millimetres of mercury (also known as torr). It can also be expressed as a fraction of normal atmospheric pressure, which is 760 torr. Vapour pressure increases with rising temperature, as surface molecules increase in kinetic energy. When the vapour pressure of liquids reaches atmospheric pressure, they boil. Solids also exert a vapour pressure, and some of them vaporize without melting ('**sublimation**'). Some examples of vapour pressures of environmental chemicals are given in table 3.1.

3.1.4 PARTITION BETWEEN DIFFERENT COMPARTMENTS OF THE ENVIRONMENT

Just as chemicals partition between immiscible liquids, so they partition also between compartments of the environment – between air and water, air and soil, etc. Here too, distribution

TABLE 3.1 *Properties of pollutants.*

Compound	Water solubility at 20–25°C (mg l^{-1})	Log K_{ow}	Vapour pressure (torr)
Sodium chloride	2.6×10^5		
Calcium chloride (6H$_2$O)	4.3×10^5		
Mercuric chloride	6.9×10^4		6.0×10^{-3}
Lead chloride	6.4×10^3		
Malathion	145	2.36	1.0×10^{-5}
Carbaryl	40		5.0×10^{-3}
Cypermethrin	< 0.2		3.9×10^{-12}
p,p'-DDT	1.2×10^{-3}	6.96	1.9×10^{-7}
Dieldrin	< 0.2		1.8×10^{-7}
2,2',4,4',5,5'-HCB	5.5×10^{-3}	6.57	8.1×10^{-7}
2,3,7,8-TCDD	8.0×10^{-6}	6.53	1.5×10^{-9}
Benzo(a)pyrene	4.0×10^{-3}	5.97	5.5×10^{-9}
Tributyl tin chloride	9.7×10^3	3.70	7.5×10^{-9}
Methyl mercuric chloride			8.5×10^{-3}

between different phases at equilibrium can be described by what are, in effect, partition coefficients, although they are usually known by other terms. **Henry's constant**, for example, relates to the distribution of a volatile chemical between air and water. A particular situation exits at the interfaces of air with solid and of water with solid, as is found in soil (see section 5.2.1). Here, chemicals may be adsorbed at the solid surface rather than being absorbed into the solid matrix. The movement of substances from one compartment to another is driven by its 'escaping tendency', or **fugacity**. The construction of models of environmental fate based on the concept of fugacity, using distribution coefficients, will be described in section 3.4.

3.1.5 MOLECULAR STABILITY AND RECALCITRANT MOLECULES

The time of residence of a chemical in the environment and, consequently, the distance it can travel is dependent upon its molecular stability. Environmental chemicals are broken down by both chemical and biochemical processes. Common methods of chemical transformation are by hydrolysis (e.g. esters such as organophosphorous and carbamate insecticides) and by oxidation and photodegradation (many types of chemicals). The stability of the chemical itself is of paramount importance, but environmental factors such as temperature, level of solar radiation, nature of adsorbing surface and pH influence the rate at which chemical degradation occurs. Many organic pollutants are readily biotransformed by the action of enzyme systems (see section 5.1.5). However, there are very large differences between groups and species, and compounds that are readily metabolized by one species may be highly persistent in others. There is particular concern about 'recalcitrant' molecules, which are highly resistant to both chemical and biochemical transformation and have long half-lives in biota as well as in soils, sediments and water. It is now clear that a number of polyhalogenated compounds have this characteristic. Examples include *p,p'*-DDE, dieldrin, some PCBs and dioxins (e.g. TCDD). The environmental problems associated with polyhalogenated compounds is a recurring theme of this text.

Although degradability is regarded as a desirable characteristic of an environmental chemical because it limits persistence, movement and biomagnification, it is necessary to strike a cautionary note. Some transformations actually lead to increased toxicity. Many carcinogens, for example, undergo metabolic activation within living organisms (see Chapter 5). Regard needs to be taken of the properties of metabolites, and the products of chemical transformation.

3.2 *Transport in water*

The pollutants present in surface waters exist in diverse states. They may be in solution and/ or in suspension. Suspended material may be in the form of droplets (e.g. oil), or particles and pollutants may be dissolved in droplets or absorbed by solid particles. All of these forms can be transported by water over considerable distances. Particulate material is liable to fall to the bottom of surface waters, e.g. where the rivers enter the sea their rate of flow is checked and the coarser transported particles fall to the bottom with estuarine deposits. Liquid droplets may rise to the surface or may be carried down by particles to the sediment, depending on their density. With oil pollution, both of these things occur – light oil rises to the surface, but 'heavy' oil residues go into sediment.

In rivers, pollutants are transported over varying distances. The distances travelled depend on factors such as the stability and physical state of the pollutants and the speed of the flow of the river. The distance travelled is likely to be greatest where stable compounds are in solution and where rivers are fast flowing. In general, the concentration of a pollutant continually falls with increasing distance below an

outfall and this may be reflected in the changing composition of the fauna and flora (figure 3.2). The importance of long-distance transport of pollutants by the rivers was clearly demonstrated when the river Rhine became polluted with the insecticide endosulfan in 1969. The initial release was evidently in the middle section of the river, near Frankfurt, but the transported compound was detected by Dutch scientists working near the Rhine estuary some 500 km downstream.

Once pollutants reach lakes or oceans, they may be transported by currents. The major oceans of the world are traversed by surface currents, so it is possible for pollutants to be moved from one continent to another. These currents are wind driven, and they move roughly at right angles away from the direction of prevailing winds. In both the Atlantic and the Pacific oceans, there are large circular patterns of currents (gyres) over most of the surface area. The movement is clockwise in the northern hemisphere, but counterclockwise in the southern hemisphere. The Gulf Stream of the North Atlantic is part of a clockwise gyre system and brings warm water to the shores of the British Isles.

The density of sea water is an important factor. Water may increase in density as the result of a fall in temperature or of an increase in salt concentration (e.g. because of evaporation). When water masses increase in density, they move towards the bottom of the ocean. Downward movements are countered by upwards movements of advecting water from the lower levels of the ocean. Deep water circulation in the oceans of the world is illustrated in figure 3.3.

It is sometimes assumed that oceans are so large that dilution will quickly reduce pollutant concentrations to such low levels that they no longer constitute a problem. One shortcoming of this argument is that the distribution of pollutants in oceans is far from uniform. The

FIGURE 3.3 *Deep water circulation in the oceans. The thick lines are major bottom currents; the thin lines represent the 'return' more general flow of water. From Turekian (1976).*

movement of particulate matter with currents, and its subsequent precipitation, ensures unequal distribution. This problem is readily seen with the precipitation of sediment in estuaries as mentioned earlier. Inshore waters tend to have substantially higher levels of pollution than the open sea.

When persistent pollutants enter marine food chains, they can be moved over large distances by migrating animals and birds. Some fish, whales and fish-eating sea birds migrate over thousands of miles, taking pollutants with them. This can lead to transfer from one ecosystem to another, e.g. where contaminated fish or birds migrate over large distances and are then eaten by vertebrate predators.

3.3 *Transport in air*

Traces of persistent pollutants such as organochlorine insecticides and PCBs have been detected in snow and in animals living in polar regions, at locations far removed from any point of release. This clearly illustrates that pollutants can be transported over very large distances by the movement by air masses. The main sink of CFCs is in the stratosphere, indicating the importance of vertical movement through the troposphere for small, stable and volatile molecules. The translocation of pollutants over large distances is dependent upon the physical state of the pollutants and upon the movement of air masses.

Consider, first of all, the physical state of the pollutants. Some air pollutants are in the gaseous state. Examples include CO, NO_x, SO_2, HF and small volatile halogenated molecules such as CFCs, trichloroethylene ($CCl_2=CHCl$) and carbon tetrachloride (CCl_4). These may move through the air by two processes – mass transport and diffusion. The question of mass transport will be returned to shortly. Diffusion of gases may be of two types. First, there is diffusion along a concentration gradient, which proceeds at rates determined by Fick's law of diffusion, which states:

$$\theta = -D(C_2 - C_1)$$

where θ is rate of diffusion, D is the diffusion constant and $(C_2 - C_1)$ is the concentration gradient. Thus, net diffusion occurs in a direction that will tend to remove the gradient. The steeper the concentration gradient the faster the rate of movement. Second, there is thermal diffusion. In situations where a thermal gradient exists, 'hot' molecules of high velocity move faster then 'cold' molecules of low velocity.

Apart from the gaseous state, pollutants exists as droplets or particles *or* in association with droplets and particles. The latter situation is commonplace. Particles of dust or soot or droplets of water can be of complex composition, containing a range of polluting substances. Apart from releases from the earth's surface, pollutants may be incorporated into rain droplets. This may happen during the course of precipitation ('wash-out') or during the formation of droplets in clouds ('rain-out'). Soluble gases such as SO_2 and NO_x tend to dissolve in rain droplets. Also, rain may bring down dust particles present in the air.

Air movements on the global scale are relatively complex. First, there is a layer of air close to the earth's surface (say up to 4 km in height) which is subject to particular turbulence and localized air flow. Pollutants released within this zone are likely to return to ground quickly, travelling only relatively short distances. On the other hand, pollutants which reach greater heights may be transported over considerable distances, carried by circulating air masses.

The part of the atmosphere relevant to this discussion extends some 35 km above the earth's surface. This is the lower atmosphere, which accounts for about 99% of the total air mass. It is divided into the troposphere (first 10–11 km) and the stratosphere, which lies above it. The troposphere is characterized by strong vertical mixing – individual molecules can move through the entire height in a matter of days. There is little vertical mixing in the stratosphere above it. The boundary between these two layers is termed the tropopause. Within the stratosphere there lies a band of relatively high ozone concentration, termed the ozone layer (ozonosphere).

Within the troposphere, there are regular patterns of air circulation, characteristic of different climatic zones (figure 3.4). Both the northern and southern hemispheres are divided into three circulation zones. First, at the equator, there are sharply rising currents of hot air to either side, drawing in flows of cooler air from the north and south respectively. In the upper part of the troposphere, the rising air cools and then moves either northward or southward. This 'poleward' air flow begins to subside and to flow towards the earth's surface at between 20 and 35 degrees latitude. Some of this air will then move in a surface flow towards the equator, thus completing the cycle. The surface winds resulting from this circulation are termed the north-east and south-east trade winds in the northern and southern hemispheres respectively. These terms illustrate the point that the air flow is not directly on a north-to-south axis – there is also a west-to-east component, which is a consequence of the influence of the earth's rotation upon air movement in the troposphere. Between approximately 30 and 60 degrees latitude, there is a reversal of this circulation pattern in both hemispheres. The surface flow is 'poleward' and not towards the equator. Finally, in the polar regions, beyond 60 degrees latitude, air flow is again reversed, as shown in figure 3.4.

It is clear from this description that air pollutants – including those associated with small particles and droplets – are liable to be transported over large distances once they enter the main air circulation a few kilometres above the earth's surface. It follows that the release of pollutants some distance above the Earth – e.g. from aeroplanes or from high chimneys – may

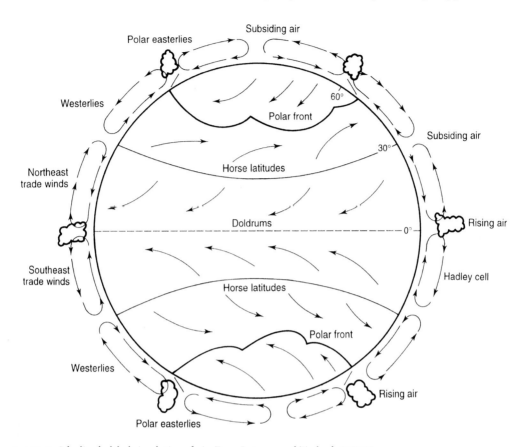

FIGURE 3.4 *Idealized global circulation of air. From Lutgens and Tarbuck (1992).*

lead to long-distance transport. Pollution problems may then become 'global' rather than local. Pollutants may be brought down by rain or snow or may be transferred to land surface by **dry deposition** at distances far removed from their original point of release. Dry deposition involves the direct absorption of gaseous components of air into surface waters or land surface. The presence of significant levels of organochlorine compounds in the arctic has been attributed to a 'cold condensation' effect. This movement of molecules from one phase of the environment to another needs to be considered when constructing models which attempt to predict the distribution of pollutants through different compartments (see section 3.3).

The discussion so far has been concerned with the movement of pollutants that are released into the atmosphere. Brief mention should also be made of molecules that are generated by chemical reactions in the atmosphere. As mentioned earlier, some of the pollutants are generated as a result of the interaction of chemicals released from internal-combustion engines, especially under conditions that give rise to photochemical smog. Ozone is generated from molecular oxygen in the stratosphere under the influence of solar radiation. Most ozone production occurs in the equatorial zone, after which it diffuses out towards the polar regions, giving rise to a more or less continuous ozone layer in the stratosphere. In the 1980s, it was discovered that a hole had appeared in the ozone layer above the South Pole. This was subsequently attributed to the destruction of ozone when it interacts with CFCs – volatile pollutants that can diffuse into the stratosphere.

3.4 *Models for environmental distribution of chemicals*

Models have been constructed which attempt to describe, in mathematical terms, the movement and distribution of chemicals between different compartments of the environment (descriptive models). Sometimes they can also be used to predict the movement and distribution of environmental chemicals (predictive models). Broadly speaking, the models are either thermodynamic or kinetic. Models of the former type do not consider the dimension of time – they are concerned with the distribution that will be found when a **thermodynamic equilibrium** is reached. Kinetic models, on the other hand, are concerned with the *rates* at which processes of transfer or transformation occur. That is to say, they include a time factor.

For the purposes of modelling, the environment can be divided into compartments physically distinct from one another and separated by phase boundaries (e.g. air–water, water–gas, etc.). At the simplest level, the distribution of a chemical between two phases **at equilibrium** is described by a partition coefficient. This represents a simple thermodynamic model in which no time factor is involved. An example of this is the octanol–water coefficient (K_{ow}) described in section 3.1.

True equilibrium states are usually found in closed systems in which the molecule(s) under consideration are not entering or leaving the system. This is not typical of the natural environment, where the systems under consideration are usually 'open' and pollutants are entering and leaving and undergoing chemical transformation and/or biotransformation. Here, the partition coefficients can describe the distribution of a pollutant between two phases, so long as the system is in a **steady state** in which the concentrations of a pollutant in the phases under consideration are constant and do not change with time. Such would be the case with the water of a river carrying a constant concentration of a lipophilic pollutant over a sediment high in organic matter. Some pollutant would partition into the sediment from the water, but an equivalent quantity would partition from the sediment into the water. Thus, the concentration of pollutant in sediment and its ambient water would remain constant. In practice, the distinction between equilibrium and steady state is not very important in the present context because, in both cases, partition coefficients effectively describe the distribution of a pollutant between two adjacent compartments.

In attempting to describe the distribution of chemicals through several compartments over relatively large areas, some success has been achieved by the use of fugacity models. These are, again, thermodynamic models. They are based on physicochemical properties of chemicals, which determine distribution, and on environmental variables such as temperature, pH, quality and quantity of light and water and air movement. The environmental variables are complex and are only predictable to a very limited degree. For this reason, fugacity models have had only very limited success when used in a predictive way. However, they have been useful as evaluative models, which describe the environmental distribution of pollutants under defined conditions (Calamari and Vighi, 1992). One virtue of this approach is that it does give some ranking order among a group of chemicals with regard to their tendency to move into air, sediments and other compartments of the environment.

Fugacity models to describe the distribution of environmental chemicals were first introduced by Mackay (1991; see also Bacci, 1993). The underlying principle is that fugacity is a measure of the tendency of a molecule to escape from one particular phase or compartment into another. It is measured by the same dimensions as pressure. When considering the distribution of a chemical through several adjoining phases, equilibrium is reached when the chemical has the same fugacity in all phases.

In any one phase,

$$f = C/Z$$

where C is the concentration of a chemical in the phase, Z is the fugacity capacity constant and f is fugacity.

Considering now a two-phase system in equilibrium:

$$f_1 = f_2$$

where f_1 and f_2 are fugacities in phase 1 and phase 2 respectively.

Thus,

$$\frac{C_1}{Z_1} = \frac{C_2}{Z_2}$$

or

$$\frac{C_1}{Z_1} = \frac{C_2}{Z_2} = K_{12}$$

where $K_{1,2}$ is the partition coefficient for the chemical between phases 1 and 2; C_1 and C_2 are concentrations in phases 1 and 2; and Z_1 and Z_2 are fugacity capacity constants in phases 1 and 2. Thus, the partition coefficient is the ratio of the fugacity capacity constant for the two compartments.

The distribution of a gas between air and water provides an example of how the model works. Here, the fugacity in air corresponds to the partial pressure of the gas (P_a). This is a measure of the 'escaping tendency'. The distribution of a gaseous pollutant between air and water is described by Henry's constant (H):

$$\frac{\text{conc. of the gas in water } \left(W\right)}{\text{partial pressure of gas } \left(P\right)} = H$$

or

$$\frac{W}{H} = P$$

Fugacity capacity constants (Z values) can be calculated for different compartments of the environment. The higher the Z values, the higher the expected concentrations in the compartment in question.

Fugacity models represent the environment as a number of compartments of known volume in which an equilibrium can be established. These compartments are described as 'units of world'. The compartments defined include 'air', 'lake water', 'soil', 'sediment' and 'biota'. Biota have been subdivided into animals and plants. An example of the distribution of environmental chemicals as described by a fugacity model is shown in figure 3.5.

It should be emphasized that these models are, at best, only of limited **predictive** value, as currently used. The inability to predict various environmental parameters such as temperature and wind speed – and the fact that chemicals are seldom at equilibrium or in the steady state – has placed a limit upon their effectiveness.

Until now this discussion has been restricted to **thermodynamic** models for systems at equilibrium or in the steady state – or when they approximate to one or other of these situations. **Kinetic** models have also been utilized and these are concerned with the **rates** at which processes of transfer or transformation occur. At the simplest level, the rate of transfer from one compartment to another follows 'first or-

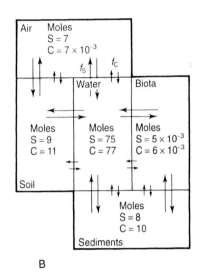

FIGURE 3.5 *The state of the different compartments of the environment is shown at two different times. In A, the two insecticides are present only in water. In B, they have moved from water into all the neighbouring compartments to achieve equilibrium. The arrows indicate direction of movement in accordance with fugacity values. Thus, in A there is only movement away from the water compartment. In B, there is equal movement in either direction on all phase boundaries because the system is in equilibrium. The number of moles (gram molecular weights) is indicated for the two insecticides. S, sulfotep; C, chlorfenvinphos. f_s and f_c refer to fugacities of sulfotep and chlorfenvinphos respectively. The long arrows are for f_s, the short ones for f_c. The major difference between the two compounds in distribution is the greater tendency for sulfotep to 'escape' from water to air. It has a higher vapour pressure (measure of fugacity). Reproduced from Calamari and Vighi (1992) with permission from John Wiley & Sons Ltd.*

der' kinetics and is described by the following equation:

$$r = -kC$$

where r is rate of transfer, k is rate constant and C is the concentration of the chemical in the phase (compartment) from which it is escaping.

Considering a pollutant that moves between compartments A and B by diffusion, a constant concentration will be reached after a certain time when equilibrium is reached. Then:

$$r_{A \to B} = r_{B \to A}$$

but

$$r_{A \to B} = -k_{AB} C_A$$

and

$$r_{B \to A} = -k_{BA} C_B$$

where k_{AB} and k_{BA} are rate constants for movement from A→B and B→A, respectively, and C_A and C_B refer to concentration in compartments A and B respectively.

It follows from this that:

$$\frac{C_A}{C_B} = K_{AB} = \frac{k_{BA}}{k_{AB}}$$

where K is the partition coefficient between compartment A and compartment B. In other words, the partition coefficient is the ratio of the rate constants at equilibrium. Much more complicated kinetic equations than these have been developed, but they lie outside the scope of this book.

To summarize, kinetic models can be used for environmental modelling. In theory, they have an advantage over thermodynamic ones: they can be used to describe the distribution of chemicals between compartments under conditions far removed from equilibrium or steady state. However, this approach has yet to be successfully developed.

3.5 *Summary*

This chapter has dealt with the long-range movements and global transport of chemicals after their release into the environment. More localized movements involving biotic factors, for example along food chains or in soil, are described in Chapter 5. The movement of chemicals depends both on their own properties and upon environmental factors. Polarity, partition coefficients, vapour pressure and molecular stability are all properties of chemicals that can influence movement and distribution in the environment. Temperature, wind speed, circulation of air masses and movements of surface waters are also critical environmental factors.

In surface waters, pollutants are transported by rivers either in the dissolved or in the particulate state and may later accumulate in lakes or in the estuaries to which they run. Ocean currents can transport pollutants over large distances. In the air, chemicals may exist in the vapour state or associated with particles or droplets. In the vapour state, movement by diffusion can be very important, as in the case of the movement of CFCs into the ozone layer. Also, pollutants can be transported over large distances by circulating air masses, leading to deposition on water or land surfaces far removed from their original point of release.

There is a growing interest in the development of evaluative and predictive models for the distribution of chemicals through the different compartments of the environment. Fugacity models are one example, depending on quantification of the 'escaping tendency' of chemicals from one environmental compartment to another.

3.6 *Further reading*

BACCI, E. (1993) *Ecotoxicology of Organic Contaminants.* Describes models for the environmental distribution of pollutants.

CALAMARI, D. and VIGHI, M. F. (1992) Describes the use of fugacity models.

CROSBY, D. G. (1998) *Environmental Toxicology and Chemistry.*

DIX, H. M. (1981) *Environmental Pollution.* Gives a wide-ranging account of the distribution and movements of pollutants in air and water.

MACKAY, D. (1991) *Multimedia Environmental Models: the Fugacity Approach.*

SCHWARZENBACH, R. P. *et al.* (1993) *Environmental Organic Chemistry.*

TUREKIAN, K. K. (1976) *Oceans,* 2nd edn. Contains a description of major ocean currents.

WAYNE, R. P. (1991) *Chemistry of Atmospheres,* 2nd edn. An authoritative and readable account of movements in the atmosphere and of photochemical reactions.

CHAPTER **4**

The fate of metals and radioactive isotopes in contaminated ecosystems

4.1 *Introduction*

Four factors control the fate of inorganic pollutants in contaminated ecosystems. These are (i) localization, (ii) persistence, (iii) bioconcentration and bioaccumulation factors and (iv) bioavailability.

4.1.1 LOCALIZATION

A pollutant is toxic when its concentration exceeds a threshold value in a particular environmental 'compartment'. The ultimate compartment is the whole planet, but compartments can be individual organisms or as small as single cells or even organelles within cells (see figure 8.1).

It has been claimed that 'the solution to pollution is dilution'. Tall chimneys operate on the 'safe dilution' approach to discharges to the environment. For example, pollution from the nickel smelting works at Sudbury, Canada, caused severe ecological disruption to the surrounding countryside. The 'solution' was to increase the height of the chimney so that metal particulates were carried further from the factory. Although the total amount of pollutants discharged remains the same, the concentration locally has markedly reduced, so that plants have begun to recolonize the vicinity of the factory, but the 'solution' has added to the acid rain of eastern North America.

At the other end of the scale, at the cellular level, organisms may compartmentalize potential toxins in insoluble deposits to prevent interference with essential biochemical reactions in the cytoplasm. For example, the epithelium of the midgut of most invertebrates contains metal-rich granules which act as intracellular sites of storage detoxification (see figures 8.4–8.6).

4.1.2 PERSISTENCE

Metals are non-biodegradable and do not break down in the environment. However, there is formation and degradation of specific compounds such as methyl mercury. Once metals get into soils or sediments, they have long residence times before they are eluted to other compartments.

Radioactive isotopes of metals decay exponentially, and persistence is dictated by the half-lives of the individual isotopes (table 1.4). High-level waste from nuclear reactors is extremely persistent as it includes isotopes with half-lives of many thousands or even millions of years.

4.1.3 BIOCONCENTRATION AND BIOACCUMULATION FACTORS

Some inorganic pollutants are assimilated by organisms to a greater extent than others. This is reflected in the **bioconcentration factor** (**BCF**), which can be expressed in the following manner:

$$BCF = \frac{\text{conc. of the chemical in the organism}}{\text{conc. in the ambient environment}}$$

For a terrestrial organism, the ambient environment is usually the soil. For an aquatic organism, it is usually the water or sediment. With inorganic chemicals, the extent of long-term **bioaccumulation** depends on the rate of excretion (see Chapter 5). Thus bioaccumulation of cadmium in animals is high relative to most other metals as it is assimilated rapidly and excreted slowly (for examples, see Chapter 11). If an organism exhibits a high bioconcentration factor for a particular substance, this may be the result of its biochemistry. For example, animals with a calcareous skeleton, exoskeleton or shell take up lead and/or stron-

tium to a greater extent than those without because these two substances follow similar biochemical pathways to calcium for which the organisms has evolved a high assimilation efficiency.

4.1.4 BIOAVAILABILITY

Another reason for a high bioconcentration factor may be that the substance in question is more bioavailable than one with a low bioconcentration factor. Methylated mercury is taken up more readily than the unmethylated form (Wolf *et al.*, 1998). pH has a marked effect on the solubility of metals in soils and water. If the pH declines (for example because of acid deposition), some metals become more soluble than others and hence more bioavailable. Aluminium is highly insoluble at normal to slightly acidic pH, but below about pH 4.5 its solubility increases dramatically and it becomes the most important factor responsible for fish kills in acidified lakes (see Chapter 14).

4.1.5 'COCKTAILS' OF INORGANIC POLLUTANTS

The subjects of synergism and antagonism between pollutants are dealt with extensively in Chapter 10. However, it is worth discussing one aspect of mixtures of pollutants which is often overlooked, i.e. the relationship between the relative toxicities of pollutants to organisms and their relative concentrations in the field (see also the discussion of risk assessment in Chapter 6). For example, negative effects of cadmium contamination of the diet of soil invertebrates on survival, growth and reproduction can be detected at about one-tenth of the concentration (by weight) at which they occur with additions of zinc to the diet. Consequently, if the concentration of zinc is 10 times that of cadmium in the diet then toxicity will be due to both metals equally. However, in regions contaminated by the two metals either from past mining activity or smelting, zinc is almost always present at about 50 times the concentration of cadmium in soils or on vegetation. Thus, zinc is responsible for toxic effects in primary consumers in these situations (Hopkin and Spurgeon, 2000). Nevertheless, because cadmium has a higher bioconcentration factor in most primary consumers than zinc, predators at the next stage in the food chain may be exposed to a zinc–cadmium ratio of less than 10 and hence be poisoned by cadmium rather than zinc.

4.2 *Terrestrial ecosystems*

4.2.1 INTRODUCTION

In terrestrial ecosystems, the soil may be contaminated with metals and radioactive isotopes as a result of previous industrial, mining or other activity, or the contamination may be due to deposition from above (figure 4.1). Concerning the latter, this can be from agricultural practices such as the application of metal-containing pesticides or metal-contaminated sewage sludge, or as wet or dry deposition from smelting activity, lead-containing car exhausts, atmospheric nuclear weapons testing or accidents such as Chernobyl.

4.2.2 METALS

Most geological deposits of metals which became exposed at the surface because of weathering were worked out in previous centuries. This, and more recent mining activity, has left a legacy of contaminated sites in which concentrations of metal can be extremely high.

Because of the long residence times of metals, mines that have been disused for many years may have a very sparse cover of vegetation (figure 4.2A). Those plants which manage to survive are often metal-tolerant strains which are generically distinct from their non-tolerant ancestors (see Chapter 13). Rehabilitation of such areas is difficult. Nowadays, the most widely used method is to 'cap' the contaminated deposit with an impermeable layer, then to cover this with top soil on which trees can be planted (figure 4.2B). Rain falling on the soil flows over the impermeable layer to the edges of the deposit rather than through the metal-contaminated material. This approach greatly reduces the flow of metal-contaminated liquid to groundwater.

Soils may be contaminated at the surface from several sources. Before the development of synthetic organic chemicals, metal-containing pesticides were widely used. In the nineteenth century, it was standard practice to spray 'Bordeaux mixture' in gardens and on crops to control pests, particularly on grape vines, as the name of the mixture would suggest. Bordeaux mixture contains copper and is still widely used in the tropics as a fungicide (Lepp and Dickinson, 1994). Arsenic, lead and chromium were used also and it is still possible to detect elevated levels of these metals in garden soils of old houses.

One method of disposal of sewage sludge is to spread the waste on agricultural fields (the source of the term 'sewage farm'). However, because drains which 'supply' sewage treatment works also take industrial waste, concentrations of metals in the sludge can be very high. The high organic matter content of sewage has a powerful binding capacity for metals which leach very slowly down the soil profile. The number of applications of sludge that can be made into farmland is restricted by the build-up of metals in soils (Alloway and Jackson, 1991).

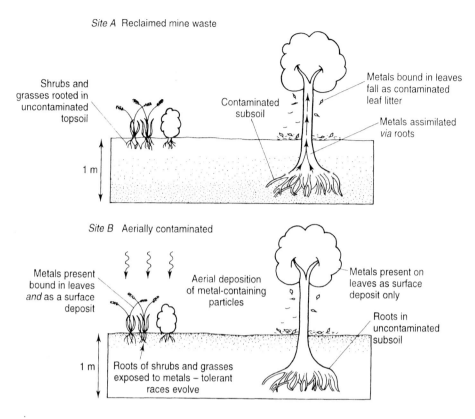

FIGURE 4.1 *Schematic diagrams comparing the distribution of metals in (A) a disused mine site 'rehabilitated' by application of uncontaminated topsoil and (B) a site subject to aerial contamination. Reproduced from Hopkin (1989) with permission from Elsevier Applied Science.*

One of the major sources of metal contamination of soils is the combustion of lead-containing petrol. In the United Kingdom, for example, leaded petrol contained about 0.4 g l^{-1} until 1985, when the maximum permitted concentration was reduced to 0.15 g l^{-1} (petrol in some less developed countries still contains as much as 3 g l^{-1} of lead). Leaded petrol was banned in the European Union from 1 January 2000 The extensive use of lead in the past has led to widespread contamination of urban soils (Culbard *et al.*, 1988). Long-range transport of the metal has occurred and elevated levels of lead can be detected in isolated regions far from industrial activity, such as Greenland (Rosman *et al.*, 1993).

The long residence time of lead in soils means that surface layers will remain contaminated with lead for several hundreds of years to come. However, the reduction in emissions to the atmosphere has been mirrored by a decline in surface deposition which, in the United Kingdom, fell rapidly after 1985 (Jones *et al.*, 1991). Lead concentrations in the air of major cities are now less than one-quarter of their values in the early 1980s. Clear evidence of the role of cars in lead contamination has been provided following the collapse of the Berlin Wall. The influx of cars from former East Germany, which ran on leaded petrol, resulted in an increase in the lead content of the moss *Polytrichum*

A

B

FIGURE 4.2 *(A) Parys Mountain, Anglesey, north Wales. During the early nineteenth century, this was the largest copper mine in the world. Mining ceased about 100 years ago, but recolonization by vegetation has been slow because of the very high concentrations of copper in surface soils. (B) Rehabilitation of mining waste at a disused copper mine in the Gusum area, Sweden. The spoil tip is being capped with an impermeable layer before landscaping with a 2-m layer of topsoil on which trees will be planted. Photographs by Steve Hopkin.*

formosum, which was monitored throughout the political change (Markert and Weckert, 1994).

In contrast to pollution from cars, deposition of metals from smelting activity tends to be fairly localized. One of the best studied sites in the world is the region surrounding a primary lead, zinc and cadmium primary smelting works at Avonmouth near Bristol, south-west England [see Hopkin (1989) and Martin and Bullock (1994) for detailed descriptions of the area]. In the close vicinity of the factory, concentrations of lead, zinc and cadmium in surface soils are at least two orders of magnitude higher than normal background levels. Significantly elevated levels of cadmium in soils can be detected up to 30 km downwind of the plant. The main effect of this heavy aerial deposition of metals is a reduction in the decomposition rate of dead vegetation which accumulates on the surface as a thick layer. Organisms such as earthworms, woodlice and millipedes, which are responsible for the initial fragmentation of leaf litter, are absent as a result of the metal contamination of their diet.

The mobility of metals in soils is dictated largely by the clay content, amount of organic matter and pH. In general, the higher the clay and/or organic matter content and pH, the more firmly bound are the metals and the longer is their residence time in soil. One of the effects of acid deposition in Europe has been 'forest dieback', which is due at least partly to nutrient deficiency (mainly magnesium). Essential elements become more mobile in acidified soils and are leached to lower soil layers to which the roots of the trees cannot penetrate (Berggren *et al.*, 1990).

In Avonmouth soils, the metals exhibit the 'classic' profile of decreased concentration with depth (figure 4.3). However, in 1976 a taller chimney was built to vent the sulphuric acid plant on the site and the pH of soils downwind decreased because of higher acid deposition. The mobility of metals increased and a 'progressive wave' of metals passed down through the soil profile (figure 4.3).

4.2.3 RADIOACTIVITY

Contamination of soils with radioactive material is a relatively recent phenomenon as most of the elements involved did not exist naturally before the development of nuclear weapons and reactors. Some regions of the world where bombs were tested, such as the Australian and Nevada deserts, are still heavily contaminated. Any 'clean up' will have to involve removal of the surface soil, but then the problem arises of what to do with the radioactive material that is removed.

The production of nuclear energy has a good safety record relative to other methods of energy production. However, there are a number of well-publicized examples of environmental contamination. Perhaps the best known is the accident at the Chernobyl complex on 26 May 1986 when one of the reactors caught fire, eventually, releasing half of its contents to the atmosphere (Edwards, 1994). Most of Europe was affected to some extent by fallout of radioactivity, which was most severe to the north-west of the site. In Byelorussia (on which 70% of the total fallout from the reactor fell; Lukashev, 1993), large areas of the country are still heavily contaminated and restrictions on certain agricultural practices remain in force (figure 4.4).

The effects of the Chernobyl fallout outside the former Soviet Union have been most persistent in Scandinavia and upland areas of north-west Europe. The vegetation in these nutrient-poor regions is adapted to retain and recycle essential elements. Metal pollutants that are deposited from the atmosphere pass down through the soil profile extremely slowly. In Cumbria, in north-west England, sheep on upland hill farms became contaminated with ra-

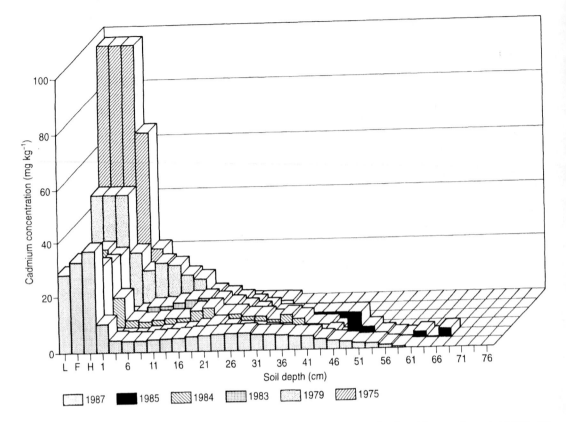

FIGURE 4.3 *Concentrations of cadmium (mg kg⁻¹ dry weight) in Hallen Wood soil profiles over the period 1975–87. Hallen Wood is 3 km north-east of a primary cadmium, lead and zinc smelting works at Avonmouth, south-west England. Each value for the mineral soil represents analysis of a block of soil collected at depths of 0–1 cm, and then at 2.5-cm intervals to the final depth. The profiles show two main features. First, a reduction over time in concentration in the litter (L), primary (F) and secondary (H) layers. Second, a progressive wave of cadmium moving down the profile. The increased mobility of cadmium was due to increased acid deposition in the woodland following construction of a tall chimney at a sulphuric acid plant at the smelting works in the mid-1970s. Reproduced from Martin and Bullock (1994) with permission from John Wiley & Sons.*

dioactive caesium for several years and were prevented from being sold for human consumption (the maximum permissible level was 1000 Bq kg⁻¹). Lambs were moved to lowland pastures before slaughter, where their radio-caesium burden was rapidly lost via faeces (Crout *et al.*, 1991). The caesium passed much more rapidly down the soil profile in those lowland fields than on the hills.

In Sweden, one of the areas most contaminated was the region occupied by the Saami community, where ¹³⁷Cs levels in reindeer reached a mean of more than 40 000 Bq kg⁻¹ (Åhman and Åhman, 1994). Since the disaster, the radioactivity of the reindeer has declined slowly and exhibited a marked seasonal fluctuation (figure 4.5). This is correlated with the change in diet from summer to winter. During summer, reindeer feed mainly on grass, herbs and leaves, which have a low radiocaesium content. During winter, lichens are an important part of the diet (up to 60%). Lichens have a radiocaesium content more than 10 times that of vascular plants from the same region. This

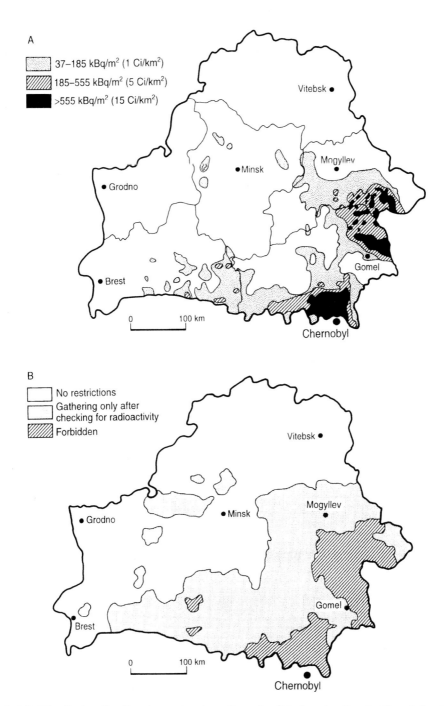

FIGURE 4.4 *(A) Distribution of radioactive caesium in surface soils of Byelorussia after the Chernobyl accident. (B) Areas of Byelorussia with restrictions for gathering mushrooms. Reproduced from Lukashev (1993) with permission from Elsevier Science Ltd, Pergamon Imprint.*

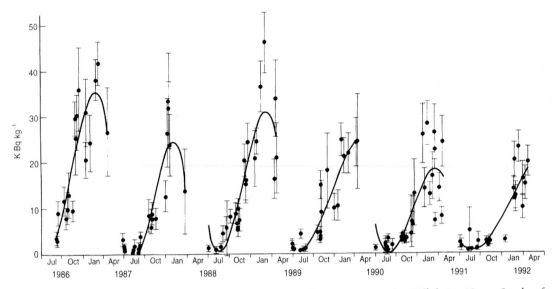

FIGURE 4.5 *Activity concentrations of* ^{137}Cs *in reindeer from the Saami community, Vilhelmina Norra, Sweden, from 1986 to 1992. Mean ± standard deviation from separate slaughter occasions (n = 10–825 animals). Reproduced from Åhman and Åhman (1994) with permission from the Health Physics Society (Lippincott, Williams & Wilkins).*

FIGURE 4.6 *Mercury (Hg) distribution in sediments of the Paraiba do Sul river, estuary and adjacent continental shelf, Rio de Janeiro State, south-east Brazil. Reproduced from Pfeiffer* et al. *(1989) with permission from Elsevier Science Publishers.*

example illustrates the importance of regular monitoring of the biota after a pollution incident as the movement and pathways of transport may be more complicated than at first thought.

4.3 *Aquatic systems*

The ultimate 'sink' for metals is the ocean.

However, because of the massive dilution of contaminants that occurs, it is difficult to prove that metals in the open sea are having a significant effect on the biota. Indeed, there is evidence that some metals, such as iron, are limiting nutrients for phytoplankton in the open ocean (Coale *et al.*, 1996). Estuaries are a different story, however, and many are grossly polluted, particularly those fed by rivers that pass through heavily industrialized regions or regions of mining activity (figure 4.6) (Grant

FIGURE 4.7 *Mercury concentrations in bed sediments of the Yatsushiro Sea (1975) and Minimata Bay (1973). Reproduced from Kudo et al. (1980) with permission from Elsevier Science Ltd, Pergamon Imprint.*

and Middleton, 1990; Bryan and Langston, 1992). When polluted fresh water reaches the sea, the flow rate slows down, suspended sediments settle on the bottom and dissolved metals are precipitated (see Chapter 3). Even if the discharges to rivers are cleaned up, the estuaries they feed may continue to be affected for many years because of remobilization of past sediment contamination. Such an effect occurred in Minimata Bay in Japan, where sediments were heavily contaminated with mercury (figure 4.7). A new quay was built in the 1970s at which much larger ships could dock. The action of their propellers has remobilized mercury-contaminated sediment, which could be detected much further out to sea than previously (Kudo *et al.*, 1980).

In water, the solubility of metals is strongly pH dependent. Streams draining mining areas are often very acidic and contain high concentrations of dissolved metals with little aquatic life. However, as the stream becomes diluted with uncontaminated water further down-stream, the pH rises and metals are precipitated onto the bed. This is the case with the stream that drains Parys mountain (figure 4.2A), where heavy deposits of iron and copper coat the submerged rocks, giving the bed of the stream a bizarre orange–brown coloration.

Acid deposition may be 'stored up' in the snow and released as a sudden 'pulse' of acidity during the spring thaw (Borg *et al.*, 1989). The resulting decline in pH of as much as one unit causes a sudden increase in the levels of soluble metals in lakes and streams (figure 4.8).

Nowadays, the deliberate release of radioactive waste into the aquatic environment is much more tightly controlled than in the past. Until the 1980s, concrete drums containing radioactive materials were routinely dumped in the ocean until the practice was banned. A bathosphere inspected some of these drums on the floor of the ocean in the early 1980s and they appeared to be intact (Sibuet *et al.*, 1985). However, their long-term integrity must remain in doubt.

FIGURE 4.8 *Variations with time in outlet water pH and concentrations of zinc in Lake 2011, Holmeshultasjön, Sweden. Note the close relationship between 'pulses' of lower pH in the lake and increased levels of dissolved zinc. Reproduced from Borg* et al. *(1989) with permission from Elsevier Science Publishers.*

The major source of radioactive contamination of the sea in Europe has been effluent from the Sellafield nuclear reprocessing plant in Cumbria, north-west England. Porpoises resident in the Irish Sea contain significantly higher concentrations of ^{137}Cs and ^{40}K than in other European waters (Berrow *et al.*, 1998). During the 1970s, discharges were high. Indeed, by far the most significant current problem is remobilization of this earlier contamination. Present day discharges are very low. Much of the radioactivity is present in sediment deposited at the time which have become covered with more recent material. Thus, by careful analysis of the different layers, it is possible to date the 'strata' and compare their radioactivity with discharge data for the year in question (Mackenzie and Scott, 1993; Mackenzie *et al.*, 1994). An example of such an approach is shown in figure 4.9.

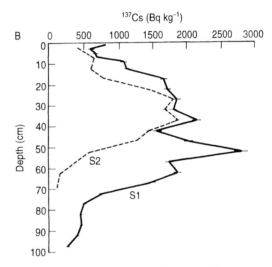

FIGURE 4.9 *(A) Variations with time in the annual quantities of ^{137}Cs discharged from the Sellafield nuclear fuel reprocessing plant. (B) ^{137}Cs concentration profiles for Solway Firth saltmarsh sediment section S1 and S2, north of Sellafield. Reproduced from Mackenzie* et al. *(1994) with permission from Elsevier Science Ltd, Pergamon Imprint.*

4.4 *Summary*

Metals are non-biodegradable pollutants, several of which have become widespread in the environment through industrial activity (mining, smelting, etc.). In aerially contaminated soils, they tend to persist for many years in the surface layers of soil. In aquatic systems, metals may become 'locked up' in bottom sediments, where they may remain for many years. However, if the pH falls, metal solubility increases and they become more mobile. One of the knock-on effects of acid rain is for metals to be transported to lower levels in the soil profile, where they may damage deep rooted plants and contaminate groundwater. Environmental contamination with radioisotopes is a relatively recent phenomenon. Fallout from the Chernobyl disaster was deposited over most of Europe. ^{137}Cs was accumulated in a wide range of organisms because of its tendency to follow the same biochemical pathways as the essential element potassium. Although present levels of release from the Sellafield reprocessing plant are low, there is concern over the remobilization of radioisotopes in sediments deposited in the 1970s, when discharge rates were high.

4.5 *Further reading*

ALLOWAY, B. J. and JACKSON, A. P. (1991) A study of metal contamination of sewage sludge.

BORG, H. *et al.* (1989) A paper which clearly demonstrates the impact on metal solubility of the seasonal dip in pH in Swedish lakes.

COALE, K. H. *et al.* (1996) Evidence for iron deficiency of phytoplankton in the open ocean.

CROUT, N. M. J. *et al.* (1991) The reasons behind the very long residence times of caesium in sheep in north-east England and measures being taken to solve the problem.

HOPKIN, S. P. (1989) *Ecophysiology of Metals in Terrestrial Invertebrates*. Contains general introductory chapters on metals and a detailed account of studies around the Avonmouth metal smelting works [see also MARTIN and BULLOCK (1994)].

JAGOE, C. H. *et al.* (1998) Demonstrates the persistent problem of severe radioactive contamination of fish near Chernobyl.

SIMKISS, K. (1993) A summary of several relevant papers on UK radioactivity after Chernobyl in this issue of the *Journal of Environmental Radioactivity*.

WOLFE, M. F. *et al.* (1998) Comprehensive review of distribution and effects of mercury in the environment.

CHAPTER 5

The fate of organic pollutants in individuals and in ecosystems

The organic pollutants discussed in this book are examples of xenobiotics. A xenobiotic is here defined as a compound which is 'foreign' to a particular organism. This means that it does not play a part in normal biochemistry. By this definition, a chemical which is normal to one organism may be foreign to another. Thus, xenobiotics may be naturally occurring as well as man-made (anthropogenic), and must have existed since early in the evolutionary history of this planet. From an evolutionary point of view, the role of naturally occurring xenobiotics as 'chemical-warfare agents', is of considerable interest. For example, there is much evidence for the evolution of detoxication mechanisms by animals to give them protection against toxic xenobiotics produced by plants. Nearly all of the organic pollutants referred to in Chapter 1 are man-made xenobiotics – which do not occur in nature. It is, however, important to remember that naturally occurring xenobiotics,

for example pyrethrins, nicotine, various mycotoxins etc., will be subject to the same toxicokinetic processes.

Toxicokinetics has relevance to ecotoxicology because it aids the understanding and prediction of the behaviour of organic pollutants within living organisms, as will shortly be explained. On the other hand, the fate of a chemical in an entire ecosystem is a more complex matter, involving movement in soils, surface waters and air and transfer along food chains. Toxicokinetic models are sometimes valuable for the prediction of the fate of chemicals in individual organisms, but more elaborate models would be required if prediction of fate in whole ecosystems were to be attempted. As discussed earlier, some success has been achieved in predicting the distribution of chemicals through major compartments of the environment (Chapter 3). However, prediction of the distribution of a chemical through the different organisms constituting an ecosystem is another matter. It may well be that the whole system is too complicated to lend itself to predictive modelling of this kind.

The general principles of toxicokinetics, as they apply to lipophilic xenobiotics in individual organisms, will now be described. The emphasis will be upon animals, with some reference to plants. Toxicokinetic models will be briefly discussed, making particular reference to their use for predicting bioconcentration and bioaccumulation. Finally, movement in ecosystems will be considered, dealing with terrestrial and aquatic systems separately.

5.1 *Fate within individual organisms*

5.1.1 GENERAL MODEL

The fate of a xenobiotic in an individual organism is represented in figure 5.1. In this figure, an integrated picture is given of the movements, interactions and biotransformations that occur after an organism has been exposed to a xenobiotic. The discussion that follows will focus, in the first place, on the situation that ex-

ists in animals, before drawing attention to some special features of plants. It should be stressed that this highly simplified model identifies those processes which are important from a toxicological point of view. The interplay between them will determine the toxic effect of a pollutant. For any particular chemical, interspecific differences in the operation of these processes will lead to corresponding differences in toxicity between species (selective toxicity).

The model identifies five types of sites – sites of uptake, metabolism, action, storage and excretion, and the arrows identify the movements of chemicals between them. The overall model will now be considered, in outline, before focusing on certain parts of it in more detail.

Once a chemical has entered an organism, four types of site which it may reach are identified, as follows.

1. Sites of (toxic) action. Here, the toxic form of a pollutant interacts with an endogenous macromolecule (e.g. protein or DNA) or structure (e.g. membrane) and this molecular interaction leads to the appearance of toxic manifestations in the whole organism. (The *chemical* acts upon the *organism*.)
2. Sites of metabolism. These are enzymes which metabolize xenobiotics. Usually

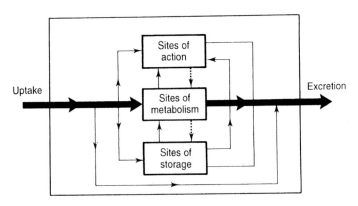

FIGURE 5.1 *General model describing the fate of lipophilic xenobiotics in living organisms. Reproduced from Walker, C. H. (Chapter 9) in Hodgson and Levi (1994) with permission from McGraw-Hill.*

metabolism causes detoxication, but in a small yet highly significant number of cases it causes activation. (The *organism* acts upon the *chemical*.)

3. Sites of storage. Here, the xenobiotic exists in an inert state from the toxicological point of view. It is not 'acting upon the organism', neither is it being 'acted upon'.

4. Sites of excretion. Excretion may be of the original pollutant, or of a biotransformation product (metabolite or conjugate). After terrestrial animals have been exposed to lipophilic xenobiotics, excretion is very largely of biotransformation products, not of original compounds (see section 5.15).

In this simple model, a single box is shown for each of the categories of sites. In reality, of course, there may be more than one type of site in any particular category – and more than one location in the body for any type of site. Thus, a xenobiotic may be stored both in fat depots and in inert membranes. Also, a target site for a neurotoxin (e.g. cholinesterase) may exist in both the central and the peripheral nervous system.

After uptake, pollutants are transported to different compartments of the body by blood and lymph (vertebrates) or haemolymph (insects). Movement into organs and tissues may be by diffusion across membranous barriers or, in the case of extremely lipophilic compounds, by transport with lipids. Uncharged molecules, which have a reasonable balance between oil and water solubility, tend to move across membranous barriers by passive diffusion. This happens if they are not too large (mol. wt < 800), and have an optimal octanol–water partition coefficient (K_{ow}) for doing so (for explanation of partition coefficient, see sections 3.3 and 5.1.2). Some very lipophilic compounds are transported 'dissolved' in lipoproteins. After partial degradation, fragments of lipoprotein are taken into cells such as hepatocytes by endocytosis, carry-

ing the associated lipophilic molecules with them. Most xenobiotics are distributed throughout the different compartments of the body after uptake. Quantitative aspects of this are described in section 5.1.7.

Many of the organic pollutants discussed in this book are highly lipophilic (hydrophobic), i.e. they have high values of K_{ow}. If not metabolized, they will be stored in fat depots or in other lipophilic sites such as membranes or lipoproteins. Such storage of potentially toxic lipophilic xenobiotics may be protective in the short term. In the long term, however, release from storage may occur, and this may lead to toxic effects in the organism. Delayed toxicity may be observed some time after initial exposure to the xenobiotic, as in the case of organochlorine insecticides such as dieldrin.

Because of their marked tendency to move into hydrophobic locations (e.g. membranes, fat depots), xenobiotics with high K_{ow} values are not *directly* excreted in the faeces or urine of terrestrial organisms to any important extent. Their efficient elimination is dependent upon biotransformation to water-soluble metabolites and conjugates (section 5.1.5), which are then readily excreted in faeces and/or urine. Thus, the thick arrow through the middle of figure 5.1 emphasizes the importance of this process for terrestrial animals. With aquatic organisms, however, loss by direct diffusion into the ambient water (e.g. across gills of fish) represents a very important mechanism of excretion for lipophilic xenobiotics.

The model can be subdivided into two parts. The processes of uptake, distribution and metabolism constitute the 'toxicokinetic' component. (In the case of drugs, this would be referred to as the 'pharmacokinetic component'.) Molecular interactions at the site of action are part of the toxicodynamic component (pharmacodynamic component in pharmacology). The operation of toxicokinetic processes determines how much of a toxic compound reaches the site of action (this may be the original xenobiotic or an active metabolite of same). By contrast, the nature and degree of interaction between the toxic compound and the site of action will determine the toxic response that is produced (toxicodynamic component). Sometimes, it is convenient to consider these two elements separately when investigating the mechanisms that underlie toxicity. A xenobiotic may be particularly toxic to a defined species for either or both of the following reasons.

1. The toxicokinetics are such that a high proportion of the active form of the xenobiotic reaches the site of action.
2. The toxicodynamics are such that a high proportion of the xenobiotic that reaches the site of action will interact there to produce a toxic response.

Conversely, another species may be insensitive to a xenobiotic because neither of these components operates in a way that favours toxicity.

Toxicokinetic aspects of the model will now be discussed in more detail. Toxicodynamic aspects will be discussed in Part 2 of the text (Chapter 7), which is concerned with effects of pollutants upon individual organisms.

5.1.2 PROCESSES OF UPTAKE

The most important routes of uptake are summarized in table 5.1.

The movement of organic molecules into the organism is usually the consequence of passive diffusion across natural barriers. This is how passage across plant or insect cuticle, vertebrate skin or membranes lining gut, lungs or tracheae usually occurs. Also, very lipophilic molecules may be absorbed from the gut in association with fat (bulk transport).

The movement of organic molecules across natural barriers by passive diffusion is dependent upon them having optimal solubility properties. To move effectively across such barriers, the molecules must, in the first place, have some affinity for the barrier itself, which is usually lipophilic in character. Also, they must have some affinity for the water which lies to the inside of the barrier. Thus, they should have a reasonable balance between lipid solubility and water solubility. This balance is indicated by the octanol–water partition coefficient (K_{ow}) value (section 3.1).

Octanol is a lipophilic (hydrophobic) solvent that is immiscible with water. For efficient movement across lipophilic barriers, K_{ow} values should not be too different from 1. Values much below 1 indicate high water solubility and very low lipid solubility. Values much higher than 1 indicate very high lipid solubility (lipophilicity) but very low water solubility. The relationship between K_{ow} and rate of movement through a lipophilic barrier is indicated in figure 5.2.

Although the K_{ow} value gives a useful general indication of the likelihood of a molecule being efficiently taken up by passive diffusion, it must be emphasized that there is no 'optimal' K_{ow} which guarantees rapid uptake in *all* situations, and other factors need to be taken into account as well. Thus, the composition and temperature of lipophilic barriers determine their state of fluidity and, therefore, the ease with which molecules can diffuse into them. At low temperatures, lipid bilayers can lose their fluidity, making diffusion through them difficult or impossible.

A further factor needs to be borne in mind with passive diffusion of pollutants which are weak acids or bases. Here, a state of equilibrium exists between charged and uncharged forms, which is determined by the pH of the ambient medium (figure 5.3).

Usually, only the uncharged form will readily cross a lipophilic barrier. Thus, the uptake of weak acids is favoured by low pH, but the uptake of many weak bases is favoured by high

TABLE 5.1 *Major routes of uptake for organic pollutants*

Type of organism	Route of uptake	Sources of pollutant
Terrestrial vertebrates	Alimentary tract	Food and ingested water
	Skin	Contaminated surfaces
	Lungs	Droplets and particles in air
		Vapour; droplets and particles in air
Terrestrial invertebrates	Alimentary tract	Food and water
	Cuticle (insects)	Contaminated surfaces
	Body wall (slugs, worms)	Contaminated environment, e.g. soil
	Tracheae	Droplets and particles in air
Fish	Gills	Pollutants in ambient water, dissolved or suspended
	Alimentary tract	Principally food
Aquatic mammals and birds	Alimentary tract	Principally food
		Small amounts from ambient water or ingested water
Aquatic amphibians	Alimentary tract	Principally food
		Small amounts from ambient water
	Skin	Pollutants in ambient water dissolved or suspended
Aquatic invertebrates	Alimentary tract	Principally food
	Respiratory surfaces	Some from ambient water
		Pollutants in ambient water, dissolved or suspended
Plants	Leaves	Pollutants in droplets or particles*
		Vapours
	Roots	Pollutants dissolved in soil water*

*Important routes of uptake for herbicides and for systemic insecticides and fungicides.

pH. Herbicides which are weak acids (e.g. 2,4-D, MCPA) penetrate plant cuticles rapidly if they are in a medium which has low pH. Within the alimentary tract of mammals, weak acids tend to be absorbed in the stomach (pH 1–2) and weak bases in the duodenum, where the pH is much higher.

Returning to table 5.1, the different types of organism will now be considered separately.

Terrestrial vertebrates and invertebrates take up lipophilic pollutants from the alimentary tract or across the skin or cuticle. Pesti-cides represent a very important category of pollutants in agricultural ecosystems where there can be substantial exposure to potentially toxic compounds by either or both of these routes. In general, uptake across the cuticle of insects is likely to be more important than up-take across the skin of vertebrates. This is be-cause insects are much smaller and have much higher ratios of surface area–body volume than do vertebrates (i.e. they have much more ab-sorbing surface per unit volume). The mobility of the organism is an important factor in deter-

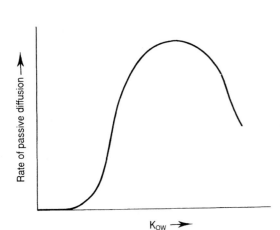

FIGURE 5.2 *Passive diffusion of xenobiotics across a biological membrane. Movement from water through membrane into water on other side.*

OCH$_2$COOH   OCH$_2$COO$^-$

2,4 D Acid   Anion of acid

FIGURE 5.3 *Equilibrium of weak acid.*

mining the rate of uptake across cuticle or skin. With invertebrates, mobile predatory species will tend to come into contact with more pesticide on soil or plant surfaces than will more sedentary species. In the case of vertebrates, movement between different locations in agricultural areas will determine the extent to which they come into contact with pesticides.

Soil organisms such as earthworms, collembola and mites may be continuously exposed to persistent pesticides present in soil. This raises the issue of the availability of compounds which are bound to clay and organic matter. While compounds dissolved in soil water are freely available for uptake, the availability of bound organic compounds is not well understood. The foregoing comments refer particularly to pesticides; however, similar considerations apply to other organic pollutants in terrestrial ecosystems.

Organic pollutants may also be absorbed via the respiratory system of vertebrates and invertebrates. Absorption by this route readily occurs with pollutants which are in the gaseous state. A more complex situation exists with pollutants associated with droplets or particles, which may be deposited in the respiratory tract. Examples include smokes and dusts emitted from factories, domestic premises or internal-combustion engines, pesticidal sprays and dust applied to agricultural land and rain which is contaminated with airborne pollutants. As yet, little is known about the extent to which terrestrial animals take up organic pollutants by this route.

In contrast to their terrestrial counterparts, **aquatic vertebrates and invertebrates** are exposed directly to many pollutants dissolved or suspended in surface waters. Uptake across respiratory surfaces or skin represents an important route of entry for many dissolved aquatic pollutants. However, as with soils, there is still little knowledge of the extent to which pollutants are taken up when they are bound to sediments or suspended particles.

Uptake from food may also be important in aquatic organisms. This is especially true of predatory birds, mammals and reptiles at the top of the marine food chain, which do not appear to take up pollutants directly from water to any important extent. They do not have gills and their skins are not thought to be very permeable to organic molecules (compare the more permeable moist skins of some amphibians).

Description of the routes of uptake of pollutants would be incomplete without a brief mention of the transfer of pollutant from parent to offspring. Pollutants may be transferred across the placenta of mammals into the developing embryo. In birds and reptiles, lipophilic compounds are transported with lipids into the egg and subsequently into the developing embryo. Also, some lipophilic pollutants are secreted into the mammalian milk, and are thus passed to offspring during suckling.

Plants can absorb pollutants across the leaf cuticle and through the roots. These processes are well characterized for translocated herbicides and for systemic insecticides and fungicides (Hassall, 1990). Gases can be absorbed through stomatal openings. Movement through leaf cuticles is by passive diffusion, and is dependent on the K_{ow} of pollutants.

5.1.3 PROCESSES OF DISTRIBUTION

In vertebrates, absorbed pollutants may travel in the blood stream and, to a lesser extent, in the lymph. Where absorption occurs from the gut, much of the absorbed pollutant will initially be taken to the liver by the hepatic portal system. Commonly, a high proportion of the circulating pollutant will then be taken into hepatocytes (first-pass effect). Entry into hepatocytes may be by diffusion across the membrane or by co-transport with lipoprotein fragments which are taken up by endocytosis. Absorption via lungs or skin may lead to a somewhat different initial pattern of distribution as the blood will travel first to tissues other than the liver. Within blood and lymph, organic molecules will be distributed between different components according to their solubility properties. Highly lipophilic compounds (high values of K_{ow}) will be associated with lipoproteins and membranes of blood cells, with little tendency to dissolve in blood water. Conversely, more polar compounds (low values of K_{ow}) will tend to dissolve more in water and will associate less with lipoproteins and membranes of blood cells.

Movement of organic molecules into the brain is of particular concern in toxicology as this is the site of action of many highly toxic substances (see Chapter 7). To enter the brain, organic molecules must cross the 'blood–brain' barrier. This consists, essentially, of membranes which lie between blood plasma and the brain. In general, lipophilic compounds can cross this barrier, but charged molecules (very low K_{ow}) cannot.

In invertebrates, movement of organic pollutants is in the haemolymph. Otherwise, distribution follows a course similar to that described for mammals.

Plants transport absorbed pollutants either in the phloem (symplastic transport) or in the transpiration stream of the xylem (apoplastic transport). Molecules entering via the roots may be transported to the aerial parts of the plant in the xylem. Molecules entering via the leaf may be taken to other parts of the plant in the phloem. (Hassall, 1990).

5.1.4 STORAGE

Xenobiotics may be located in positions where they are not able to interact with their sites of action and they are not subject to metabolism. Of particular importance are lipophilic (hydrophobic) environments, especially fat depots, but also lipoprotein micelles and cell membranes which lack sites of action or enzymes that can metabolize the xenobiotic in question. (It should be emphasized that no general rule applies here – an inert membrane for one xenobiotic may contain a site of action or a detoxifying enzyme for another.) Many lipophilic xenobiotics can be stored in depot fat and some are stored as a consequence of binding to proteins. An example of the latter type of storage is the binding of the rodenticide warfarin to serum albumin.

The amount of depot fat in vertebrates is subject to considerable variation. At times, when food is plentiful, fat depots may be built up – in some cases to the point where they account for 20% or more of the total body weight. An example of this is puffin (*Fratercula arctica*) chicks which are 'balls of fat' when they leave the nesting burrows and find their way to the sea. At other times – when food is scarce, during illness, egg laying or migration – fat depots are run down to provide energy. The rapid mobilization of depot fat will bring a rapid release of stored pollutants into the bloodstream which will then find their way to sites of action and metabolism. Thus, in the short term, storage of lipophilic pollutants in fat depots may minimize their toxic effect. Conversely, longer-term release from storage may lead to toxic effects.

The problem of delayed toxicity was well illustrated in studies on the effects of the insecticide dieldrin on Eider ducks (*Somateria mollissima*) in the Netherlands during the 1960s (Koeman and van Genderen, 1972). During the breeding season, female Eider ducks died of dieldrin poisoning, males did not. At the onset of the breeding season, females build up considerable fat depots, which are then utilized during the course of egg laying (large clutches of eggs are produced). In the present example, significant quantities of dieldrin were laid down in fat depots. Because of the size of the fat depots, the concentrations in other tissues were not particularly high. The dieldrin concentrations in fat were 10–20 times greater than in tissues such as liver or brain. With the mobilization of fat depots, however, blood dieldrin levels rose rapidly from about 0.02 μg ml^{-1} to 0.5 μg ml^{-1}, and birds were poisoned because of the corresponding increase of concentrations of the insecticide in the brain. Similar effects are to be anticipated in other situations where mobilization of fat depots leads to the relatively rapid release of stored lipophilic compounds of high toxicity. Other organochlorine insecticide residues (e.g. *p,p'*-DDT, heptachlor epoxide), organomercury compounds, polychlorodibenzodioxins (PCDDs) and polychlorinated biphenyls (PCBs) can be redistributed in a similar fashion. Also, such compounds may cause delayed toxicity during illness, starvation or migration.

The importance of storage of persistent lipophilic compounds in depot fat is evident from many analyses of vertebrate samples from terrestrial, marine and freshwater ecosystems. A wide range of organochlorine compounds can be identified using techniques such as capillary gas chromatography. Concentrations are particularly high in predators at the top of food pyramids (see later sections of this chapter).

5.1.5 METABOLISM

The enzymic metabolism of most lipophilic xenobiotics occurs in two phases (see Timbrell, 1999):

The initial (phase I) biotransformation involves oxidation, hydrolysis, hydration or reduction in the great majority of cases and normally leads to the production of metabolites which contain hydroxyl groups. The hydroxyl group introduced in this initial step is necessary for most, but not all, of the subsequent conjugation reactions which constitute the second stage (phase II) of biotransformation. These two phases lead to a progressive increase in water solubility, moving from a lipophilic xenobiotic to a more polar metabolite and then to an even more polar conjugate. Most conjugates are negatively charged (anions), have appreciable water solubility and are readily excreted in bile and/or urine. After exposure of vertebrates and insects to lipophilic xenobiotics, most of the excreted products are conjugated.

The scheme shown above represents a simplification of the real situation. Phase I may involve more than one step. Also, some xenobiotics (especially if they already possess hydroxyl groups) may undergo conjugation directly. The relationship between metabolism and excretion will be explained in the next section (section 5.1.6).

In the great majority of cases, biotransformation leads to a loss of toxicity (detoxication) and is protective to the organism. However, in a small – but highly significant – number of cases, metabolism leads to an increase in toxicity (activation). In particular, oxidation in phase I leads to the production of reactive metabolites which can bind to cellular macromolecules. Oxidation of organophosphorous insecticides such as dimethoate, diazinon, malathion, disyston, chlorpyriphos and many others leads to the production of reactive metabolites (oxons) which can phosphorylate, and thereby inhibit, acetylcholinesterase of the nervous system (Chapter 7). Oxidation of carcinogens such as benzo(*a*)pyrene, aflatoxin and vinyl chloride leads to the formation of reactive metabolites which can bind to DNA (Chapter 7). Thus, relatively inert molecules, which are not themselves able to cause toxic effects, are converted to reactive metabolites having very short biological half-lives, which can cause cellular damage. One of the curious features of biochemical toxicology is that many of the most destructive types of molecules are reactive metabolites which are difficult or impossible to detect because of their short biological half-lives. Proof of their existence and the damage that they cause often depends on identification of the modifications that they cause to cellular macromolecules, e.g. inhibited acetylcholinesterase or damaged DNA (see Chapter 10).

The remainder of this section will be devoted to a brief description of the major enzymes concerned with biotransformation, using organic pollutants as examples.

The major classes of enzyme which metabolize xenobiotics in vertebrates and invertebrates are presented in table 5.2 (Gibson and Skett, 1986). Many of the enzymes responsible for phase I biotransformations are located in the endoplasmic reticulum – notably that of the liver (vertebrates), hepatopancreas, fat body and gut (invertebrates). Lipophilic xenobiotics tend to move into the endoplasmic reticulum, but their more polar biotransformation products tend to partition out into the cytosol. Conjugating enzymes such as sulphotransferases and glutathione-S-transferases located in the cytosol can then conjugate the metabolites. The conjugates are readily excreted in urine and/or bile. Thus, the increase in polarity which results from the sequential biotransformation shown above leads to movement first from membrane to cytosol and then from cytosol to urine and/or bile. Although many of the types of enzymes shown in table 5.2 are also found in *plants*, the activities are usually low in comparison with those found in animals.

Of the enzymes responsible for phase I biotransformations, the **microsomal monooxygenases (mixed function oxidases)** are the most versatile (Waterman and Johnston, 1991; Lewis, 1996). They are found in all vertebrates and invertebrates and are present in the endoplasmic reticulum of a variety of tissues. Low levels have been found in plants. Unlike other phase I enzymes, the biotransformations that they catalyse are not restricted to any particular functional group. They are able to metabolize the great majority of lipophilic xenobiotics – so long as they are not too large (large molecules would not be able to fit into the available space adjacent to the catalytic site). The main exception to this rule is presented by highly halogenated compounds such as certain PCBs and PBBs – microsomal monooxygenase does not work effectively against C–halogen bonds (i.e. C–Cl, C–Br, etc.).

Oxidations by the microsomal mono-

TABLE 5.2 *Enzymes that metabolize lipophilic xenobiotics (phase I)*

Name	Principal location	Cofactors	Substrates
Microsomal monooxygenases (mixed function oxidases)	Endoplasmic reticulum of many animal tissues, especially liver of vertebrates, hepatopancreas, fat body and gut of invertebrates	NADPH (NADH) O_2	Most lipophilic xenobiotics of molecular weight < 800
Carboxyl esterases	Endoplasmic reticulum of many animal tissues; also found in cytosol and in serum/plasma of vertebrates	None known	Lipophilic carboxyl esters
'A' esterases	Endoplasmic reticulum of certain cell types of vertebrates; mammalian serum/plasma (associated with high-density lipoprotein)	Ca^{2+}	Organophosphate esters
Epoxide hydrolases	Principally endoplasmic reticulum of animal cells; some in cytosol	None known	Organic epoxides
Reductases (a number of different enzymes can express this activity at low levels of dissolved oxygen, including certain flavoproteins and haemproteins)	Endoplasmic reticulum and cytosol of a number of types of animal cell	NADH/ NADPH	Organonitro-compounds, some organo-halogen compounds, e.g. *p,p'*-DDT

oxygenase system depend upon the activation of molecular oxygen (O_2) after it has been bound to an associated haemprotein, cytochrome P_{450}. Activation is accomplished by the transfer of electrons to bound O_2, which then splits, one atom being used to oxidize the substrate (e.g. organic pollutant) and the other being used to form water. Electrons for this purpose come from NADPH (sometimes from NADH). The overall reaction can be written thus:

$$NADPH_2 + X + O_2 \rightarrow XO + H_2O + NADP$$

where X = substrate.

The haemprotein cytochrome P_{450} exists in many different forms which have contrasting, yet overlapping, substrate specificities (Lewis, 1996; Livingstone and Stegeman, 1998; Nelson. 1998). Some forms of cytochrome P_{450} are readily *inducible*. The appearance of a lipophilic xenobiotic inside the body can trigger the synthesis of more cytochrome P_{450}, leading to more rapid biotransformation of xenobiotics. This is normally protective, enabling the organism to eliminate more rapidly the xenobiotic which originally caused induction. (There are, however, cases where the chemical causing induction is not metabolized by the cytochrome P_{450} that is induced, e.g. highly chlorinated PCBs.)

Although oxidation by cytochrome P_{450} causes detoxication in the great majority of cases, often introducing hydroxyl groups to facilitate conjugation, there are some very important exceptions to this rule. The oxidative desulphuration of organophosphorous insecticides which possess a thion group (e.g. diazinon, dimethoate, disyston and malathion) leads to the formation of oxons, which are active anticholinesterases (see figure 5.4). Some organochlorine insecticides of the cyclodiene group are converted to stable and toxic epoxides (figure 5.4). Thus, aldrin is converted to dieldrin, and heptachlor to heptachlor epoxide. Also, a number of carcinogens are activated by the same system. Polycyclic aromatic hydrocarbons such as benzo(*a*)pyrene (figure 5.4) and aflatoxin B are converted into epoxides which are strongly electrophilic and can bind to DNA. Nitrosamines are converted to methyl radicals and other reactive species when oxidized by cytochrome P_{450}. Vinyl chloride is also activated by cytochrome P_{450}.

Hepatic microsomal monooxygenase (HMO) activity towards xenobiotic substrates varies substantially among different groups, species and strains. In omnivorous and herbivorous mammals, activity is inversely related to body weight (figure 5.5), small mammals tending to have more activity per unit body weight than large mammals. Fish have relatively low HMO activities, with no clear relationship to body weight. Birds have variable activities. Omnivorous and herbivorous species have HMO activities similar to mammals of similar body size (on average a little lower), but fish-eating birds and specialized predators (e.g. sparrowhawk, *Accipiter nisus*) have activities similar to fish (Walker, 1980, 1998a; Ronis and Walker, 1989). However, birds generally, and fish-eating birds in particular, show the same relationships between HMO activity and body weight observed in mammals (Walker, 1980).

These differences are explicable in terms of the detoxifying function of HMO. Terrestrial vertebrates depend upon HMO for the effective elimination of lipophilic xenobiotics, but fish have far less dependence on metabolic detoxication because they 'excrete' uncharged lipophilic molecules into ambient water by passive diffusion. (The same argument can be applied to amphibia which lose xenobiotics by diffusion across skin.) Small mammals have much higher surface area/body volume ratios than large mammals. Consequently, they take

in food – and associated xenobiotics – much more rapidly in order to obtain sufficient metabolic energy to maintain body temperature. Thus, they need higher levels of detoxifying enzymes than large mammals because they take in xenobiotics more rapidly. This argument does not apply to poikilotherms, so it is not surprising that the small amount of HMO possessed by fish is not obviously related to body size - in contrast to the situation in mammals and birds. With specialized predators such as fish-eating birds and raptors such as the sparrowhawk (very largely bird eating), the need for detoxication is small because their food does not contain many xenobiotics. Indeed, the food is similar in composition to the predators themselves, especially with bird-eating predators such as the sparrowhawk and peregrine falcon. Plants, by contrast, contain many compounds which are xenobiotics to animals. Some of these compounds have the function of protecting plants against 'grazing' by animals. Thus, the variations in HMO activity shown in figure 5.5 can be explained in terms of the requirements of different species for detoxication of liposoluble xenobiotics.

Turning now to **esterases**. Aldridge (1953) distinguished between 'A' esterases which hydrolyse organophosphates and 'B' esterases which are inhibited by them.

Carboxyl esterases are examples of 'B' esterases and constitute an important group of detoxifying enzymes which hydrolyse lipophilic carboxyl esters to form acids and alcohols (figure 5.4). They are widely distributed in nature, being found in membranes (especially endoplasmic reticulum) from a variety of tissues in all animals so far investigated. They are also found in cytosol and in vertebrate plasma–serum. A number of different forms have been recognized, showing contrasting yet overlapping substrate specificities. Purified carboxyl esterases from mammalian liver endoplasmic reticulum metabolize both xenobiotic esters and endogenous

FIGURE 5.4 *(i) Phase I biotransformations. MO, microsomal monooxygenase.*

Diazinon

Permethrin

Diazoxon

Benzo(a)pyrene 7,8 oxide

Nitropyrene

FIGURE 5.4 (ii) (Continued).

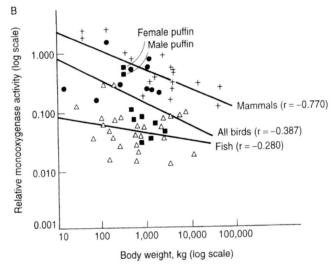

FIGURE 5.5 *Monooxygenase activities of mammals, birds and fish. (A) Mammals and birds. Reproduced from Ronis and Walker (1989) with permission from Academic Press. (B) Mammals, birds and fish. Activities are of hepatic microsomal monooxygenases to a range of substrates, expressed in relation to body weight. Each point represents one species (males and females are sometimes entered separately).*

ones. In mammals, the range of different forms of carboxyl esterases differs considerably between tissues. Liver, for example, has a wide range of forms, but blood and muscle have a smaller number of forms.

'A' **esterases** are enzymes which can hydrolyse organophosphorous triesters and diesters which possess an oxon group (P=O) (figure 5.4ii). These esters are lipophilic in character

and this type of esterase does not appear to hydrolyse organophosphates which are ionized. The known substrates are mainly organophosphorous insecticides, although some forms of 'A' esterase can hydrolyse nerve gases (e.g. soman and tabun). In vertebrates, 'A' esterase activity is found in the endoplasmic reticulum of liver and other tissues. In mammals, activity is also found in serum/plasma, some of it in as-

sociation with high-density lipoprotein. 'A' esterase exists in a number of different forms.
Those forms which hydrolyse organophosphorous insecticides are calcium dependent and
lose their activity in the presence of chelating
agents which bind calcium.

There are some marked differences with regard to 'A' esterases. In contrast to mammals,
birds have little or no serum–plasma 'A' esterase (Walker and Thompson, 1991; Walker *et
al.*, 1991b). Some species of insects have no 'A'
esterase activity.

Epoxide hydrolases are found, principally, in
the endoplasmic reticulum of animals, mammalian liver being a particularly rich source. A
different form of the enzyme is found in
cystosol. Epoxide hydrolases hydrate a wide
range of aromatic and aliphatic epoxides to form

trans diols and do not require cofactors (figure
5.4). Epoxide hydrolases of the endoplasmic
reticulum hydrate epoxides generated by microsomal monooxygenases. In some cases, this
represents a protective function because the
epoxides are strong electrophiles which can form
adducts with cellular macromolecules (see
Chapter 7). Epoxide hydrolases can metabolize
endogenous as well as xenobiotic substrates.
Epoxides of steroids and of insect juvenile hormone are examples of endogenous substrates.
As with monooxygenases and esterases, this
enzyme generates metabolites with hydroxyl
groups, which are then available for conjugation.

Reductases are enzymes which can catalyse
the transfer of electrons to organic molecules
such as nitroaromatic and organohalogen com-

FIGURE 5.6 *Phase II biotransformations.*

pounds (figure 5.4). Whether enzymes operate as reductases or not often depends upon the availability of oxygen. Where oxygen is freely available, it can act as an electron acceptor (i.e. electrons will flow to oxygen rather than to the organic molecule). Both flavoproteins and haemproteins (e.g. cytochrome P_{450}) have been shown to act as reductases when oxygen levels are low. The flow of electrons to oxygen leads to the formation of oxygen radicals (e.g. superoxide anion and hydroxyl radical) which can cause cellular damage. This question will be discussed later. When nitroaromatic compounds are reduced, they are converted into amines. Reduction of halogenated compounds leads to dehalogenation (e.g. conversion of *p,p'*-DDT to *p,p'*-DDD).

Of the phase II enzymes which are responsible for conjugation, **glucuronyl transferases** are largely confined to membranes – especially the endoplasmic reticulum of vertebrate liver. These enzymes catalyse the interaction between lipophilic organic molecules (xenobiotic and endogenous) with labile hydrogen atoms (usually of hydroxyl groups) and glucuronic acid to form glucuronides (see figure 5.6). The glucuronic acid is supplied by a nucleotide cofactor, uridine diphosphate glucuronic acid (UDPGA), which is synthesized in the cytosol. Thus, the glucuronyl transferase enzyme and its associated lipophilic substrate receive the polar cofactor from the hydrophilic environment of the cytosol. Glucuronides exist very largely as anions at cellular pH (approximately 7.4), and move away from the membrane and into the cytosol when they are formed (table 5.3).

Glucuronides are subject to hydrolysis by enzymes termed glucuronidases (e.g. in the alimentary tract), thus leading to a reversal of the reaction shown in figure 5.6 and a release of the hydroxyl compound that was originally conjugated. The consequences of this will be discussed later when discussing excretion. Glucuronyl transferases, like other enzymes concerned with xenobiotic metabolism, exist in a number of different forms.

Sulphotransferases exist in the cytosol of various cell types (especially liver) in a number of different forms. They catalyse the transfer of the sulphate group from the phosphoadenine phosphosulphate (PAPS) to xenobiotic and endogenous substrates that possess a free hydroxyl group (figure 5.6). The cofactor PAPS is generated in the cytosol. The resulting sulphate conjugates exist as anions with appreciable water

TABLE 5.3 *Enzymes that metabolize xenobiotics (phase II)*

Name	Principal location	Cofactors	Substrates
Glucuronyl transferases	Endoplasmic reticulum of many animal cells	UDP-glucuronic acid	Especially organic compounds with free OH groups; also some organic compounds with free –SH or NH_2
Sulphotransferases	Cytosol of many animal cells	Phosphoadenine phosphosulphate	Many organic compounds with free OH groups
Glutathione-S-transferases	Mainly in cytosol of many types of animal cell; sometimes a little in endoplasmic reticulum	Reduced glutathione	Many foreign electrophiles, including some organohalogen compounds and many organic epoxides

solubility. Like glucuronides, sulphate conjugates are subject to enzyme hydrolysis – in this case catalysed by sulphatases.

Glutathione-S-transferases are found in the cytosol of many cell types, especially vertebrate liver. A number of different forms are known in vertebrates and invertebrates. These enzymes catalyse the conjugation of reduced glutathione to a variety of xenobiotics which are of electrophilic character (figure 5.6). This conjugation would proceed naturally in the absence of the enzyme, but very slowly. The enzyme speeds up the reaction by binding the reduced glutathione in close proximity to the xenobiotic.

Conjugations with glutathione are not dependent upon the presence of hydroxyl groups. Important substrates include certain organohalogen compounds and organophosphorous insecticides. The glutathione conjugates so formed often undergo further modification before excretion. In vertebrates, further metabolism of the glutathione moiety leads to the formation of mercapturic acid conjugates, which are usually the predominant excreted forms. Both glutathione conjugates and the mercapturic acids derived from them are anionic in character.

Conjugation in phase II metabolism promotes excretion and, with very few exceptions, has a detoxifying function (however, see section 5.1.6 for further discussion). The conjugating enzymes discussed here are very important in vertebrates and invertebrates. However, a number of other types of conjugates are known, which are often 'group specific'. This is particularly true of peptide conjugation.

Before leaving the subject of metabolism, mention should be made of oxyradical formation. As discussed earlier when considering reductases, electrons can be passed on to molecular oxygen, causing the generation of highly reactive oxyradicals. Examples include the superoxide anion ($O_2 \cdot^-$), the hydroxyl radical (OH·) and hydrogen peroxide (H_2O_2). A description of the routes by which these are formed lies outside the scope of this book. The significant point, in the context of ecotoxicology, is that pollutants may promote the formation of such oxyradicals and that this may lead to cellular damage. Thus, some organic pollutants (e.g. paraquat herbicide and nitroaromatic compounds) may pass electrons on to oxygen to form oxyradicals. Additionally, certain refractory compounds may stabilize superoxide (e.g. PCBs, which combine with cytochrome P_{450} but are not themselves metabolized).

Cells possess enzyme systems which detoxify oxyradicals, e.g. superoxide dismutase, catalase and peroxidase. However, they are not necessarily able to cope with increased rates of formation of oxyradicals caused by the action of organic pollutants. This is an important but difficult area of ecotoxicology, about which far too little is known.

5.1.6 SITES OF EXCRETION

The excretion of xenobiotics and their metabolites and conjugates has been studied in some detail in vertebrates. Far less is known about these processes in invertebrates, insects having received more attention than other groups. This account will be mainly concerned with the situation in vertebrates, with some reference to invertebrates.

Many aquatic vertebrates can 'excrete' lipophilic xenobiotics by diffusion into ambient water. Fish can do this across gills and amphibia, such as frogs, across permeable skin. Aquatic birds, on the other hand, do not have permeable membranes in contact with water; skin and feathers appear not to be readily permeable to pollutants. With aquatic mammals (whales, porpoises and seals), the skin also seems to be relatively impermeable to such compounds.

With terrestrial vertebrates and invertebrates, the effective elimination of lipophilic xenobiotics depends upon converting them into water-soluble metabolites and conjugates which can be excreted [some highly lipophilic compounds are excreted to a limited extent in milk (mammals) or eggs (birds and reptiles and invertebrates)]. Vertebrates excrete these biotransformation products in bile and/or urine.

Considering excretion in urine first; soluble metabolites and conjugates are removed from blood in the glomerular filtrate, which then passes down the proximal and distal tubules of the kidney. During the movement of the filtrate down the renal tubules, some xenobiotics pass into the tubular lumen from the plasma by passive diffusion. Also, there may be some active transport of weak acids and bases across the walls of the proximal tubules into the lumen. There may also be some reabsorption of xenobiotics from the tubular lumen into the plasma. Eventually, urine collects in the bladder and is voided independently of the faeces in the case of mammals, but is combined with faeces in the case of birds, reptiles and amphibia.

Conjugates and metabolites may also be excreted via bile. In the first place, these move across the plasma membrane of the hepatocytes into bile canaliculi (figure 5.7). Bile then passes into the main bile duct and is eventually released into the duodenum. Once in the duodenum, conjugates may pass completely through the alimentary tract to be voided with faeces. As conjugates are usually polar (in most cases they are anions), they are not readily reabsorbed by passive diffusion across the wall of the alimentary tract. If, on the other hand, they are broken down in the gut, e.g. by the action of enzymes such as glucuronidases and sulphatases, the metabolites (or, sometimes, original xenobiotics) so released may be reabsorbed into the blood stream by passive diffusion. This is because they are uncharged and can, therefore, readily cross membrane barriers. The reabsorbed compounds are then returned to the

FIGURE 5.7 *Excretion in bile – transverse section of liver cell showing bile canaliculus.*

liver, reconjugated and the cycle repeated. This process is termed 'enterohepatic circulation'. Some xenobiotic metabolites may be recycled many times before they finally appear in faeces. The process of enterohepatic circulation may have toxicological consequences. Recycled metabolites may have toxic effects. Also, metabolites which are themselves of low toxicity may be transformed into toxic secondary metabolites in the gut (e.g. by microbial action). In mammals, excretion via urine does not raise the problem of enterohepatic circulation or of further biotransformation within the gut.

The extent to which excretion occurs in bile or in urine depends upon the molecular weight of the xenobiotic and the species in question. As noted above, the majority of excreted conjugates are organic anions. When their molecular weights are below 300, excretion is predominantly in the urine. Above molecular weight 600, it is predominantly in the bile. Between these limits, the preferred rate of excretion depends upon the species and upon the molecular weight of the anionic conjugate (figure 5.8). Threshold molecular weights have been proposed for anionic conjugates in different species (figure 5.8). These are weights above which there is likely to be appreciable ($< 10\%$) total excretion in bile. The following figures have been proposed: rat 325 ($\pm$ 50), guinea pig 425 ($\pm$ 50) and rabbit 475 ($\pm$ 50). Thus, rats

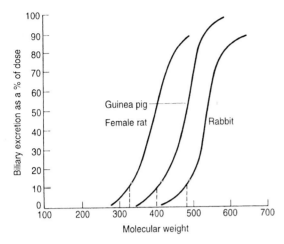

FIGURE 5.8 *Excretion routes of anionic conjugates. Reproduced from Moriarty (1975) with permission from Academic Press.*

show a greater tendency than rabbits to excrete via bile and may, therefore, be more susceptible to the toxic action of certain compounds for the reasons stated above.

The situation in terrestrial invertebrates appears to be similar to that for vertebrates, except that the hepatopancreas or fat body serves the function of the liver.

5.1.7 TOXICOKINETIC MODELS

The processes by which pollutants are taken up, distributed, stored, metabolized and excreted have already been discussed and are summarized in figure 5.1. From a toxicological point of view, a living organism may be divided into different sites, and these may be ascribed different functions – uptake, storage, action, metabolism and excretion. This simple model is purely descriptive and not quantitative. It does not define the rates at which the processes of transfer or biotransformation occur. In toxicokinetics, the central issue is the rates of these processes, the determination of which can lead to the development of both descriptive and

predictive models. In toxicokinetic terms, the organism can be divided into compartments which usually represent particular organs and tissues. Each of these compartments contains a discrete quantity of xenobiotic which is subject to particular rates of transfer and biotransformation. These compartments may or may not correspond to the sites defined in figure 5.1.

The development of multicompartmental models represents an ideal which is seldom realized in practice. Although it is desirable to have maximum information about the kinetics of particular xenobiotics in those compartments of the body that are of toxicological interest, a large amount of work is involved in determining kinetic constants even for one pollutant and one species. In ecotoxicology, such an approach is clearly of little value because concern is about a large range of organisms and organic pollutants. Interest is largely restricted to simple models which treat the whole organism as a simple compartment. The following account will be concerned principally with these.

If an organism is continuously exposed to a constant level of an organic pollutant, the concentration of the pollutant in the whole organism will increase over a period of time (figure 5.9). An aquatic organism may be exposed to a pollutant dissolved in ambient water; a terrestrial organism to a pollutant in its food. Initially, the rate of increase in the tissue concentration of the pollutant will be rapid, but will then begin to tail off and will eventually reach a plateau, so long as a lethal concentration is not reached beforehand. When the pollutant concentration reaches a plateau, the system is said to be in a 'steady state'. The rate at which a pollutant is being taken up is balanced by the rate at which it is being lost. (Loss may be due to metabolism and/or to direct excretion of the original pollutant.) If exposure to the pollutant now ceases, the tissue concentration of the pollutant will begin to fall. In the simplest situation, where the kinetics of loss

FIGURE 5.9 *Kinetics of uptake and loss. When an animal is continually dosed with a chemical, the log concentration increases with time until a steady state is reached. When dosing of the chemical stops, the log concentration falls linearly with time if first-order kinetics apply (as in a one-compartment model).*

can be described by a single rate constant (one compartment model), the rate of loss is proportional to tissue concentration. First-order kinetics apply and this situation can be described by the equation:

$$-\frac{dC}{dt} = KC \text{ or } C_t = C_0 e^{-Kt}$$

where C is the concentration in whole organism, C_0 is the concentration at the time when exposure ceases, C_t is the concentration at time t, K is the rate constant for loss and $-dC/dt$ is the rate of loss of pollutant from the organism. This negative exponential decline in concentration is illustrated in figure 5.10B.

It follows that the log concentration of the xenobiotic (log C) is linear with respect to time (figure 5.10C). The biological half-life (t_{50}) is the time that it will take for the concentration to fall by one-half, and this can be readily determined from this plot of log C versus time.

Sometimes the whole organism does not behave as a single compartment, in which case more complex equations are needed to describe the rate of loss after cessation of exposure. Sometimes the rate of loss is biphasic, and here the loss can be described by a more complex equation derived from a two-compartment model.

Figure 5.9 represents a simplified one-compartment situation, which assumes that the rate of uptake of xenobiotic is constant and that the state of the organism remains the same. The

organism is continuously exposed to a xenobiotic until a steady rate is reached, after which dosing is discontinued. In practice, the rate of uptake is not absolutely constant. For example, if the source of the xenobiotic is food, there is likely to be diurnal variation in the rate of ingestion pattern. Sometimes the activity of enzymes will change because of induction or inhibition, which thus can change the rate at which a xenobiotic is lost – and ultimately the steady-state concentration for any defined rate of uptake of xenobiotic. In reality, the change of xenobiotic concentration with time is usually more complex than the simple situation shown in figure 5.9.

5.1.8 TOXICOKINETIC MODELS FOR BIOCONCENTRATION AND BIOACCUMULATION

Toxicokinetic models have been developed which describe – and most importantly, predict – the degree of bioconcentration or bioaccumulation of organic pollutants by animals continuously exposed to xenobiotics, when in the **steady state** (Moriarty, 1975, 1999; Moriarty and Walker, 1987; Walker, 1990a). It is important to emphasize the value of steady-state models. The tissue concentrations in the steady state represent the maximum levels that are to be expected, given a particular level of exposure. Further, they are not time depend-

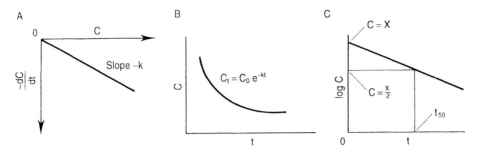

FIGURE 5.10 *Kinetics of loss. C, tissue concentration; t, time (minutes or hours); t$_{50}$, half-life; x, initial tissue concentration. (A) Rate of loss of chemical is proportional to tissue concentration. (B) The tissue concentration of chemical falls exponentially with time. (C) The log of the tissue concentration of a pollutant falls linearly with time.*

ent. Bioconcentration or bioaccumulation factors determined before the steady state is reached are of little value as they only apply to particular periods of exposure and do not indicate the maximum concentration that may be reached.

In this account, the following definitions will be used for biocentration factor (BCF) and bioaccumulation factor (BAF):

$$BCF = \frac{\text{conc. in organism}}{\text{conc. in ambient medium}}$$

$$BAF = \frac{\text{conc. in organism}}{\text{conc. in food (or ingested water)}}$$

For reasons stated above, these factors should be determined when the system is in the steady state. Bioconcentration factors are important in aquatic ecotoxicology, where the ambient medium is a major source of organic pollutants. Bioaccumulation factors are critical in terrestrial ecotoxicology, where food (or, in some cases, ingested water) represents a major source of organic pollutants.

For aquatic organisms (e.g. the edible mussel, *Mytilus edulis*), a close relationship has often been demonstrated between bioconcentration factors for lipophilic organic pollutants and their K_{ow} values (figure 5.11). Here, a simple situation

applies. Most of the uptake and loss of xenobiotic is accomplished by passive diffusion, with metabolism playing only a minor role (exchange diffusion). At the steady state, the BCF is determined by the partition between water and the hydrophobic components of the organism. In the case of non-polar xenobiotics of high lipophilicity, the partition lies very much in the direction of the organism. In other words, high BCF values are associated with high values for K_{ow}. (Sometimes, K_{ow} values are transformed to pK_{ow} values by converting them to $\log_{10}$ values.)

In more complex situations – where other sources of uptake and loss become more important – this simple model will break down. Thus, with fish, BCF values for certain pollutants fall well below the values predicted by a plot such as that shown in figure 5.11 because they are rapidly metabolized, i.e. they are less than would be expected from K_{ow} values. The rate of loss is then greater than would be expected from passive diffusion alone.

For terrestrial organisms, a different approach is required to predict BAFs. In the simplest situation, both uptake and loss may be ascribed to single processes, each of which will be governed by one rate constant (figure 5.9). With strongly lipophilic compounds, food may represent the main source of pollution, whereas metabolism may represent the main source of

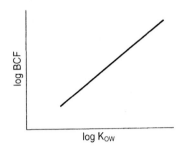

FIGURE 5.11 *Bioconcentration factors and* K_{ow} *values.*

account in the development of predictive models. To date, the only model which meets these criteria is the one which uses K_{ow} values to predict bioconcentration by certain aquatic organisms.

5.2 *Organic pollutants in terrestrial ecosystems*

loss (table 5.4) (Walker 1987, 1990a). Here, it should be possible to develop predictive models for bioaccumulation in the steady state using rate constants for these two processes, so long as metabolism is simple. As yet, however, no such models have been validated.

There is much interest in predictive models which will enable reasonable assessments to be made of the extent to which various organic pollutants may be bioconcentrated or bioaccumulated by different organisms. Such models could be of great assistance in the process of risk assessment for environmental chemicals (Chapter 6). They would need to be simple and cheap to operate. The major routes of uptake and loss of organic pollutants by organisms from terrestrial and aquatic habitats are summarized in table 5.4, and these need to be taken into

The routes by which the pollutants reach the land surface and terrestrial animals and plants were described in Chapter 2. Pollutants may remain on the land surface, enter terrestrial food chains or be transferred to air or water. Contamination of the land surface occurs where there is dumping, for example at landfill sites or in the vicinity of industrial premises, and in soils. Agricultural soils are of particular interest and concern because they receive treatments with pesticides which, by definition, have high toxicity to certain types of organism. Once pesticides enter soils, questions arise about residues in crops, contamination of drinking water and possible effects on soil organisms and soil fertility. The fate of organic pollutants in soil will be discussed before dealing with their movement through terrestrial food chains.

TABLE 5.4 *Model system for bioaccumulation of lipophilic pollutants**†

Type of organism	Routes of uptake			Routes of loss	
	Diffusion	Food	Water	Diffusion	Metabolism
Aquatic					
Molluscs	+++++	(+)		+++++	
Fish with substantial enzyme activity	++++	+→+++		++++	+→+++
Terrestrial					
e.g. predators		++++	(+)		+++++

*The importance of routes of uptake and loss are indicated by a scoring system on the scale + to +++++.
†After Walker (1975).

5.2.1 FATE IN SOILS

Pesticides are the most important pollutants in agricultural soils, and these may reach the soil directly or by transfer of residues from plants to which they have been applied. Pesticides are applied as sprays, granules or dust. Some pesticides are sufficiently soluble to be fully dissolved in spray water, but in most cases they are formulated as emulsifiable concentrates (dissolved in an oily liquid) or wettable powders (fine particles mixed with inert diluent) because of their low water solubility. In the latter, droplets (emulsifiable concentrate) or particles (wettable powder) are suspended in water. The availability of the pesticide is dependent on formulation; on rates of release from particles, droplets or granular formulations. Apart from pesticides, soils are sometimes contaminated by hydrocarbons, PCBs and other industrial chemicals. Such compounds are usually far less toxic than pesticides, and pollution problems associated with them are more localized. The fate of pesticides and other organic pollutants in soils will now be discussed before considering their transfer along food chains.

Soils are complex associations among living organisms, mineral particles and organic matter. The clay fraction of the minerals and the so-called humus of the organic matter are colloids (diameter < 0.002 mm), which on account of their small size present a large surface area per unit volume. When organic compounds enter soils, they become distributed between soil water, soil air and the available surfaces of soil minerals and organic matter. Where they are in the form of – or associated with – particles or droplets, some time will elapse before the individual molecules are distributed between these compartments of the soil. This will be the case, for example, when pesticides formulated as wettable powders are applied as sprays.

The distribution of organic compounds in soil is dependent upon their physical properties – especially solubility (e.g. K_{ow} values), vapour pressure and chemical stability (figure 5.12). Taking chemical stability first, organic pollutants are broken down by hydrolysis, oxidation, isomerization and, if on the surface, the action of light (photochemical breakdown). Usually, breakdown leads to a loss of toxicity,

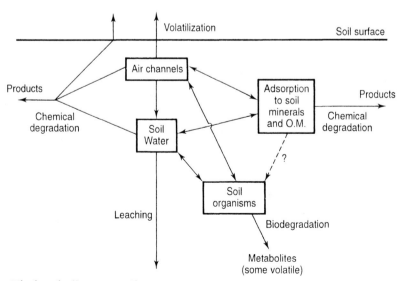

FIGURE 5.12 *The fate of pollutants in soil.*

but occasionally the products are themselves highly toxic (e.g. the isomerization of the organophosphorous compound malathion to isomalathion). Polar compounds (hydrophilic compounds of low K_{ow}) tend to dissolve in water and are adsorbed to soil colloids only to a limited degree. (Sometimes, organic compounds which exist as ions provide an exception to this rule because they become strongly associated with sites on soil colloids which bear an opposite charge. Thus, the herbicide paraquat exists as a cation and binds strongly to negatively charged sites of clay minerals.) Conversely, compounds of low water solubility (high K_{ow}) tend to become strongly adsorbed to the surfaces of clay and soil organic matter, but exist only at very low concentrations in soil water. Compounds of high vapour pressure tend to volatilize into the soil air or directly from the soil surface into the atmosphere. If they enter soil air, they may be retained within the soil for some time, but will eventually pass into the atmosphere.

The binding of organic molecules to soil colloids restricts their movement in the soil and their availability to soil organisms. Thus, compounds of high K_{ow} (i.e. lipophilic compounds) applied to soil show little tendency to be leached down through the soil profile by water. Some soil-acting herbicides (e.g. simazine) exhibit depth selection in their herbicidal action because of this phenomenon. They are only toxic to surface-rooting weeds because they are not carried far enough down the profile to be taken up by deeper rooting ones. Limitation of availability to soil organisms restricts the rate of biotransformation and the toxicity of lipophilic compounds. Thus, chemically stable lipophilic compounds often have long half-lives in soil because they are tightly bound to clay and/or organic matter and are metabolized very slowly.

By contrast, hydrophilic compounds (e.g. soluble herbicides such as 2,4-D and MCPA), which are not strongly bound to soil clay or organic matter, move more freely in soil and are readily available to soil organisms. They tend to be carried down the soil profile by water. Also, they tend not to be very persistent because soil organisms metabolize them relatively rapidly.

Active forms of organochlorine insecticides such as p,p'-DDT and dieldrin provide examples of lipophilic compounds which are only slowly metabolized. Even when freely available, they are metabolized only slowly because they are poor substrates for detoxifying enzymes. The loss of compounds such as these from soils is biphasic (figure 5.13). Immediately after application, there is a period of relatively rapid loss when newly applied insecticide, which is present as particles or dissolved in an oily solvent, is volatilized or simply blown away as dust (Edwards, 1976). During this period, molecules of the insecticide become adsorbed to soil colloids and the period of initial rapid loss is succeeded by a period of slow exponential loss. This is because the adsorbed insecticide only slowly becomes available for loss by evaporation or metabolism. During the period of slow loss, the half-lives for these compounds can run into years – in some cases even tens of years (see table 5.5). The half-lives depend on the compound, the type of soil and the climate. Compounds with higher vapour pressure tend to be lost more rapidly than compounds of lower vapour pressure. Compounds which are biotransformed rapidly tend to be lost more rapidly than compounds which resist biotransformation. Persistence tends to be greatest in heavy soils which are high in clay and/or organic matter. Persistence is also favoured by low temperatures. Under tropical conditions, rates of loss due to volatilization, chemical breakdown and biotransformation tend to be faster than under cooler more temperate conditions.

Hydrophilic compounds, such as the herbicides 2,4-D and MCPA and the insecticide carbofuran, show a fundamentally different

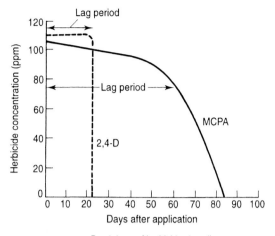

Breakdown of herbicides in soil

Disappearance of persistent organochlorine insecticides from soils

FIGURE 5.13 *Loss of pesticides from soil. From Walker (1975).*

TABLE 5.5 *Persistence of organochlorine insecticides in soils**†

Compound	Time for 50% loss (years)	Time for 95% loss (years)
Dieldrin	0.5–4	4–30
p,p'-DDT	2.5–5	5–25
Lindane (γ-BHC)	(Approx. 1.5)	3–10

*After Walker (1975).
†As the rate of loss from soils falls into different phases, true half-lives cannot be determined. However, estimates have been made of the time taken for a particular percentage of an applied dose to disappear.

pattern of loss from that just described (figure 5.13). Immediately after application to soil, the rate of loss is relatively slow. However, after a period of time (lag phase), the rate becomes much more rapid and the compound is quickly lost. Typically, the chemical disappears within a period of days or weeks. If more of the compound is immediately added to the soil, then this also quickly disappears. The soil has become 'enriched'. Strains of microorganisms have developed in the soil which can metabolize the organic compound. If no more compound is added to the soil, this 'enrichment' will be lost and the soil will revert to its original state. This enrichment phenomenon is sometimes a problem in agriculture; certain pesticides lose their effectiveness if they are used too often (carbofuran is a case in point).

The fate of organic compounds in soil – their movement and their decay curves – have been predicted with some success using models which incorporate parameters such as K_{ow} and vapour pressure.

5.2.2 TRANSFER ALONG TERRESTRIAL FOOD CHAINS

Organisms in terrestrial ecosystems may take up pollutants from their food and ingested water or directly from air, water or solid surfaces with which they come into contact. Uptake from ambient water is not such an important route as it is for aquatic organisms, although it should be borne in mind that soil organisms (e.g. earthworms) may acquire pollutants in this way. Uptake from ingested food or water is a very important route of uptake – often the major route for terrestrial vertebrates. Here, passage of organic pollutants or their stable biotransformation products along food chains is a matter of great importance. Where compounds have long biological half-lives, their passage along a food chain may lead to

biomagnification at some or all of the stages. Typically, the highest concentrations of pollutant will be found in predators of the higher trophic levels of the food pyramid. Also, because of the mobility of some animals (especially birds), they may be transported to areas far removed from the point where they were originally released, e.g. between Africa and Europe. For a discussion of the problem, see Balk and Koeman (1984).

The existence of relatively high levels of persistent lipophilic pollutants in terrestrial predators has often been reported. Examples include organochlorine insecticides (figure 5.14), PCBs and methyl mercury. There is a shortage of reliable data on the concentrations of persistent pollutants in organisms of different trophic levels of the same terrestrial ecosystem at the same time.

Levels of persistent organochlorine insecticides were determined in organisms from different trophic levels of terrestrial ecosystems in Britain during the 1960s. Like data from the marine environment (figure 5.15 and section 5.3), these point to a general upward trend in residues of compounds such as p,p'-DDE (metabolite of p,p'-DDT) and dieldrin with movement along the food chain. However, there is a danger of oversimplifying the complex situation that existed in the field at a time when there were very large temporal and spatial variations in exposure to these compounds. For example, grain-eating birds acquired lethal doses of dieldrin in some areas (e.g. agricultural areas of eastern England), where they could feed on grain that had been dressed with it (Turtle *et al.*, 1963; Moore and Walker, 1964); in nearby areas where the chemical was not used, the exposure of the same species was almost zero. Also, biomagnification did not necessarily occur at each step in the food chain, even for highly persistent compounds. Dressed seed might have a dieldrin residue of $100 \ \mu g \ g^{-1}$, but a grain-eating bird or mammal could not nor-

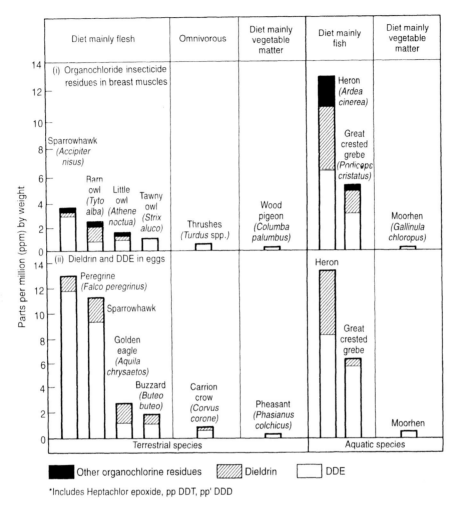

FIGURE 5.14 *Organochlorine insecticides in British birds. From Walker (1975).*

mally build up such a concentration in its tissues because lethal dieldrin poisoning would normally occur at levels well below this. Further, species such as the sparrowhawk and peregrine tend to feed selectively. Thus, in the situation under consideration, they may have tended to select prey with the highest residue concentrations in their tissues, i.e. individuals showing sublethal symptoms of poisoning. They may have been exposed not to average concentrations of pollutant in the tissue of the prey species but to the highest concentrations in surviving members of that species.

One important issue is the tendency of predators to bioaccumulate persistent pollutants with long biological half-lives present in prey, when exposed to them over long periods. Available data suggest that predatory birds exposed to compounds such as dieldrin or *p,p′*-DDE can achieve bioaccumulation factors of 5- to 15-fold in relation to their prey if a steady state is reached. As mentioned earlier, predatory birds are thought to be particularly efficient bioaccumulators of lipophilic xenobiotics because they have poorly developed oxidative detoxication systems.

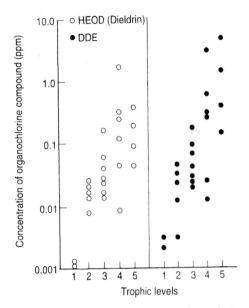

FIGURE 5.15 *Organochlorine insecticides in the Farne Island ecosystem. From Walker (1975).*

In summary, certain persistent lipophilic pollutants have been shown to be transferred along terrestrial food chains, sometimes reaching their highest concentrations in predators of the highest trophic levels. There has been a major problem with certain pesticides used as seed dressings (e.g. aldrin, dieldrin, methyl mercury), where the concentration at the first step of the food chain is already high and grain-eating species can quickly acquire lethal doses if they feed on dressed grain. The persistence of most of these compounds is due to a combination of high lipophilicity (high K_{ow}) and resistance to metabolic detoxication (especially in species such as specialized predators, which are deficient in detoxifying enzymes anyway). Lipophilic compounds that are readily metabolized tend not to be persistent in terrestrial animals. If they have toxic effects, these seem to be 'acute' in nature – of short duration and limited to the species immediately exposed to them and to the area in which they are released.

5.3 *Organic pollutants in aquatic ecosystems*

As in the case of soils, the fate of organic pollutants entering aquatic ecosystems depends on their physical properties, especially lipophilicity, vapour pressure and chemical stability. Compounds which lack stability, for example compounds that tend to be hydrolysed, present few problems as aquatic pollutants unless their transformation products happen to be toxic. Also, volatile compounds tend not to persist in aquatic ecosystems. Polarity is once again important in determining distribution and persistence. Hydrophilic compounds tend to be dissolved in and distributed throughout surface water. Lipophilic compounds, on the other hand, tend to become associated with particulate matter, notably of sediments. Sometimes they may also exist on the surface of water, e.g. dissolved in surface oil films.

5.3.1 POLLUTANTS IN SEDIMENTS

Like soils, the sediments of rivers, lakes and seas are associations between organic and inorganic particles and living organisms. Organic pollutants may be adsorbed to the particles of sediments, and this limits their mobility and their availability to bottom-dwelling organisms. It is often uncertain to what extent pollutants can be taken up directly by animals from the adsorbed state or whether they need first to move into the aqueous medium before they become available. The extent to which an adsorbed pollutant can be taken up directly will depend upon the nature of the chemical, the nature of the surfaces to which it is bound, the strength of the binding, the species which is taking it up and – in some cases at least – the temperature, pH and oxygen content of the ambient water.

The oxygen content of water can be important in determining the nature and the rate of both chemical and biochemical transformations. As the oxygen content declines, so there will be a tendency for oxidative transformations to be replaced by reductive ones (section 5.1.5). Sediments can differ widely with regard to their oxygen content. At the bottom of deep seas, conditions are anaerobic, whereas in shallow fast-moving streams oxygen levels are relatively high. Sediments of intertidal zones along the sea-shore experience fluctuating oxygen levels in accordance with tidal movements.

5.3.2 TRANSFER ALONG AQUATIC FOOD CHAINS

The same general considerations apply as in the case of terrestrial food chains (section 5.2.2), except that exchange diffusion between organisms and ambient water is a complicating factor in aquatic food chains. This makes more difficult the interpretation of data such as those shown in figure 5.15. The residue levels shown for the persistent organochlorine compounds dieldrin and p,p'-DDE (metabolite of p,p'-DDT) were measured in organisms sampled from different trophic levels of the marine ecosystem around the Farne Islands, Northumberland, UK, during the period 1962–4 (Robinson *et al.*, 1967). This shows a strong relationship between the log concentrations of the residues and the trophic levels of the organisms in which they were determined. The lowest levels are in the plants (brown algae) in trophic level 1, the highest in vertebrate predators of trophic levels 4 and 5. It is interesting that p,p'-DDE is subject to a steeper gradient than dieldrin. p,p'-DDE is metabolized more slowly and has a longer biological half-life than dieldrin in the animals that have been studied. It appears as if biomagnification occurred at every stage of the food chain. However, aquatic invertebrates and fish in trophic levels 2–4 obtained an unknown proportion of their residue burden directly from water and/or sediment and not from food. Also with predators (as mentioned previously), there may be selective predation. The prey that they eat may contain higher organochlorine residues than the average value shown in figure 5.15. These points aside, the predators appear to be achieving a substantial biomagnification (i.e. a BAF considerably greater than 1), assuming that most of the pollutant burden comes from food. The apparent BAF calculated from the data given in figure 5.15 lies between 50 and 60 for the fish-eating shag (*Phalacrocorax aristotelis*) (Walker, 1990b). The variation in residue levels in the principal prey species, the sand eel (*Ammodytes lanceolatus*), was not large. Even if they were feeding on the most contaminated individuals, a BF of several-fold is still suggested.

The fish-eating birds shown to bioaccumulate substantial levels of persistent organochlorine compounds in this and other studies have low activities of the monooxygenase system, which is mainly responsible for their detoxication (Ronis and Walker, 1989; Walker, 1990b; figure 5.5). The species in question include the shag, cormorant (*Phalacrocorax carbo*), guillemot (*Uria aalge*) and razorbill (*Alca torda*). In general, a deficiency in their detoxication system favours bioaccumulation.

There is a general concern about the build-up of relatively high levels of persistent pollutants in predators at the top of aquatic food chains (for a recent review of the question, see Walker and Livingstone, 1992). Apart from the organochlorine insecticides just mentioned, persistent polychlorinated biphenyls (PBBs) and polychlorinated dibenzodioxins (PCDDs) have also given rise to concern. In addition to sea birds, marine mammals such as seals and cetaceans (porpoises, dolphins and whales) have been shown to bioaccumulate these compounds – and to pass them to their offspring via milk.

Like sea birds, predatory marine mammals have poorly developed monooxygenase detoxication systems. The above trend has also been observed in freshwater ecosystems, where relatively high levels of persistent organochlorine compounds have been found in predatory birds [e.g. heron (*Ardea cinerea*)] and mammals [e.g. otter (*Lutra lutra*)] in Western Europe and in predatory birds [e.g. double-crested cormorant (*Phalacrocorax auritus*)] of the Great Lakes of Canada and the USA.

Lipophilic pollutants which have relatively short half-lives (e.g. polycyclic aromatic hydrocarbons; PAHs) do not show the same tendency to pass along food chains and to be biomagnified. Some invertebrates of the lower trophic levels (e.g. the mussel *Mytilus edulis*) bioconcentrate and/or bioaccumulate them because they have little ability to metabolize them. On the other hand, fish, birds and mammals metabolize them rapidly by monooxygenase attack, so there is little tendency for them to reach the higher trophic levels let alone to be bioaccumulated there. Although these compounds do not raise the problem of biomagnification, it must be emphasized that some PAHs and other readily degradable compounds are subject to metabolic activation (section 5.1.5).

5.4 *Summary*

Organic pollutants are examples of xenobiotics – compounds which are 'foreign' to particular organisms. A model is presented to describe the fate of xenobiotics in living organisms, as seen from the toxicological point of view. In this model, five types of site are identified with which the xenobiotic can interact. These are: sites of uptake, sites of metabolism, sites of storage, sites of action and sites of excretion. In animals, xenobiotics are circulated in blood and lymph (vertebrates) or in haemolymph (invertebrates). In plants, they move in the phloem and/or xylem. In this way, xenobiotics are distributed around different organs and tissues.

Lipophilic xenobiotics move across membranous barriers by simple diffusion or, in some cases, by transport with certain macromolecules (e.g. lipoprotein fragments) which can traverse membranes. Metabolism of lipophilic xenobiotics by animals occurs in two phases. Phase I, which can be oxidation, hydrolysis, hydration or reduction, yields metabolites of greater polarity than the parent compound. Phase II involves conjugation to produce polar conjugates, which are usually anions and are readily excreted. A variety of monooxyygenases and hydrolases are responsible for phase I biotransformations. In phase II biotransformations, conjugation is usually with cellular anions such as glucuronide, sulphate or glutathione. The excretion of water-soluble metabolites and conjugates completes the process of metabolic detoxication. Sometimes, phase I metabolism causes activation rather than detoxication, as in the case of certain oxidations of benzo-(*a*)pyrene and organophosphorous insecticides.

Persistent lipophilic pollutants with long biological half-lives can undergo biomagnification when they pass along terrestrial or aquatic food chains, and reach especially high concentrations in vertebrate predators at the top of food chains. Examples include organochlorine insecticides such as dieldrin and DDT, many PCBs, dioxin (TCDD), methyl mercury and tributyl tin. Specialist predators, for example fish-eating birds and bird-eating birds, have poor oxidative detoxication systems and efficiently bioaccumulate organochlorine compounds. Compounds of this type are often highly persistent in soil. Water-soluble and readily biodegradable compounds do not usually cause problems of persistence and bioaccumulation.

5.5 *Further reading*

CROSBY, D. G. (1998) *Environmental Toxicology and Chemistry*. Good account of environmental chemistry.

ENVIRONMENTAL HEALTH CRITERIA. A range of monographs on particular pollutants published by the World Health Organization (over 160 titles). Some of them give detailed reviews of the fate of particular pollutants. Selected titles are in the references at the end of this book.

HODGSON, E. and LEVI, P. (1994) *Introduction to Biochemical Toxicology*, 2nd edn. A multiauthor text covering the main aspects of biochemical toxicology in reasonable depth. Covers toxicokinetics of xenobiotics and mode of action of 'poisons'.

MORIARTY, F. (1975) *Organochlorine Insecticides: Persistent Organic Pollutants*. Probably the best researched of all the persistent organic pollutants. Much information on their fate and environmental behaviour.

SCHÜÜRMANN, G. and MARKERT, B. (eds) (1998) *Exotoxicology*. A text that deals in depth with many topics in the first section of the book.

TIMBRELL, J. A. (1999) *Principles of Biochemical Toxicology*, 3rd edn. A single-author text giving a very clear and readable account of the basic principles of the subject.

WALKER, C. H. and LIVINGSTONE, D. R. (1992) *Persistent Pollutants in Marine Ecosystems*. Gives a detailed account of the fate and levels of persistent organic compounds in marine food chains.

PART 2

Effects of pollutants on individual organisms

CHAPTER 6

Toxicity testing

The first five chapters have been concerned with the fate of chemicals in the environment and within living organisms. This chapter begins the second part of the text, in which emphasis shifts to questions about the effects that chemicals have upon individual organisms. Ecotoxicology is concerned with the harmful effects of chemicals. This chapter will address the questions of what constitutes harm and how is toxicity measured? It is important to do this before moving on to the more complex issues of Part 3, in which the relationship between toxic effects upon individuals and consequent adverse effects at the levels of population, community or ecosystem becomes a crucial question.

6.1 General principles

Of central importance in both toxicology and ecotoxicology is the relationship between the *quantity* of chemical to which an organism is exposed and the nature and degree of consequent harmful (toxic) effects. Dose–response relationships provide the basis for assessment of *hazards* and *risks* presented by environmental chemicals (more later). This simple basic concept immediately raises questions about the definition of poisons because everything depends on dose. Paracelsus (1493–1541) recognized the dilemma and stated: 'All substances are poisons; there is none that is not a poison. The right dose differentiates a poison and a remedy' – or, in other words, 'the dose maketh the poison'. Thus, no chemical is poisonous if the dose is low enough, whereas all chemicals are poisonous if the dose is high enough (even apparently harmless substances such as sugar and salt can be toxic to animals at high doses). It is advisable to remember this principle when reading sensational popular articles on the subject, in which reference is often made to 'poisonous' or 'toxic' substances in the environment without pointing out that the levels are far too small to have any toxic effect.

There are many different ways in which toxicity can be measured. Most commonly, the measure ('end-point') is death, although there is a growing interest in the use of more sophisticated indices. The desire to minimize lethal toxicity testing with vertebrate animals is a significant driving force here. Biochemical, physiological, reproductive and behavioural effects can all provide measures of toxicity. Many toxicity tests provide an estimate of the dose (or the concentration in food, air or water) which will cause a toxic response at the 50% level, e.g. the median lethal dose is the dose that will kill 50% of a population. It is also possible – and this approach is gaining in popularity – to establish the highest concentration or dose that will *not* cause an effect.

Several terms used in relation to toxicity testing require definition. First, in lethal toxicity testing, LD_{50} represents the median lethal dose, whereas LC_{50} represents the median lethal concentration. In toxicity tests which determine these values, it is also possible to determine the highest doses or concentrations which cause no toxicity – the no observed effect dose (NOED) and no observed effect concentration (NOEC) respectively. These values can only be determined in situations where a higher dose or concentration has produced an effect in the same toxicity test. These points are illustrated in figure 6.1, which refers to the determination of an LC_{50} after 96 h exposure. If a test is carried out where the end-point is an adverse response other than death, then an EC_{50} or ED_{50} is determined. Here, the concentration or dose producing the effect in 50% of the population is determined. As with lethal toxicity testing, it is possible to determine NOEC or NOED following this approach. However, values for NOEC or NOED are only meaningful in a test in which a higher dose has been shown to produce an effect.

Apart from toxicity tests involving the use of live animals which are the principal subject of this chapter, there are other ways of evaluating the toxic properties of chemicals. In the main, these stem from an understanding of the mode of action of chemicals. For example, bacterial mutagenicity assays (e.g. the Ames test) aid the identification of substances which can act as carcinogens or mutagens in mammals. Also, the study of the relationship between structure and toxicity ('quantitative structure–activity relationships' or QSARs) can aid the identification of toxic molecules. Further examples are given in section 6.7.2. These approaches become more viable as molecular mechanisms of toxicity come to be better known and they can lead to an understanding of the *molecular characteristics* which cause a chemical to interact adversely with cellular macromolecules.

An environmental chemical can enter a living organism by one or more routes of uptake. Depending on the chemical, the species and the environmental conditions, one route of uptake may be dominant or more than one may be significant. Both the efficiency of uptake and the degree of toxic effect differ between these routes, e.g. for most lipophilic compounds absorption from the gut of vertebrates is faster than absorption from the skin. Consequently, toxicity is usually higher with oral administration than with topical application.

When tests are performed on terrestrial animals, it is common to apply single measured

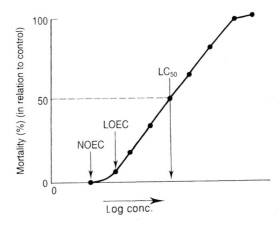

FIGURE 6.1 *Toxicity after 96 h exposure in an aquatic toxicity test. It should be noted that NOEC can be determined only where LOEC is known – otherwise there would be no indication of a concentration that can be toxic. NOEC, no observed effect concentration; LOEC, lowest observed effect concentration; LC_{50}, median lethal concentration at 96 h.*

doses, e.g. orally, topically (i.e. applied to skin or cuticle) or by injection into tissues or body fluids. There may also be continuous exposure to a constant concentration of chemical in food, in air or in soil (soil organisms). With aquatic organisms, direct uptake from water is predominantly from water or food, and toxicity testing usually involves continuous exposure to defined concentrations rather than the administration of single doses. In selecting the route of administration of chemicals in toxicity testing, account needs to be taken of the major route(s) of uptake in nature.

There can be very large differences between groups of organisms and between species in their susceptibility to the toxic action of chemicals; there can also be large differences between different strains of the same species – as in the case of strains of insects which have become resistant to insecticides (see section 13.6). The selective toxicity ratio (SER) is expressed in terms of the median lethal dose (MLD) in the following way:

$$SER = \frac{\text{MLD (or concentration) for species A}}{\text{MLD (or concentration) for species B}}$$

Selectivity ratios between beneficial organisms and pests are considered when evaluating the environmental safety of pesticides (table 6.1). Some examples of selectivity are discussed in Hodgson and Levi (1994) and Walker (1983).

The fact that chemicals can be so selective makes it difficult to extrapolate toxicity data from one species to another. For practical reasons, it is only possible to perform toxicity tests on a very limited number of species. Thus, toxicity tests are very rarely carried out on those species thought to be at risk in the field. When regulatory authorities take decisions about the release of chemicals into the environment, they base their decisions on toxicity data obtained with surrogate species, which may be very different from field species in their susceptibility

to particular chemicals. For example, estimation of the toxicity of a chemical to wild birds in a risk assessment exercise will be based on data from two or three laboratory species at most, although globally there are over 9000 species of great diversity. In another example, tests on the pH preferences of eight species of springtails (Insecta: *Collembola*) by Van Straalen and Verhoef (1997) showed that some species had a much greater tendency to settle in strongly acidified soils than did the standard test collembolan *Folsomia candida*, which is clearly not representative of all springtail species. This problem of extrapolating toxicity data obtained with one species to another is also a feature of human toxicology. For humans, the surrogate species are rats and mice. In general, the difficult practice of extrapolating between species becomes easier with a better understanding of the mechanisms responsible for toxicity. Such understanding facilitates interspecies comparisons and can aid the development of models to predict toxicity to individual species. Data obtained in *in vitro* studies can contribute to these models, e.g. metabolic rate constants (Chapter 5) may be used for models that estimate tissue concentrations and half-lives of organic chemicals.

Multiple species tests represent a more sophisticated approach to ecotoxicity testing than single species tests, but they are at an early stage of development. It is likely to be some years before they become routine because of the complexity of the interactions between species. Tests using a simple predator–prey system, for example with mites preying on Collembola (Axelsen *et al.*, 1997; Hamers and Krogh, 1997), have provided useful information However, results from tests with three or more species are difficult to relate to field conditions.

The following account will describe toxicity tests which have been widely used to aid the process of risk assessment, which will be considered later in section 6.4. It should be

TABLE 6.1 *Selective toxicity of some pesticides**

Compound	Acute oral LD$_{50}$ (mg kg^{-1})			Topical/dermal LD$_{50}$ (mg kg^{-1})		
	Rat	Birds	SER	Rat	Insect	SER
Organophosphorous insecticides					Housefly	
Dimethoate	250	26 (4)	9.9	925	0.20	4.6 × 10^3
Fenitrothion	462	332 (4)	1.4	> 3000	5.7	526
Dichloros	27	9.6 (2)	2.8	488	0.80	610
Diazinon	450	4.5 (4)	100	850	1.9	447
Malathion	1650	685 (3)	2.4	> 4000	17.4	> 230
Pirimiphos-methyl	1400	162 (3)	8.6			
Pirimiphos-ethyl	138	6.5 (2)	21			
Organochlorine insecticides					Housefly	
DDT	400	923 (3)	0.43	2500	14.0	179
Dieldrin	40	91 (7)	75	75	1.0	75
γ-HCH	200	118†	1.7	750	3	250
Carbamate insecticides					Housefly	
Carbaryl	500	990 (5)	0.5	> 4000	> 500	
Baygon (Propoxur)	135	26 (6)	5.2	> 2400	25	96
Carbofuran	6	2.4 (4)	2.5		7	
Aldicarb	6	3.6 (4)	1.6	1	6	0.16
Zectran	39	12 (6)	3.2	3.2		
Pyrethroids					Bee	
Permethrin	500	> 13 000 (4)	< 0.04	> 2500	0.017	> 1.47 × 10^5
Cypermethrin	250	> 10 000‡	< 0.25	> 4800	0.11	> 4.4 × 10^4
Fenvalerate	451	> 4000 (3)	< 0.11	4000	0.21	2.4 × 10^4
Deltamethryn	129	4000‡	< 0.03	> 800	0.035	2.3 × 10^4

For birds, an average value of LD$_{50}$ is usually given. A number in brackets ( ) indicates the number of different species used to obtain the average.

Where ranges of LD$_{50}$ values are given in the original source, an average has been calculated to simplify presentation.

Selectivity ratio (SER) = (LD$_{50}$ rat)/(LD$_{50}$ other species)

*From Hodgson and Levi (1994).

†Value for pheasant.

‡Value for duck.

emphasized that the values obtained by toxicity testing (e.g. LD$_{50}$, LC$_{50}$) are very dependent upon the conditions under which tests are performed. Factors such as formulation of chemical, route of dosing, feeding regimen, temperature, humidity, state of health of animals etc. can all influence the values that are obtained. The interpretation of toxicity data needs to be carried out with caution because (i) they are determined under a well-defined set of conditions, which may differ considerably from those prevailing in the natural environment, and (ii) they are not even very reproducible because it is very difficult to control all the variables mentioned above. At best, standard toxicity data give a measure of toxicity under closely defined sets of operating conditions. There has sometimes been a tendency to accord toxicity data with more significance than they actually have.

The transformation of percentage kills by statistical procedures requires brief discussion. The underlying assumption is that the data can be fitted to a statistical distribution which will give a straight line when plotted against the log of the dose (or concentration) of the chemical. This is illustrated in figure 6.2, which shows a normal gaussian distribution in a toxicity test in relation to the dose of chemical given. If the median concentration is ascribed the value of 0, then values to either side of it can be measured as normal equivalent deviates (NEDs). Comparisons can then be made between measured doses or concentrations producing a particular response and those that would be expected if the data followed normal distribution. In practice, the data from toxicity tests usually depart from normal distribution when approaching the extremes of 0% and 100% (as in figure 6.1). It is possible to correct for these deviations by converting percentage kills into NEDs or the related probit units. When these values are plotted against dose or concentration, a straight line is obtained, as in figure 6.8. Probit values for percentage responses can be looked up in tables.

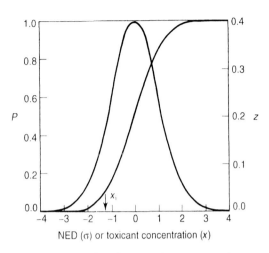

FIGURE 6.2 *Plot of standardized normal frequency distribution (P) with its integral, the cumulative normal frequency function (z).*

of by-products of the manufacturing process. Furthermore, the formulation actually marketed will contain additives such as carrier solvents, emulsifiers and stabilizers, all of which may have some effect on its toxicity. In environmental risk assessment, tests need to be carried out on the products actually released into the environment if there is to be a realistic estimation of toxic impact.

Matters can become more complicated still when considering the pollution that actually exists in the environment. Sewage, factory or pulp mill effluents released into surface waters often contain complex mixtures of pollutants. So too do contaminated sediments and soils, where the situation can be complicated by the presence of highly persistent lipophilic compounds with long biological half-lives (see discussion in Chapter 5). Although toxicity is usually additive, there is the possibility of potentiation of toxicity when animals or plants are exposed to mixtures; the toxicity of a mixture can greatly exceed the summation of the toxicities of its component chemicals (see Chapter 9).

6.2 *Determination of the toxicity of mixtures*

Much toxicity testing is performed upon single compounds. This is a necessary part of the environmental risk assessment of pesticides and a wide range of industrial chemicals. Sometimes testing is carried out on relatively pure samples of the chemicals. Quite often, however, the materials tested contain appreciable quantities of other compounds. In the case of pesticides, for instance, the technical product used in formulations may contain appreciable quantities

Tests performed on environmental samples such as water, sediment, soil and air can measure the toxicity of mixtures. If these are accompanied by chemical analysis, the measured toxicity can sometimes be compared with toxicity predicted from the detected residues of chemicals (the toxicities of individual chemicals determined by analysis are calculated and are then summated to give the predicted toxicity for the whole mixture). Not infrequently, the measured toxicity differs markedly from the predicted toxicity. There are several possible causes for such a discrepancy. Apart from the questions of potentiation or antagonism, the chemical analysis may be incomplete, overlooking the presence of certain toxic molecules. This can be a particularly difficult problem with complex mixtures of organic pollutants at low concentrations. In soils and sediments, there is also the question of availability. To what extent is a strongly adsorbed chemical available to aquatic organisms in contact with sediments, or to terrestrial organisms dwelling in soil? In general, relating chemical analysis to toxicity is relatively easy when only a small number of chemicals (say up to three) express toxicity.

The assays described in the following account may be used to evaluate the toxicity of mixtures both in industrial products and in environmental samples. It should, however, be emphasized that it can be very difficult to attribute the toxicity that is measured to particular compounds, or combinations of compounds in the case of environmental samples. When dealing with the regular monitoring of effluents containing complex mixtures of chemicals, the composition of which may change radically with time, it is desirable to analyse large numbers of samples. Rapid, inexpensive tests, including bioassays (see section 6.7.2), have an important role here.

6.3 *Toxicity testing with terrestrial organisms*

6.3.1 INTRODUCTION

It is sometimes stated that toxicity testing with terrestrial animals is simpler than with aquatic animals as only one route of exposure via the gut has to be considered. Although this is the case with widely utilized tests using rats and mice, this is certainly not the case with more recently developed tests with invertebrates in which exposure via the external medium (air, soil) is important also. In this section, ecotoxicological tests with invertebrates, plants and birds will be described.

6.3.2 INVERTEBRATES

Two ecotoxicological tests using earthworms and bees are in widespread use and have been developed to test the effects of chemicals on 'representative' terrestrial invertebrates. Several other tests are at various stages of formulation at the time of writing. The methods of testing on four types of organism will be covered here: widely used standard tests with earthworms, a test currently undergoing international validation with springtails, tests in the early stages of development with woodlice and a standard test using bees. The earthworm tests mainly measure effects of chemicals that pass across the body surface. The springtail test measures effects on reproduction, again mainly through contact poisoning of the adults, their eggs or juveniles that hatch from the eggs. The isopod test measures the effects of chemicals on feeding rates, mainly through feeding repellence. The bee test is designed to assess the effects of chemicals on 'beneficial insects'. For further details of these and other tests using enchytraeid worms, mites, beetle, centipedes, millipedes and nematodes, see Løkke and Van Gestel (1998).

Earthworms

The OECD earthworm test is the best established method and has reached the stage of being a legislative requirement in some countries before new chemicals can be released into the environment. Several approaches have been adopted (described in detail in Greig-Smith *et al.*, 1992a; Sheppard *et al.*, 1998). The most widely used species is *Eisenia fetida* (figure 6.3A), which is easy to culture in the laboratory and has a reproductive cycle of about 6 weeks at

A

B

FIGURE 6.3 *(A) Group of earthworms* (Eisenia fetida) *in standard OECD soil. Adult worms are approximately 8 cm in length. (B) Juvenile* Eisenia fetida *emerging from a cocoon of approximately 2 mm in diameter. Photographs by Steve Hopkin.*

20°C. However, this is a species native to the southern Mediterranean which survives only in compost or manure heaps or in heated glasshouses during the winter in northern latitudes. Thus, care should be taken in extrapolating the results of tests with *Eisenia* to other species of earthworm.

One problem with the standard OECD method is that the worms have no food during the experiment and tend to lose weight. More recent studies have included a small pellet of cow or horse manure on the surface of the artificial soil which allows the worms to put on weight. Fed worms also have a higher reproductive rate than starved specimens. The bioconcentration factor (concentrations of the chemical in worms/concentration of the chemical in the soil) can also be determined from such experiments and can be used to predict the exposure of earthworm predators to potential toxins in the field.

When the results of toxicity tests using laboratory soils have been compared with those in 'natural' soils brought in from the field, it is clear that chemicals are invariably much more 'bioavailable' to worms in the artificial medium. Indeed for metals such as zinc, worms are affected at concentrations 5–10 times lower in artificial soils than in field soils (Spurgeon *et al.*,

1994). Uptake of metals by worms is more closely related to the concentrations in soil pore water than to total content (Janssen *et al.*, 1997), and accumulation rates tend to be relatively high in soils with low pH and low organic matter (Spurgeon and Hopkin, 1996). The artificial soil has a lower organic matter content than many field soils, and the clay used has a relatively low cation exchange capacity. In terms of 'environmental protection', therefore, the test includes an in-built 'safety factor' which should be borne in mind when results of laboratory tests are being extrapolated to the field.

A test currently under development assesses the integrity of membranes of **lysosomes** in earthworm coelomocytes (Weeks and Svendsen, 1996). When worms are stressed, the lysosomal membranes become 'leaky'. The extent of membrane disruption is measured by the time taken for a dye, neutral red, to diffuse out of the lysosomes into the cytoplasm of the coelomocytes in coelomic fluid removed from the coelom and spread onto a microscopic slide. The neutral red retention time has been successfully used as a biomarker assay to measure the impact of pollution on earthworms in the aftermath of a plastics fire in England (Svendsen *et al.*, 1996).

BOX 6.1 *OECD earthworm toxicity test.*

The basic test uses an artificial soil made up from 70% sand, 20% kaolin clay and 10% *Sphagnum* peat (by weight). Water is added to constitute 35% (by weight) of the final soil and the pH is adjusted to 6.3 with calcium carbonate. Ten adult worms are added to each of the four replicates of a control and an ascending series of concentrations of the chemicals under test. The worms are left for 14 days, after which the number of survivors are counted. An LC_{50} for survival can then be calculated. Some workers have run their experiments for a longer period to allow the worms to reproduce (e.g. Spurgeon et al., 1994). The reproductive rate is easy to access by counting the number of cocoons produced (figure 6.3B). Effects on reproduction are detected at lower concentrations of metal in soils than effects on growth and survival and it is thought that the former approach is more ecologically relevant in trying to access potential effects in the field (figure 6.4). Interestingly, there is also evidence that the artificial soil may be deficient in copper (figure 6.4C), although further work is required to substantiate this suggestion.

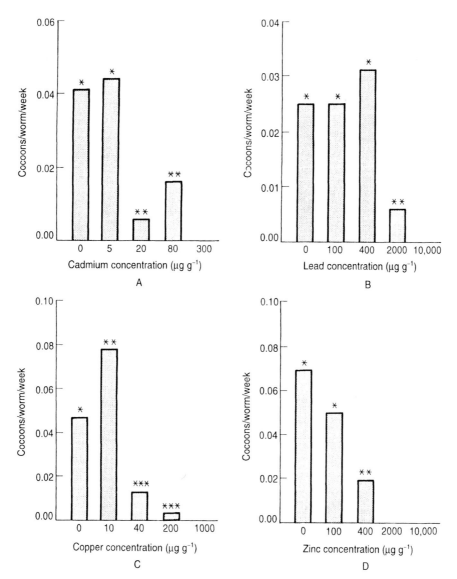

FIGURE 6.4 *Rate of cocoon production by* Eisenia fetida *exposed to (A) cadmium, (B) lead, (C) copper and (D) zinc (μg g⁻¹ dry weight). Bars with the same number of asterisks (*) were not significantly different at* P < 0.05. *Reproduced from Spurgeon* et al. *(1994) with permission from Elsevier Science.*

Springtails

Springtails (Insecta: *Collembola*) are among the most abundant of all soil invertebrates (Hopkin, 1997). Although the springtail test has not reached such an advanced legislative stage as that for earthworms, many laboratories in the world use this species to assess the effects of chemicals on non-target soil arthropods (Wiles and Frampton, 1996; Crouau *et al.*, 1999). Comparison of a range of test organisms by Ronday and Houx (1996) concluded that *Folsomia candida* (figure 6.5) was the most suitable animal for soil quality assessment.

BOX 6.2 *Test on springtails.*

The most widely used species is a parthenogenetic strain of *Folsomia candida* (figure 6.5) which has a reproductive cycle of 3–4 weeks at 20°C. A similar experimental design to that used in the earthworm test is constructed with four replicates each of a control, and an ascending series of concentrations of the test chemical. The chemical is mixed in the same formulation of artificial soil as described for the earthworm test above. Ten adult springtails of the same age are placed in a small container about the same size as a plastic vending-machine cup (about 10 cm in height, 6 cm in diameter), together with a small amount of dried yeast for food. They are left for a minimum of 3 weeks, then the soil is flooded. All the springtails, including any offspring produced during the experiment, float to the surface of the water. Each container is photographed on film or captured digitally and the images are projected onto a large screen on which the number of progeny are counted. Developments in image analysis software mean that it is now possible to count and measure the length of all animals automatically (Martikainen and Krogh, 1999). The concentration which causes 50% (or other percentage) reduction in reproduction compared with controls can then be calculated. If a larger number of replicates is prepared, then a proportion of the containers can be flooded at intervals to produce a time response (figure 6.6).

As for earthworms, chemicals tend to be more toxic in the artificial soil than their toxic equivalents in the field. Much closer laboratory–field comparability can be obtained by 'ageing' soils after application of contaminants and percolating soils with water before conducting the test (Smit and Van Gestel, 1998).

Woodlice

Tests with woodlice (Crustacea: *Isopoda, Oniscidea*) are still at an early stage of development. One problem with species such as *Oniscus asellus* (figure 6.7B) and *Porcellio scaber* is their low growth rate and long reproductive cycle relative to *Eisenia fetida* and *Folsomia candida*. Even at 25°C, it still takes a minimum of 6 months for newly hatched juvenile woodlice to reach reproductive age. However, woodlice are potentially important test animals as they are common and widespread and perform an important role in physically breaking down leaf material into smaller particles which are more easily decomposed by microbes (for a review, see Drobne, 1997).

Woodlice are extremely resilient to starvation and can survive for many weeks without food. However, one parameter that can be measured with relative ease is the feeding rate. Woodlice eat leaf litter and convert this to faecal pellets with very consistent shape and weight

(figure 6.7B). Thus, by counting the number of faecal pellets produced during specified time intervals, it is possible to determine whether the presence of a chemical on the leaves is deterring the woodlice from feeding. Reduced feeding rates will have knock-on effects such as slower growth rate and lengthened reproductive period. However, more experiments need to be conducted before woodlice can be considered eligible for 'standard toxicity test' status (Hopkin *et al.*, 1986).

A recent development is the ability to quantify aspects of the behaviour of woodlice in response to pollution by automated video tracking. Woodlice collected from leaf litter contaminated with residues of a plastics fire spent significantly less time actually walking and showed less tendency to turn than did controls (Sørensen *et al.*, 1999). Such behavioural effects often occur at concentrations well below those at which effects on growth and reproduction can be detected.

FIGURE 6.5 *Adults and juveniles of* Folsomia candida, *a parthenogenetic springtail. The largest specimen is 2 mm in length. Photograph by Steve Hopkin.*

FIGURE 6.6 *Total numbers of* Folsomia candida *individuals at different levels of cadmium in artificial soil. Blank* (□), *34.8* (●), *71.3* (△), *148* (■), *326* (○), *707* (▲) *and 1491* (▽) *μg Cd g^{-1} dry weight. Reproduced from Crommentuijn* et al. *(1993) with permission from Academic Press.*

Beneficial arthropods

Toxicity tests upon insects have been widely used to establish the effectiveness of insecticides against pest species, many of which belong to the orders *Lepidoptera, Diptera, Hemiptera* and *Coleoptera*. These have been widely used by pes-

ticide manufacturers when screening new compounds. Such tests generally use the lethal endpoint and have also been valuable in identifying resistance to pesticides in the field. Similar testing procedures have also been used to establish toxicity to beneficial insects, including natural parasites and predators of insect pests. Toxicity testing procedures for insects will now be briefly reviewed.

A common procedure is to apply a solution of the test chemical directly to the outside of an insect, using a microsyringe. In this way, a known dose is administered per insect and a topical LD_{50} can be estimated, using the same procedure as described later for vertebrates (see 6.3.3). Commonly, the chemical is dissolved in an organic solvent, e.g. acetone. Tests of this kind are useful for ranking the toxicity of chemicals and for identifying resistance, but the exposure is unrealistic and does not reflect that which occurs in the field. With insecticides and other pesticides, insects are exposed to formulations – sprays, granules and dusts – under field conditions. Thus, in more realistic testing protocols, insects are exposed to insecticidal

A

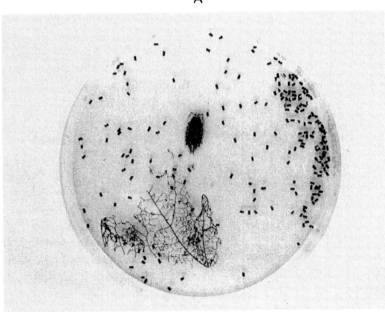

B

FIGURE 6.7 *(A) Half a field maple leaf* (Acer campestre) *in a Petri dish of 9 cm in diameter before addition of a woodlouse. (B) Three weeks later, the leaf has been reduced to faecal pellets by the feeding activity of* Oniscus asellus. *Photographs by Steve Hopkin.*

deposits on the surfaces of the dishes in which they are contained or on their food. After such exposures, LC_{50}s can be calculated (i.e. median lethal concentrations, which will be encountered again later in the context of aquatic toxicity testing). The German regulatory authority for pesticides has published details of formal laboratory tests for **beneficial insects** other than bees, and these are described by Jepson in Calow (1993).

Following concern as to the effects of pesticides on non-target insects, a number of laboratories test new chemicals on honey bees (Gough *et al.*, 1994).

6.3.3 VERTEBRATES

The toxicity of chemicals to mammals, birds and other vertebrates has commonly been measured as a median lethal dose (LD_{50}). In routine toxicity testing, a single dose is given orally to obtain an estimate of acute oral LD_{50}. The procedures for doing this are described in some detail in many standard texts (for example see Calow, 1993) and will only be given in outline here. To do this, groups of animals are given

doses of the test chemical over a range of values which centre on a rough estimate of LD_{50} obtained in a preliminary range of test. The percentage of animals which die in each group over a fixed period following dosing is then plotted against the log of the dose (mg kg^{-1}). To obtain a straight line relationship between dose and mortality, it is necessary to transform percentage kills into normal equivalent deviates (NEDs) or probit values (probit analysis) (figure 6.8).

Values of LD_{50} are sometimes obtained using other methods of dosing. Chemicals may be injected into the blood, into the muscle or into the peritoneal cavity, or they may be applied directly to the skin (dermal LD_{50}). Sometimes they are continuously administered in the food or water over a fixed period of time. In this case, toxicity is expressed as a 'median' lethal concentration in food or water over a stipulated period, e.g. a 5-day oral LC_{50}.

With birds, reproductive toxicity tests are sometimes carried out. The end-points here may be clutch size, shell thickness, hatchability of eggs, embryotoxicity and viability of chicks. For further details on avian toxicity tests, see Walker in Calow (1993).

BOX 6.3 *Toxicity tests with bees.*

The standard method is to collect workers from a hive, anaesthetize them with carbon dioxide and then maintain the bees in cylinders of wire mesh (ten in each container) with three replicates of each test concentration plus controls. Each container is supplied with a sugar solution in a feeding bottle containing a known concentration of a chemical being tested. The cages are kept at constant temperature (typically 25°C) and are checked at 1, 2, 4, 24 and 48 h for morality of bees. In an additional test, contact toxicity is measured by applying the chemical directly to the thorax of bees. However, it is not possible to uncritically apply results of such tests to all species of bees because of their differing life styles (Thompson and Hunt, 1999).

A field-based method can also be used, but this requires considerably more space. A beehive is maintained in a large enclosed tunnel of polythene or netting and a pollen- and/or nectar-bearing crop is sprayed with the test chemical. Bees which return to the hive but then die inside during the night are ejected in the morning by healthy workers. The level of mortality can thus be simply determined by counting the number of dead bees in a tray placed under the hive entrance.

FIGURE 6.8 *Determination of LD$_{50}$. For details, see text.*

During the course of this type of toxicity testing, it is possible to establish values for 'no observed effect dose' or 'no observed effect concentration', i.e. the highest dose or concentration given that produces no lethal effect. Nowadays, there is a movement away from the use of the LD$_{50}$ tests on ethical grounds. The British Toxicity Society has proposed an alternative procedure which requires the use of far

fewer animals. This would merely classify compounds as 'harmful', 'toxic' or 'very toxic' (for further details, see Timbrell, 1995).

6.3.4 PLANTS

A wide variety of methods have been developed for assessing the toxicity of chemicals to plants. In this section, tests to assess the effects of metals will be covered in most detail as this is where most work has been concentrated. However, the basic principles of these tests are applicable to other chemicals and in tests to assess the effects of acid rain and gaseous pollutants.

Plants which naturally contain very high concentrations of specific metals are known as **metallophytes** (Baker and Proctor, 1990). Some metallophytes are able to grow naturally on metal-contaminated soils. Other non-metallophytes have evolved genetically distinct strains which are resistant to metals and are able to grow much better in contaminated situations than non-resistant plants of the same species (Schat and Bookum, 1992). Resistant plants may also grow more successfully in nutrient-poor conditions which often exist in acidic mine waste (Ernst *et al.*, 1992) (see section 13.6). There has been considerable interest in using tolerant plants to revegetate old mine sites (Plenderleith and Bell, 1990).

A number of techniques have been developed for quantifying metal tolerance on plants (Baker and Walker, 1989). The degree of tolerance is measured by comparing the response with tolerant plants (from contaminated sites) and expressing this as a percentage of the performance of non-tolerant plants (collected from 'clean' areas). The principal parameters measured are seedling survival, biomass, shoot growth and root growth.

Seedling survival is the number of plants surviving from seed after a specified time period. It is a better measure then straightforward germination. Many non-resistant plants will successfully germinate in metal-contaminated soil but subsequently they fail to grow and remain characteristically stunted.

Rates of dry matter production and the biomass yield of resistant plants are generally found to be lower that their non-resistant counterparts when grown in uncontaminated soil. This reduction is believed to be due to the energy expenditure in metal-tolerance mechanisms, such as compartmentalization of metals in intracellular 'compartments' (Vasquez *et al.*, 1994) (see section 13.6.2). When grown in contaminated soil or nutrient solutions, the growth of resistant plants exceeds that of non-resistant ones. Other ways of measuring this difference are shoot and root length. Shoot length can be compared in plants grown in soil but measuring differences in root length is much easier if the plants are reared in nutrient solutions in clear containers. Regular monitoring of the root length can be conducted without disturbing the plants. This approach has been used in numerous demonstrations of resistance as it allows the composition of the liquid medium to be closely controlled (table 6.2).

Deleterious effects on plants of gaseous pollutants such as ozone can be detected by methods in addition to those reported above. Yellowing of leaves, **chlorosis**, is a characteristic feature of stress. The critical air concentrations can be determined by exposing replicates to different levels of the gas under test in separate chambers, although such experiments are expensive to conduct. The degree of chlorosis can be measured by counting the number of leaves affected or by determining the area of leaf exhibiting the yellow coloration (Mehlhorn *et al.*, 1991). Different varieties of the same species may exhibit different degrees of chlorosis at the same chemical concentration and may be useful for biological monitoring of pollution (e.g. tobacco; see Chapter 11).

6.4 *Toxicity testing with aquatic organisms*

The basic principles of aquatic toxicity testing are similar to those already described for terrestrial organisms. However, there are particular questions about the main routes of uptake which influence some aspects of the design of tests.

With aquatic organisms, direct uptake from water is a route of major importance (e.g. uptake across the gills of a fish or across the permeable skin of amphibia). Uptake can also occur from food during its passage through the alimentary system and bottom-dwelling organisms are exposed to residues in sediment. The

TABLE 6.2 *Percentage tolerance of three species of grass collected from Hallen Wood (contaminated by aerial deposition from a smelting works) and Midger Wood (uncontaminated)**

	Hallen plants	**Midger plants**
Dactylus glomerata	105.9	57.4
Holcus lanatus	113.8	28.1
Deschampsia	82.2	37.5

*The tolerance was calculated from the mean length of roots in 2 p.p.m. Cd per mean length of roots with no Cd. The measurements were made after 14 days in full-strength Hoaglands culture solution. From Martin and Bullock (1994).

relative importance of these routes of uptake differs between organisms and between chemicals and depends on environmental conditions. In some cases, all of these routes may operate in one organism at one time.

6.4.1 TESTS CONCERNED WITH DIRECT ABSORPTION FROM WATER

Much of the toxicity testing carried out with aquatic organisms (e.g. fish, *Daphnia*, *Gammarus pulex*) is concerned with direct absorption of chemicals from water. The chemicals may be in solution, in suspension or both. Organisms are exposed to different concentrations of the chemical in water to determine values for median lethal concentration. Although absorption is primarily direct from water, it is not possible to completely exclude contamination of food, and thus some uptake from this source.

One difficulty in aquatic toxicity testing is maintaining a constant concentration of chemical in water. Chemicals are lost in water because of (i) absorption and metabolism by the test organism and (ii) volatilization, degradation and adsorption from water. Where the rate of loss is relatively low, tests may be performed using **static** or *semistatic* systems. With static systems, the water is not changed for the duration of the test. With the semistatic systems, the water is replaced at regular intervals (usually every 24 h). A better, but more complex and expensive, method for renewing test solutions is provided by a **continuous flow** (*flow through*) system. With systems of this type, test solution is continuously renewed, thus ensuring a constant concentration of the test chemical and preventing the build-up of contamination from faeces, algae, mucus, etc. If organisms are exposed to a chemical for a sufficiently long time, steady-state concentrations will be reached in the tissues (Chapter 5).

The toxic effect of a chemical depends upon the concentration present in the tissue(s) where the site of action is located (Chapters 5 and 7). This, in turn, depends upon the concentration of the chemical in water and the period of time over which exposure has occurred. Thus, the median lethal concentration (LC_{50}) is related to the exposure period (figure 6.9). With increasing exposure time, the LC_{50} becomes less until the median lethal threshold concentration (threshold LC_{50}) is reached. At this point, further increases in exposure period cause no change in mortality. It may reasonably be supposed that when the threshold LC_{50} is reached the system is in a steady state, i.e. the tissue concentration is no longer increasing with time.

In performing aquatic toxicity tests, preliminary ranging tests are necessary to obtain a rough estimate of toxicity. Here, small groups of test organisms (typically two or three individuals per group) are exposed to a wide range of concentrations of the test chemical on a log scale. The results of the ranging test can be used to plan a full toxicity test, in which larger numbers of test animals are exposed to a narrower range of concentrations, centring on the LC_{50} estimated from the ranging test. The percentage mortality in each of the test groups is recorded over various time intervals for the duration of the test. The data obtained can be used to determine the LC_{50} at different exposure times.

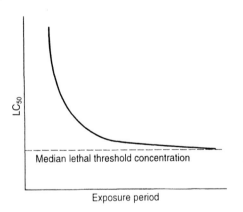

FIGURE 6.9 *Relationship of LC_{50} to the exposure period.*

Initially, the data can be plotted in two ways.

1. At one fixed exposure concentration, a series of chosen survival times can be plotted against the percentage of the sample surviving at each of them (e.g. 25% of the sample survive for 4 days). This can then be repeated for each of the remaining concentrations to obtain values for the median survival periods (median periods of response).

2. At one fixed exposure period, the percentage mortality can be related to the exposure concentration. From this, the median lethal concentration can be calculated for each of the individual exposure periods (figure 6.10).

From these data, two further plots can be produced. Using data from 1, median survival time can be plotted against concentration of chemical. Using data from 2, the median lethal concentration can be related to exposure period, and from this LC_{50} values can be estimated for particular exposure periods, e.g. LC_{50} at 96 h.

Up to this point, the discussion has been restricted to lethal toxicity testing. It should be emphasized, however, that end-points other than lethality may be used. More generally, median effective concentrations (EC_{50}) can be determined for a variety of non-lethal endpoints.

As the toxic effect of a chemical is highly dependent upon the period of exposure, two important issues are: (i) over what period should a toxicity test be conducted? and (ii) what exposure periods should be chosen for the estimation of EC_{50} or LC_{50}? The answer to both of these questions depends on the organism be-

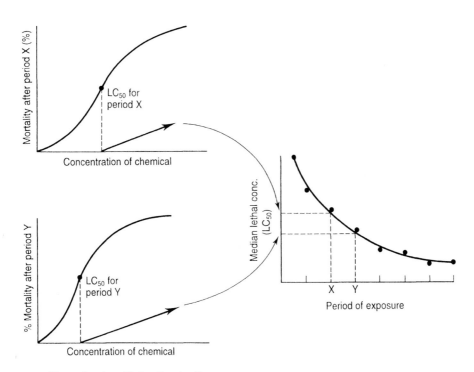

FIGURE 6.10 *Determination of* LC_{50}. *For details, see text.*

ing tested and the purpose of the test. Tests with *Daphnia* are commonly of only 24–48 h duration in static systems. By contrast, fish toxicity tests are usually of longer duration. Simple screening tests commonly require relatively short exposure periods. Typical measures of toxicity are LC_{50} at 48 h for *Daphnia* and LC_{50} at 96 h for fish. Such data give an indication of the toxicity of one chemical in relation to others (i.e. a ranking of toxicity). On the other hand, longer exposure periods may be required in which toxicity data are needed for the evaluation of water quality. Here, exposure may be continued until a median threshold concentration has been established (figure 6.9).

Some aquatic toxicity tests work to non-lethal end-points, and examples of this are given in boxes 6.4 and 6.5.

A number of factors influence the values obtained in aquatic toxicity testing, apart from the inherent properties of the chemical and of the test organism. A critical question is the availability of the chemical to the test organism. Availability may be limited because of ad-sorption to particulate matter or to the surface of the tank.

6.4.2 SEDIMENT TOXICITY TESTS

Sediments represent an important source of pollutants to aquatic organisms. Lipophilic organic pollutants and some metals are strongly retained by sediments in which they have long half-lives. An example of a sediment toxicity test is given in box 6.6.

The toxicity of chemicals associated with sediments is difficult to assess because it is uncertain how much of the total quantity is available to the test organism. Strongly adsorbed molecules which exist at low concentrations in water may nevertheless be directly absorbed by aquatic animals with passage of sediment particles through the alimentary tract or across respiratory surfaces. The development of improved sediment toxicity tests is being actively pursued.

BOX 6.4 *Daphnia reproduction test (figure 6.11A).*

Neonates of **Daphnia magna** less than 24 h old are used in this test. One neonate is placed into 60–80 ml of test solution, contained within a 100 ml beaker. Usually, ten replicates are used for five test concentrations plus a control. Thus, sixty neonates are used per test. The test is performed at 20°C, using a 16:8-h L/D photoperiod. Every 2 or 3 days, the number of surviving organisms and the number of young are determined (static renewal system). The adults are then moved to fresh test solution and the young are discarded. *Daphnia* are fed on microalgae (typically *Chlorella* or *Scenedesmus*). There is no generally agreed standardization of the food supply and variations in the quantity and quality can affect the outcome of tests. For example, deficiencies of certain trace elements in food may not become apparent until several generations have been cultured (see Caffrey and Keating, 1997). Usually, the first brood is observed after 8–10 days, with subsequent broods appearing at approximately 2-day intervals. The duration of the test is 21 days (five broods). The reproduction data obtained are used to calculate lowest observed effect concentration (LOEC) and NOEC data. Comparison is made between the number of offspring per surviving female in the toxicant treatments and the reproductive output of controls.

Care should be taken to standardize the clone used for the test. Baird *et al.* (1990) found a more than 100-fold difference in the LC_{50} for cadmium in *D. magna* between two clones.

FIGURE 6.11 *Organisms used in aquatic ecotoxicity testing. (A)* Daphnia magna *of 2 mm length. (B) Chironomid larvae of 1 cm length. Photographs by Steve Hopkin.*

Box 6.5 *Algal toxicity test.*

The essential feature of an algal toxicity test is to determine the effects of a chemical on a green algal population growing exponentially in a nutrient-enriched medium for 72 or 96 h (Lewis, 1990). Cell density is determined either using a direct measurement (microscopic counting) or one of several indirect techniques (spectrophotometric method or electronic particle counter). Controlled experimental factors include the initial cell density, test temperature, pH, light quality and intensity. Typically, an EC_{50} value and the NOEC are calculated based on growth inhibition. The EC_{50} value is the concentration of the test substance that causes a 50% reduction of the effect parameter.

The most frequently used test species is the freshwater green coccoid *Selenastrum capricornutum*, which is the only species recommended by the US Environmental Protection Agency for monitoring the toxicity effects of effluents. At least 10 other species have been used. However, Lewis (1990) concluded that these are too few data to support the environmental relevance of results using the standard method in their present format. The main problem is that interspecific variation in response to chemicals is significant and the field validation of most laboratory-derived results is lacking (a problem common to most tests). The way forward is probably to conduct simultaneous multiple species tests to assess the relative toxicities of different chemicals to different species.

Box 6.6 *Sediment toxicity test using the marine amphipod* Rhephoxynius abronius.

Sediment samples are taken in the field, to a depth of 2–5 cm below the sediment–water interface. After mixing sediments, subsamples are removed and are taken through 0.5-mm sieves to remove biota, before adding to 1 l flasks containing aerated seawater to a depth of 2 cm. Samples of contaminated sediments may be diluted with varying proportions of clean sediments when assessing their toxicity. Controls are run with samples of clean sediment. Sediment samples may also be spiked with different concentrations of chemicals whose sediment toxicity is to be assessed. Samples of *R. abronius* are obtained from uncontaminated sites and are acclimated at the test density for 4 days, before adding 20 test organisms to each flask. The duration of this static test is usually 10 days. Survival is the most commonly used end-point. Where fivefold replication was used for this test, there was a 75% certainty of detecting a mean survival change of 15%.

The same general principle has been followed with **chironomid** larvae (figure 6.11B).

6.5 *Risk assessment*

The toxicity data obtained by the testing procedures described here are eventually used to make assessments of hazard and risk (Walker *et al*, 1991a; Calow, 1993, 1994; Suter, 1993). For the purposes of this discussion, the following definitions will be used:

hazard = the potential to cause harm

risk = the probability that harm will be caused

Risk assessment depends on making a comparison between two things:

1. the toxicity of a compound expressed as a concentration (EC_{50}, LC_{50} or NOEC can be used);
2. the anticipated exposure of an organism to the same chemical, expressed in the

same units as 1 (a concentration in water, food or soil to which the organism is exposed).

In hazard assessment, a toxicity test can give a plot which relates the frequency of a toxic effect (e.g. mortality) to the dose that is given (figure 6.1). From this, NOEC and an EC_{50} can be estimated. This can be compared with a putative 'high' environmental concentration to decide whether a hazard exists. A ranking of compounds according to their toxicity is important at this stage. If toxicity is very low, then a compound is not regarded as being hazardous.

In risk assessment, further calculations are carried out to obtain values for the predicted environmental concentration (PEC) and the predicted environmental no effect concentration (PNEC). The details of these calculations lie beyond the scope of the book. In the case of PEC, calculations are based on known rates of release and dilution factors in the environment. If, for example, a chemical is used on an industrial process, the level of the industrial effluent is measured or calculated. This figure is then divided by the dilution that occurs in receiving waters (e.g. river or lake) to obtain a value for PEC. The PNEC can be estimated by dividing LC_{50} or EC_{50} for the most sensitive species tested in the laboratory by an arbitrary safety factor (often 1000). This is to allow for the great uncertainty in extrapolating from laboratory toxicity data for one species to expected field toxicity to other species.

$$\frac{PEC}{PNEC} = \text{risk quotient}$$

If this value is well below 1, the risk is low. If it is 1 or above, there is a substantial risk.

Clearly, calculations such as these provide only rough estimates of risk. It is necessary to include a large safety factor to allow for uncertainties about environmental toxicity. Apart from the uncertainty over toxicity to particular organisms in the field, there is the further problem of estimating environmental exposure. In terrestrial ecosystems, it is often very difficult to obtain a realistic value for PEC. If a mobile species (e.g. bird, mammal or insect) obtains a chemical mainly from its food – and the chemical is distributed very unevenly throughout the ecosystem – then it is not possible to estimate exposure with any degree of accuracy. A grain-eating bird may not consume grain treated with a pesticide; a mammal may or may not be in a field when it is sprayed with a pesticide.

Considerations such as these strengthen the case for the development of new strategies using biomarkers for the purpose of risk assessment. Biomarker assays can provide measures of exposure and sometimes of toxic effect under actual field conditions, e.g. in field trials with pesticides. This issue will be discussed further in Chapter 15.

6.6 *Toxicity testing in the field*

If the normal procedures of risk assessment raise uncertainties about the safety of a chemical, further testing may be necessary before a decision can finally be taken about permitted release into the environment. In the case of pesticides, field trials are sometimes required. In a full-scale field trial, a pesticide is likely to be applied at a dose, and under conditions which are most likely to produce toxic effects ('worst case scenario') (Somerville and Walker, 1990). A variety of measurements may be made, depending on the chemical, the method of application, the habitat, the climate and agricultural system. These may include looking for dead animals and birds ('corpse counting'), estimating population numbers and breeding success, measuring residues in soil, crop and animals and, occasionally, biomarker assays (e.g. cholinesterase inhibition, eggshell thinning in birds). Trials such as these are expensive and are not undertaken lightly. An example of a field study concerned with the toxic effects of a radionuclide is given in box 6.7.

Of growing interest are 'mesocosms' – small, carefully controlled systems which provide some simulation of the conditions in the natural environment. Experimental ponds provide an example of mesocosms. Ponds of standard size can be established, which become colonized by plants, insects and vertebrates. The effect of chemicals on the pond communities can then be tested. Such mesocosms provide a half-way house between closely controlled laboratory tests and extensive, only loosely controlled, full-scale field trials. One of their advantages is that they do allow replication, which is not usually possible in full-scale field trials (for reviews of mesocosms, see Ramade, 1992; Crossland, 1994).

6.7 Alternative methods in ecotoxicity testing

In recent years, there has been mounting op-

BOX 6.7 *Field study to assess environmental effects of ^{137}Cs on birds.*

One interesting recent study (Lowe, 1991) has demonstrated the importance of laboratory testing for field extrapolation if one is to prove or disprove a causative relationship between levels of an environmental pollutant and an ecological effect. This concerns an area in the vicinity of the Sellafield nuclear waste reprocessing complex in North-west England.

Since 1983, concern had been expressed about the apparent decline in numbers of birds in the Ravenglass estuary in West Cumbria, particularly of the black-headed gull colony on the Drigg dunes. Suggestions had been made that the decline might be due to excessive radiation in the birds' food and their general environment. In Lowe's (1991) study, 12 species of marine invertebrate from Ravenglass were analysed. Most of them are known to be important food for birds but none of them showed excessive contamination with radionuclides.

Analysis of a sample of carcasses from the area showed that oystercatchers (*Haematopus ostralegus*) and shelduck (*Tadorna tadorna*) had some of the highest concentrations of ^{137}Cs of all birds in their tissues, yet their breeding success and populations were unaffected.

Black-headed gulls, on the other hand, were found to be feeding mainly inland and were the least contaminated with radionuclides of all the birds at Ravenglass, yet this species and its breeding success were in decline. Calculations of the total dose equivalent rate to the whole body of the most contaminated black-headed gull amounted to 9.8×10^{-4} mSv h^{-1} (about 8.4×10^{-4} mGy h^{-1}, whole body absorbed dose rate), and the background exposure dose was of the order of 8.3×10^{-4} mGy h^{-1}. As a minimum chronic dose of 1000 mGy day^{-1} has been found necessary to retard growth of nestling birds, and 9600 mGy over 20 days of incubation to cause the death of 50% of embryos in black-headed gulls' eggs in laboratory experiments, the concentrations of radionuclides in the food, body tissues and general environment were at least three orders of magnitude too low to have any effect.

The most likely cause of the desertion of the gullery was the combination of an uncontrolled fox population, the severest outbreak of myxomatosis among the rabbits since 1954 and the driest May–July period on record, all in the same year (1984).

position to the use of animals in testing procedures which cause suffering to them. In particular, toxicity tests upon vertebrates which use lethality as the end-point have been heavily criticized (Balls *et al.*, 1991; Van Zutphen and Balls, 1997; Walker, 1998b; Walker *et al.*, 1998). These objections have been raised to tests for human risk assessment as well as ecotoxicity tests. Quite apart from ethical considerations such as these, there have also been strong criticisms of existing practices in ecotoxicity testing on scientific grounds. Organizations actively involved in the quest for alternative methods include **Fund for the Replacement on Animals in Medical Experiments (FRAME)** and the **European Centre for the Validation of Alternative Methods (ECVAM)**.

The limited value of estimates of environmental risk based on data from standard toxicity tests was explained in the previous section. The ultimate concern in ecotoxicology is about the effects that pollutants have at the levels of population, community and ecosystem. Estimates of the probability that certain individuals may experience toxic effects tell us virtually nothing about this. Indirect effects, such as the decline of the grey partridge on agricultural land as a consequence of the use of herbicides (see Chapter 12), are not predictable by ecotoxicity testing! Indeed, the herbicides normally used on agricultural land have very low toxicity to partridges and other farmland birds and would appear very safe in a risk assessment exercise. The development of alternative methods for ecotoxicity testing could serve two purposes – improving the quality of the science and reducing suffering caused to animals.

Alternative testing procedures can be divided into two major categories:

1. alternative methods for estimating toxicity to vertebrates;
2. alternative methods and strategies which work to more ecologically relevant end-points.

Although the first of these addresses most directly the immediate question of replacing tests that cause suffering to vertebrate animals, the latter deals with the longer-term issue of the development of a different and more ecological approach to environmental risk assessment. If strategies and tests working to more ecological end-points are successfully developed, there may well be a case for redirecting limited resources away from current testing practices towards others that provide better assessments of environmental risk. These two different approaches will now be briefly described.

6.7.1 ALTERNATIVE METHODS FOR ESTIMATING TOXICITY TO VERTEBRATES

Toxicity tests that use live vertebrates

It is frequently asserted that, with our present state of knowledge, it is important to continue with some toxicity tests upon live vertebrates because of the uncertainties of estimating vertebrate toxicity by other means. Even in this situation, however, steps can be taken to reduce suffering. The three major aims of FRAME, ECVAM and related organizations are the replacement, reduction and refinement of toxicity tests upon animals (the three Rs). The last two can still be followed where vertebrate toxicity testing continues. Included here are testing procedures which work to a lethal end-point, but reduce the number of animals used. The so-called 'fixed dose procedure' provides an example of this approach (see 6.3.3). Of greater long-term interest is the use of other end-points, including biomarker responses in tissues such as blood, which can be obtained by non-destructive sampling. Biomarker responses can give early warning of the operation of a toxic mechanism or process before overt symptoms of intoxication are shown. At present, this approach is only feasible for a limited number of chemicals whose mode of action is well understood. In time, however, it should become more widely applicable, as we learn more about the mode of action of toxic chemicals.

Toxicity tests on non-vertebrates

Sometimes, correlations have been shown between the toxicity of a group of related chemicals to vertebrates and their toxicity to a non-vertebrate species, e.g. the toxicity to fish sometimes correlates well with toxicity to *Daphnia*. Such correlations are likely where there is a similar mode of action in the vertebrate to that in the non-vertebrate. However, toxicity is dependent on toxicokinetic factors (e.g. metabolism), which differ greatly between vertebrates and other groups, and the actual toxicity expressed as LD_{50} or LC_{50} can vary greatly between vertebrates and invertebrates even when considering only compounds with a similar mode of action. Over a wide range of environmental chemicals of differing modes of action, tests with non-vertebrate species do not give a reliable guide to vertebrate toxicity.

Tests using cellular systems

Vertebrate cells, for example hepatocytes of mammals, fish and birds, are sometimes used to measure toxicity. Sometimes, these contain reporter genes, which mediate a characteristic response when there is exposure to chemicals that operate a toxic mechanism. One example of this is the use of genetically modified mouse hepatocytes, which emit light when dioxins or planar PCBs interact with the Ah receptor (**CALUX** system) (Murk *et al.*, 1997). Such systems can be very useful in screening procedures to show the presence of chemicals operating a particular toxic mechanism in a vertebrate species. However, they do not provide reliable estimates of median lethal doses that would be found in a normal toxicity test.

Predictive models

The best known examples are quantitative structure–activity relationships (**QSARs**), which have been used to predict toxicity from physicochemical properties of environmental pollutants. At the present stage of development, QSARs cannot be regarded as workable alternatives to toxicity tests, although they can give valuable information and are likely to be of greater predictive value in future, as the technology improves. A more promising approach, in the longer term, may be the incorporation of *in vitro* data obtained from cellular systems (see above) into more sophisticated models that incorporate toxicokinetic parameters. Such an approach is now being considered to obtain better prediction of the human toxicity of drugs and environmental chemicals utilizing data obtained from human *in vitro* systems. In theory, it should help to overcome the serious problem of making interspecies comparisons when evaluating toxicity; a problem that is more difficult in ecotoxicology than in human toxicology! Many species can be studied *in vitro* which are not available for normal toxicity testing; practical, ethical and economic factors all limit the number and the variety of species that can be used for ecotoxicity testing.

6.7.2 ALTERNATIVE METHODS AND STRATEGIES THAT WORK TO MORE ECOLOGICAL END-POINTS

The ultimate concern in ecotoxicology is about effects that chemicals have at the level of population and above in the field. The whole question of effects at these higher levels is the subject of the last part of the present text. In the context of toxicity testing, however, the possibility of using data that relate to such effects in the field instead of, or in addition to, toxicity data obtained in the laboratory for individual species will now be briefly considered.

Field studies

'Field studies' is a very broad title, encompassing a wide range of activities from various types of monitoring to field trials with pesticides. In theory, the study of the effects of chemicals on individuals, populations and communities in the field addresses directly the basic issues of ecotoxicology. The overriding problem with field studies is the very limited opportunity to control variables such as temperature, rainfall, movements of water and air and migration of animals, all of which can influence the effects of environmental chemicals. This can be carried out to a limited extent in field trials, where pesticides, for example, are applied in a controlled way and effects are monitored, making comparison with control areas. The use of biomarkers to measure responses to the chemical in individual organisms can provide a causal link between exposure to a chemical and a change at the population level (e.g. population decline, decline in reproductive success or increased mortality rate), as will be explained further in Chapter 15. It is also possible to bring contaminated material, for example soil from the field, into the laboratory and then to expose animals under more controlled conditions.

Many field studies do not involve controlled release of chemicals, but rather the investigation of an ongoing pollution problem (see examples given in Chapters 12 and 15). Some degree of experimental control is still possible; for example, the deployment of 'clean' organisms along pollution gradients. Residues of chemicals detected by analysis can be related to biomarker responses that they are known to cause, and these responses, in turn, can be related to population changes. Changes in

populations and communities caused by chemicals may be identified by 'biotic indices', such as river invertebrate prediction and classification (RIVPACS) (Wright *et al.*, 1993) or structural analysis of communities (see Chapter 14). The development of resistance to chemicals may involve genetic changes that can be measured and related to their original cause.

Microcosms and mesocosms

Model populations or communities can be established in the laboratory or the field (see section 6.6) which simulate but do not exactly reproduce 'the real world'. They can be used to run controlled experiments with adequate replication and can demonstrate the effects of chemicals on ecological processes such as the carbon and nitrogen cycles (see Chapter 14). The major problem is in interpreting data coming from them and relating it to 'the real world'.

Theoretical models

Data from biomarker studies, field studies and microcosms/mesocosms can be incorporated into mathematical models which attempt to predict effects of chemicals at the level of population or above. This approach is at an early stage of development and will not be discussed further here, but will be mentioned in the final section of the text.

6.8 *Summary*

Concern over the possible environmental impact of 'new' and existing chemicals has led to the development of a range of methods for testing their biological effects. End-points of such tests include mortality, reproduction and, more recently, behaviour. The most widely used organisms are algae, earthworms, springtails, honeybees, *Daphnia*, fish and birds. Several other animals have been proposed in recent years, and alternative testing methods are under active development. There are, however, problems in uncritically extrapolating the results of such tests to field conditions. Tests provide a useful 'safety net' for screening for the most toxic chemicals, but are unlikely ever to predict ecological effects on all species.

6.9 *Further reading*

BALLS, M. *et al.* (eds) (1991) *Animals and Alternatives in Toxicology*. A broad view of alternative tests, including ecotoxicity tests.

CALOW, P (ed.) (1993, 1994) *Handbook of Ecotoxicology*, Vols 1 and 2. A comprehensive text, describing all the main types of ecotoxicity testing and the scientific bases for them. Useful description of risk assessment.

HUDSON, R. H. *et al.* (1984) *Handbook of Toxicity of Pesticides to Wildlife*. Many examples of toxicity of pesticides to wildlife.

LEWIS, M. A. (1990) A review of algal toxicity testing.

LØKKE, H. and VAN GESTEL. K. (eds) (1998) *Handbook of Soil Invertebrate Toxicity Tests*. A comprehensive manual.

SUTER, G. W. (1993) *Ecological Risk Assessment*. Gives an ecological view of risk assessment.

TIMBRELL, J. (1995) *Introduction to Toxicology*, 2nd edn. A straightforward account of toxicity and toxicity testing.

WALKER, C. H. (1998b) and WALKER, C. H. *et al.* (1998) Review alternative methods for ecotoxicity testing.

CHAPTER **7**

Biochemical effects of pollutants

7.1 Introduction

When pollutants enter living organisms, they cause a variety of changes (for a general account, see Guthrie and Perry, 1980; Hodgson and Levi, 1994; Timbrell, 1995). These changes (bio-effects) are, broadly speaking, of two kinds: those which serve to protect the organisms against the harmful effects of the chemical and those which do not. Some examples of both types are given in table 7.1.

We consider, first, **protective responses**. Some protective mechanisms work by reducing the concentration of free pollutants in the cell, thereby preventing or limiting interactions with cellular components which may be harmful to the organism. Organic pollutants often cause the induction of enzymes that can metabolize them (see also Chapter 5). One of the most important of these enzyme systems is the monooxygenase system, whose function is to increase the rate of production of water-solu-

TABLE 7.1 *Protective and non-protective responses to chemicals*

Type of effect	Example	Consequences
Protective	Induction of monooxygenases	Increase in rate of metabolism of pollutant to more water-soluble metabolite and thus increase in rate of excretion (Chapter 5)
	Induction of metallothionein	Increases the rate of binding with metals to decrease bioavailability
Non-protective (may or may not lead to toxic manifestations)	Inhibition of AChE	Toxic effects seen above 50% inhibition
	Formation of DNA adducts	May cause harmful effects if leading to mutation

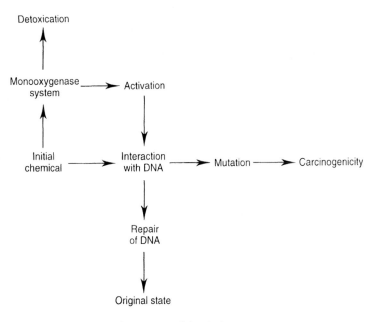

FIGURE 7.1 *Pathways for activation and detoxication of chemicals.*

ble metabolites and conjugates of low toxicity which can be rapidly excreted. In this case, metabolism causes detoxification. However, in a small, yet important, number of cases, metabolism leads to the production of active metabolites (e.g. of carcinogens or organophosphorous insecticides) that can cause more damage to the cell than the original compounds. An outline of these reactions is shown in figure 7.1.

Another mechanism whereby the bioavailability of pollutants is reduced is by binding to another molecule. This can lead to excretion or storage. The metallothioneins are examples of proteins which can bind metal ions and which become induced when there is exposure to high metal concentrations. There are also inducible proteins which can bind organic pollutants; this mechanism provides the basis of resistance to some drugs by removing them from the cell.

In addition to protective mechanisms which control the levels of free pollutants, there are other responses which are concerned with the repair of damage caused by pollutants. The release of stress proteins falls into this category. When organisms are exposed to chemical insult or heat shock, stress proteins are released which have the function of repairing cellular damage. Similarly, if pollutants cause damage to DNA, repair mechanisms can come into play (figure 7.2). There are many examples of homeostatic mechanisms such as these which restore cellular systems to their normal functional state after toxic damage has been caused by chemicals.

Pollutants also cause changes *unrelated to any protective function* (table 7.1). These include alterations in the level of function or enzymes, receptors and reactive intermediates and changes in DNA. In many cases, such changes cause no obvious harm to the organism and appear to be unrelated to toxicity, although they can seldom be regarded as beneficial; indeed, as with protective mechanisms, there will be an energy cost associated with them (Chapters 8 and 13). In others, molecular changes have serious consequences at the level of the cell or

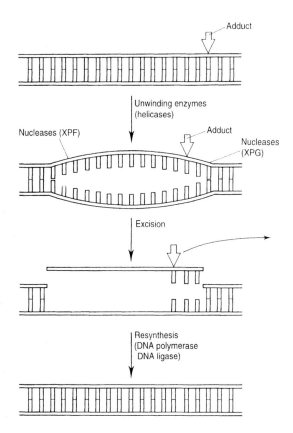

FIGURE 7.2 *Mechanism of DNA repair after adduct formation to remove covalently bound adduct.*

the whole animal. The health of the organism is adversely affected, sometimes so seriously that death will ensue. Here, the biochemical changes are at the site of action and represent the molecular basis of toxicity. The question of the relationship between biochemical changes of this kind and toxic manifestations at higher levels of organization will be returned to later (Chapter 10). The changes caused by pollutants entering living organisms may be specific for a particular type of chemical or they may be nonspecific. Some examples, in decreasing order of specificity, are given in table 7.2. The use of these changes to assess the impact of chemicals is considered in the chapter on biomarkers (Chapter 10) (Walker, 1992).

Similarly, responses to any particular pollutant may be specific to certain groups of organisms or may occur in nearly all organisms. The degree of induction of monooxygenases in response to a particular pollutant shows considerable phylogenetic variation between species. Thus, some inducing agents such as barbiturates produce marked responses in mammals, but no responses in fish or in some birds. In the following account, two categories of biochemical effects will be considered. First, those which have a protective function, and, second, those which represent the molecular mode of action of pollutants. Consideration will be given to the consequences of these effects at the levels of the cell and of the whole organism.

7.2 *Protective biochemical responses*

When the concentration of a xenobiotic or inorganic ion exceeds a certain level in the cell, it may trigger off responses designed to protect the organism against potential toxic effects. Very commonly, this response is an increase in the *quantity* of a protein which can facilitate the removal of the molecule or ion. In the case of lipophilic xenobiotics, a number of enzymes are *induced* which can increase the rate of biotransformation of the molecule to water-soluble and readily excretable metabolites and conjugates. (**Induction** involves an increase in the activity of an enzyme as a consequence of an increase in its cellular concentration produced in response to a chemical.) Prominent among these enzymes are the monooxygenases of the endoplasmic reticulum of vertebrates and invertebrates (Chapter 5), which have cytochrome P_{450} as their catalytic centre. A number of inducible forms exist in the livers of vertebrate animals. Induction involves an increase in both cytochrome P_{450} and the enzyme activities associated with it. In mammals, a

TABLE 7.2 *Specificity of response*

Biological response (biomarker)	Chemicals	Comment
Inhibition of ALAD	Lead	Specific for lead
Inhibition of vitamin K cycle	Anticoagulant rodenticides, e.g. warfarin	These compounds are antagonists for vitamin K
Inhibition of AChE	Organophosphorous compounds and carbamates	Specific to these two classes, OPs can be separated from carbamates by strength of binding to AChE
Induction of monooxygenases	Organochlorines, polynuclear aromatics	A wide variety of man-made and natural chemicals cause induction; particular P_{450} isoforms may be induced by certain types of pollutant (Chapter 5)

number of inducible forms of cytochrome P_{450} belong to a single gene family – family 2 (Nebert and Gonzalez, 1987).

Many lipophilic xenobiotics are both inducers and substrates for this type of cytochrome P_{450}. Another group of enzymes – cytochrome P_{450} family 1 – have a much more restricted range of inducers and substrates. These enzymes interact particularly with flat (co-planar) molecules, such as polycyclic aromatic hydrocarbons (PAHs), co-planar PCBs and dioxins. Although metabolism by a cytochrome P_{450} usually causes detoxication, there are exceptions. In particular, induction of a cytochrome P_{450} of family 1 can cause increased activation of carcinogens (e.g. certain PAHs) and of co-planar PCBs which can act as thyroxine antagonists. The induction of different types of cytochrome P_{450} in diverse groups of animals is reviewed in Livingstone and Stegeman (1998).

When certain metal ions exceed a critical cellular level, another type of protein is induced. Metallothioneins can increase in concentration after exposure to various metals. These are binding proteins, rich in SH groups, which can lower cellular concentrations of metal ions such as Cu^{2+}, Cd^{2+} and Hg^{2+} by sequestrating them (Hamer, 1986).

The foregoing responses are concerned with prevention of toxic damage by the simple strategy of removing potentially harmful xenobiotics and ions before they interact to a significant degree with their sites of action. Other responses correct damage after it has occurred. Important examples are the production of stress proteins and the operation of DNA repair mechanisms. Chemical and heat shock can cause the release of stress proteins into the cell, which have the function of repairing cellular damage. When chemicals cause damage to DNA by forming adducts, there is an increase in the operation of DNA repair mechanisms (figure 7.2).

7.3 Molecular mechanisms of toxicity

An understanding of mechanisms of toxicity at the molecular level is important for several reasons. First, it can pave the way for developing antidotes to the adverse effects of the pollutant. Second, the understanding of the mechanism can lead to the development of assays which can be used to demonstrate and measure the deleterious effects of the chemical. Both of these are important in human medicine, which has been the driving force for much of this work. An example of this is shown for the inhibition of acetylcholinesterase by the active forms of an organophosphorous insecticide (OP) in figure 7.3. The OP interacts with a hydroxyl group which is a functional part of the enzyme. The phosphorylated enzyme produced has no activity, i.e. it cannot hydrolyse the natural substrate acetylcholine. Another influential factor has been the interest in developing new biocides. If the site of action of a pesticide is known, a molecular model of it can be constructed. Other molecules can then be identified which will fit the same site and will interact there.

Molecules which fit the model can then be synthesized and tested as novel biocides.

Some of the most toxic compounds known are highly reactive and readily form covalent bonds with their sites of action. Frequently, reactive molecules of this type are generated by enzymatic attack upon compounds which are, in themselves, unreactive and 'non-toxic'. This is the case with many chemical carcinogens [e.g. benzo(*a*)pyrene (BaP), dibenz(ah)anthracene, aflatoxin, acetylaminofluorene and others]. The enzymes involved (often the monooxygenase system that usually protects the organism) can sometimes generate the reactive compounds that cause cellular damage. This phenomenon is considered in more detail in section 7.4.1 on genotoxic compounds.

FIGURE 7.3 *Mechanism of action of AChE. Under normal conditions, acetylcholine binds to acetylcholinesterase and is then broken down (hydrolysed) to yield acetic acid and choline which break away from the enzyme. Organosphosphates bind to a hydroxyl group belonging to the amino acid serine which is part of the binding site shown on the right-hand side of the enzyme surface. When this happens, the enzyme is inhibited and can no longer hydrolyse acetylcholine. There is currently dispute over the nature of the binding site for acetylcholine shown on the left-hand side of the diagram.*

Toxic effects are sometimes due to molecular interactions which do not lead to the formation of covalent bonds. The reversible binding of a molecule to a site on a cellular macromolecule can lead to toxicity. The target may be a receptor site for a chemical messenger [e.g. a receptor for acetylcholine or GABA (gamma aminobutyric acid)] or it may be a pore through a membrane which normally allows the passage of ions (e.g. the blockage of Na^+ channels by tetrodotoxin).

To gain a full understanding of the toxic effects of chemicals, it is necessary to link initial molecular interactions to consequent effects at higher levels of organization. The extent to which such a molecular interaction occurs is, in a general way, related to the dose of the chemical received – although the relationship is rarely a simple one. There are several reasons why many dose–response relationships are not simple. Low-level exposure, up to a particular threshold, may produce no measurable interaction at all. In some cases, protective mechanisms remove the chemical before it can reach its site of action. For this reason, there are often major differences between *in vitro* and *in vivo* experiments. In other cases when the dose exceeds a particular threshold level, protective mechanisms may come into play which reduce the amount of chemical reaching its active site. An example of this is the induction of metallothionein which decreases the bioavailability of toxic metals such as cadmium. Another reason for lack of effect at low levels of exposure is that many systems have reserve capacity. The enzyme carbonic anhydrase, which catalyses the conversion of carbon dioxide to carbonic acid, has to be inhibited by more than 50% before physiological effects are seen. Similarly, the inhibition of brain acetylcholinesterase must usually exceed 50% before consequent physiological disturbances are seen (table 7.1). In other cases, especially carcinogenicity, it has been argued that no minimum

safe level exists. In theory, a single molecular interaction can initiate the whole process that leads to the development of cancer.

Thus, dose–response relationships may be complex with regard to the molecular interactions which underlie toxicity. Clearly, there can be further complications when relating dose to toxic manifestations at higher levels of organization (e.g. toxic effects seen at the level of cell function or the function of the whole organism.). Here, it is not just a question of the relationship between dose and the degree of molecular interaction (e.g. the percentage inhibition of an enzyme); there may also be a complex relationship between the degree of molecular interaction and consequent higher level effects, e.g. because of the intervention of protective mechanisms such as the induction of stress proteins. In studying responses to toxic molecules, it is very important to construct appropriate dose–response curves of the range of exposures that are likely to be experienced. It would be unwise to assume that a dose–response curve is a simple straight line relationship.

7.4 Examples of molecular mechanisms of toxicity

Molecular interactions between xenobiotics and sites of action, which lead to toxic manifestations, may be highly specific for certain types of organism or very non-specific. In the simplest case, molecules are highly selective between two species because one species has a certain type of site of action which does not occur in the other. For example, the pesticide dimilin acts on the site of formation of chitin and thus affects only those arthropods that utilize this material to form their exoskeleton. The organophosphorous insecticides which act on the nervous system are toxic to all animals, but have little or no toxicity towards plants. Animals have a site of action in the nervous system (in the example given, a form of acetylcholinesterase, which is discussed in more detail in section 7.4.2), whereas plants have no nervous systems and no similar site of action. Some compounds show little selective toxicity and may be regarded as general biocides. An example is dinitroorthocresol (DNOC) and related compounds, which act upon mitochondrial membranes and cause the uncoupling of oxidative phosphorylation. This system is found in all eukaryotes, and 'uncouplers' of this type can run down the gradient of protons across mitochondrial membranes in general, thereby inhibiting or preventing the synthesis of ATP.

Some of most subtle selectivity occurs in resistant strains, a subject dealt with in more detail in Chapter 12. Here, two strains of the same species can have different forms of the same site of action – one of them susceptible to a toxic molecule the other non-susceptible, e.g. some strains of insects resistant to organophosphorous insecticide have forms of acetylcholinesterase which differ from those of susceptible strains of the same species. The difference between the 'susceptible' and 'resistant' forms of acetylcholinesterase may be just a single amino acid in the entire molecule – and thus a very small difference in the structure of a site of action can cause a large difference in toxicity. An organophosphorous insecticide may show little tendency to interact with the 'resistant' enzyme.

In the following account, animals and plants will be considered separately.

7.4.1 GENOTOXIC COMPOUNDS

Many compounds which act as carcinogens are known to cause damage to DNA, i.e. they are genotoxic, and it is strongly suspected that this is a causal relationship. When cells with damaged DNA divide, mutant cells can be produced. Some mutant cells are tumour cells which will follow uncoordinated growth patterns and may migrate within the organisms to produce secondary growths (metastasis) in other locations.

The relationship between DNA changes and harm to the organism is complex. Although adduct formation (covalent binding of the pollutant to DNA) is a good index of exposure, the relationship of adduct formation to harm to the organism is less well defined. For the most part, these DNA adducts are short lived. DNA repair mechanisms quickly excise the adducted structures and replace them with the original moiety (figure 7.2). Sometimes, however, they are relatively stable, and when the cell divides the adducted element is mistranslated and a mutant cell is produced. Thus, whereas there are good data to relate the number of DNA–BaP adducts and the extent of cigarette smoking, the relationship between DNA–BaP adducts and lung cancer is less well established.

Certain polycyclic aromatic hydrocarbons or 'PAHs' (e.g. BaP and dibenz(ah)anthracene), acetylaminofluorene, aflatoxin and vinyl chloride are all examples of genotoxic pollutants. In all of them, it is not the original compound that interacts with DNA. Indeed, the original compounds are relatively stable and unreactive. Enzymatic metabolism (mainly oxidative metabolism by one or more forms of monooxygenase) produces highly reactive and short-lived metabolites which can bind to DNA (figure 7.1).

This type of molecular interaction would appear to be common and widespread. A significant number of pollutants are known to have mutagenic properties – certain PAHs being a case in point. PAHs are present in smoke and soot and in crude oil. They are, therefore, present in urban areas, wherever there is traffic, smoke and soot, and in marine locations where there have been oil spillages.

7.4.2 NEUROTOXIC COMPOUNDS

The nervous system of both vertebrates and invertebrates is very sensitive to the toxic effects of chemicals and there are many examples of neurotoxins – both naturally occurring and man-made. Among the 'natural' neurotoxins should be included tetrodotoxin (from the puffer fish), botulinum toxin (from the anaerobic bacterium *Clostridium botulinum*), atropine (from deadly nightshade, *Atropa belladonna*), the natural insecticides nicotine (from the wild tobacco *Nicotiana tabacum*), pyrethrin (from the flowering heads of *Chrysanthemum* sp.) and many more. Among man-made compounds, it is interesting to note that the four major groups of insecticide – organochlorine, organophosphorous, carbamates and pyrethroid insecticides – all act as nerve poisons.

All of the examples given disturb, in some way, the normal transmission of impulses along nerves and/or across synapses (i.e. junctions between nerves or between nerve endings and muscle and gland cells). However, a distinction can be made between compounds which act directly upon receptors or pores situated in the nerve membrane and those which inhibit the acetylcholinesterase (AChE) of synapses. These two groups will now be considered separately.

The passage of an action potential along a nerve is dependent upon the flow of Na^+ and K^+ across the nerve membrane. During the normal passage of an action potential, Na^+ channels (figure 7.4) are open for a brief instant, allowing the inward flow of Na^+ ions. They are then closed to terminate the Na^+ flow. Pyrethroid insecticides, natural pyrethrins and DDT all interact with Na^+ channels to disturb this function. The usual consequence of their interaction is retarded closure of the channels. This can cause a prolongation of the flow of Na^+ ions across the membrane, which leads to disturbance of the normal passage of the action potential. This type of poisoning causes uncontrolled repetitive spontaneous discharges along the nerve. Several action potentials are generated in response to a single stimulus instead of just one. Uncoordinated muscular tremors and twitches are characteristic symptoms of this type of poisoning. The interaction of these hydrophobic 'water-insoluble' compounds with Na^+ channels is reversible and does not appear to involve the formation of covalent bonds. It is likely that these compounds first dissolve in the lipids of the nerve membrane before interacting with some site on the Na^+ channel, which spans the membrane.

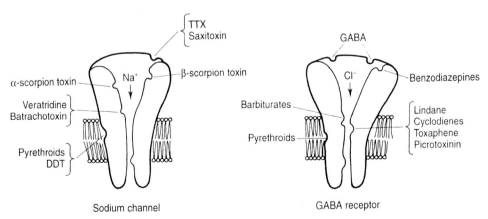

FIGURE 7.4 *Sodium channels and GABA receptors. Reproduced after Eldefrawi and Eldefrawi (1990) with permission from Marcel Dekker Inc., New York.*

Another site of action for insecticides and other neurotoxins is the GABA receptor, which functions as a Cl⁻ channel through the nerve membrane (figure 7.4). Chlorinated cyclodiene insecticides, or their active metabolites (e.g. dieldrin, endrin, heptachlor epoxide), act as GABA antagonists. By binding to the receptors, they reduce the flow of Cl⁻ ions. So, too, does the insecticide γ-HCH (lindane) and naturally occurring picrotoxinin. In vertebrates, these compounds act upon GABA receptors of the brain. Convulsions are typical symptoms of this type of poisoning.

Receptors for acetylcholine, which are located on post-synaptic membranes (i.e. on the other side of the synapse from nerve endings), represent the site of action for a number of chemicals. Thus, nicotine can act as a partial mimic (agonist) of acetylcholine at what are termed 'nicotinic receptors' for acetylcholine. Similarly, atropine can act upon 'muscarinic' receptors for acetylcholine. Nicotinic receptors and muscarinic receptors differ from one another in their structure, their response to toxic chemicals and their location within the nervous system. In vertebrates, nicotinic receptors are found especially at the neuromuscular junction and on many synapses of autonomic ganglia. By contrast, muscarinic receptors are especially found on synapses at the endings of parasympathetic nerves – typically on the membranes of smooth muscle or gland.

Compounds that inhibit AChE represent one of the most toxic groups of compounds to vertebrates and invertebrates. As noted earlier (Chapter 1), OPs are particularly important here. Some were developed for chemical warfare, many others have been developed as insecticides. Another group of anticholinesterases are the carbamates which have insecticidal action. (These should not be confused with other carbamates used as herbicides or fungicides, which do not act as anticholinesterases).

The toxic action of these compounds is indi-rect in that their primary effect is to inhibit the action of AChEs, which have the function of breaking down acetylcholine released into the synapse. Acetylcholine release occurs from the endings of cholinergic nerves and it functions as a chemical messenger. When an impulse reaches a nerve ending, acetylcholine is released and carries the signal across the synaptic cleft to a receptor on the post-synaptic membrane (which may be of a nerve cell, a muscle cell or a gland cell) (see section 8.4.1). When acetylcholine interacts with its receptor, a signal is generated on the post-synaptic membrane, so that the impulse (message) is carried on. For effective neuronal control, it is essential that this signal is rapidly terminated. To achieve this, acetylcholine must be rapidly broken down by AChE in the vicinity of the receptor (figure 7.3). Anticholinesterases have the effect of reducing or preventing altogether the breakdown of acetylcholine. As a consequence, acetylcholine builds up in the synapse, leading to 'over-stimulation' of the receptor and the continued production of a signal after this should have stopped. If this situation continues, the signalling system will eventually run down, resulting in synaptic block. At this point, it will no longer be possible for acetylcholine to relay signals across the synapse. In the case of neuromuscular junctions so affected, tetanus will result, with the muscle in a fixed state, unable to contract or relax in response to nerve stimulation.

7.4.3 MITOCHONDRIAL POISONS

Mitochondria have a vital role in energy transformation and are found in all eukaryotes. It is not, therefore, surprising that some of the most dangerous non-selective biocides act upon mitochondrial systems.

Uncouplers of oxidative phosphorylation such as 2,4-dinitrophenol fall into this category. When a mitochondrion is functioning normally, it has a gradient (electrochemical gradient) of proteins across its inner membrane (figure 7.5). In fact, there is an excess of protons on the outside of the inner membrane and a deficiency on the inside. The maintenance of this gradient depends upon the mitochondrial membrane being essentially impermeable to protons. The production of ATP by mitochondria is driven by energy stored in this proton gradient. If the proton gradient is lost, ATP production will cease. The compounds in question can eliminate this gradient, dissipating in the form of heat the energy associated with it and thus preventing the energy stored in a proton gradient from being used to biosynthesize ATP. Other mitochondrial poisons, such as the naturally occurring insecticide rotenone and cyanide ions, can inhibit the operation of the electron transport chain of the inner membrane of the mitochondrion, thus preventing the production of the proton gradient referred to above. They do so by interacting with components of the electron-carrier system.

7.4.4 VITAMIN K ANTAGONISTS

Warfarin and certain related rodenticides are toxic to vertebrates because they act as antagonists of vitamin K. Vitamin K plays an essential role in the synthesis of clotting proteins in the liver. It undergoes a cyclical series of changes (vitamin K cycle) during the course of which the clotting proteins become carboxylated. After carboxylation has occurred, the clotting proteins are released into the blood, where they play a vital role in the process of clotting, which occurs when there is damage to blood vessels. Warfarin and related compounds have a structural resemblance to vitamin K and strongly compete with it for binding sites, even at low concentrations. This leads to inhibition of the vitamin K cycle and, consequently, to the incomplete synthesis of clotting proteins. Under these conditions, the levels of clotting proteins fall in blood, and the blood loses its ability to clot. Failure of clotting results in haemorrhage.

Flocoumafen and related compounds, referred to as second-generation anticoagulant

FIGURE 7.5 *Mitochondrial poisons. The figure shows a diagrammatic cross-section of a mitochondrion. Protons (H^+) are actively transported from the matrix across the inner membrane as a result of electron flow along the electron transport chain. These protons can then flow back to the matrix, via the enzyme ATP synthetase, where energy associated with them is used to synthesize ATP. 'Uncouplers' such as 2,4-dinitrophenol can carry these protons back to the matrix before ATP synthesis occurs. Other poisons, for example rotenone and CN^-, can inhibit the flow of electrons along the electron transport chain.*

rodenticides, were developed to circumvent the problem of resistance to warfarin. These rodenticides become lethal when the receptor sites in the liver are saturated. There is a dramatic difference in the toxicity of flocoumafen to rats and to quail (LD_{50} 0.25 mg kg^{-1} and 300 mg kg^{-1} respectively.). An important factor in these species difference is that the rodenticide is metabolized more rapidly by quail than by rat (Huckle *et al.*, 1989).

7.4.5 THYROXINE ANTAGONISTS

The thyroid gland produces two hormones which have marked effects upon metabolic processes in many tissues. One of these, thyroxine (1-3,3,5,5-tetraiodothyronine or T_4), binds to the protein transthyretin (TTR) which is part of a transport protein complex found in blood. The other part of the complex consists of a second protein to which is bound retinal (vitamin A) (figure 7.6). Thus, transthyretin–thyroxine is attached to the retinal-binding protein (RBP). Certain hydroxy metabolites of the PCB congener 3,3'-4,4'-tetrachlorobiphenyl (3,3,'-4,4'-TCB) compete with thyroxine for its binding site on TTR. In particular, the metabolite 4'-hydroxy-3,3',4-5'tetrachlorobiphenyl binds very strongly and effectively competes with thyroxine (Brouwer *et al.*, 1990). These metabolites are formed as a result of metabolism of the original PCBs congener by a particular cytochrome P_{450} form of the monooxygenase system (P_{450} 1A1).

When binding occurs, there are two consequences. First, thyroxine is displaced and is lost from the blood. Second, the retinol complex breaks away and retinol is lost from the blood. Thus, the levels of thyroxine and retinol will fall as a consequence of the production of certain PCB metabolites.

7.4.6 INHIBITION OF ATPASES

The ATPases (adenosine triphosphatases) are a family of enzymes (Na$^+$, K$^+$-ATPase, Ca^{2+}-ATPase, etc.) involved in the transport of ions. These enzymes are involved in the osmoregulation of a variety of organisms and the effects of a number of organochlorines of this process have been investigated. The avian salt gland which enables pelagic seabirds to maintain their salt balance in a marine environment is also dependent on ATPases. Another ATPase-dependent organ is the avian oviduct. The inhibition of Ca^{2+}-ATPase, which is involved in the transport of calcium, by DDE (the persistent metabolite of DDT) is considered to be the basis of DDE-induced eggshell thinning. This phenomenon is considered in some detail in Chapter 15.

7.4.7 ENVIRONMENTAL OESTROGENS AND ANDROGENS

Although endocrine disruptors have been known for many years, the topic has recently attracted more interest and is now one of the most controversial environmental issues. A good review has recently been published by the Institute of Forestry and Nature Research in The Netherlands (Janssen *et al.*, 1998). An oestrogenic chemical is one which can imitate an oestrogen by binding to the oestrogen receptor. One mechanism is for the pseudo-oestrogen to bind to the oestrogen receptor, thus stimulating the transcription activity, resulting in the induction of oestrogenic processes. Another mechanism is for the compound to act as an anti-oestrogen, binding strongly to the oestrogen receptor and thus blocking the effects of endogenous oestrogens. This can result in masculinization of the organism.

Normal

RBP ——Retinol (r)——→ RBPr ——Transthyretin (TTR)——→ TTR – RBPr ——→

T$_4$

In presence of TCB

3,3',4,4'-TCB ——Monooxygenase——→ TCBOH   TTR – TCBOH

RBP ——Retinol (r)——→ RBPr ——————→ TTR ——→ RBPr

T$_4$ RBBr   T$_4$

L-Thyroxine (T$_4$)

3,3',4,4'-TCB

4'-OH-3,4,3',5'-TCB (TCBOH)

FIGURE 7.6 *Mechanism of toxicity of a polychlorinated biphenyl. Retinol (r) binds to retinol-binding protein (RBP) which is then attached to transthyretin (TTR). Thyroxine (T4) binds to TTR and is transported via the blood in this form. The coplanar PCB 3,3',4,4'-tetrachlorobiphenyl (3,3',4,4'-TCB) is converted into hydroxymetabolites by the inducible cytochrome P$_{450}$, called P$_{450}$ 1A1. The metabolite 4'-OH-3,3',4,5'-tetrachlorobiphenyl (TCBOH) is structurally similar to thyroxine and strongly competes for thyroxine binding sites. The consequences are loss of thyroxine from TTR, the fragmentation of the TTR–RBPr complex and loss of both thyroxine and retinol from blood.*

A considerable number of man-made chemicals are oestrogenic. These include some of the common organochlorine insecticides, tributyl tin, phthalates and nonylphenols. The oestrogenic activity of these compounds varies widely. For example, the *o,p'* isomers of DDT and DDE are much more active than the *p,p'* isomers. In addition to man-made chemicals, there are many natural oestrogenic compounds. Over 300 plants belonging to more than 16 different families have been shown to contain substances with oestrogenic activity. The best known are isoflavonids and the coumestans. The resorcyclic acid lactones are found in fungi. The

interest in endocrine-disrupting chemicals has led to the development of rapid screening techniques, such as transcription of an oestrogen-responsive reporter gene in transfected cells, production of vitellogenin in a hepatocyte culture of rainbow trout cells and in recombination yeast cell cultures containing the human oestrogen receptor gene.

The best documented cases of endocrine disruption caused by pollutants are the induction of **vitellogenin** in male fish and masculinization of female gastropods. Vitellogenin is a protein synthesized in the liver of female fish and transported to oocytes to form the yolk of the eggs.

It is well established that the induction of vitellogenin is under oestrogenic control. Elevated levels of vitellogenin were found in fish that had been exposed to sewage treatment effluent after hermaphrodite fish had been reported in treatment lagoons. This phenomenon is considered in more detail in Chapter 10.

Pollutants with androgenic activity can cause masculinization of female gastropods. Imposex (females developing male characteristics such as penis and vas deferens) has been widely reported in marine gastropods associated with marinas. Detailed studies have been carried out at the Plymouth Marine Laboratory since the first finding of imposex in the dog whelk (*Nucella lapillus*) in Plymouth Sound in 1969. Laboratory experiments and *in situ* transfer experiments have shown that imposex is initiated in dog whelks by tributyl tin at very low concentrations. This phenomenon is considered in more detail in Chapter 15.

In both this and the preceding example, the underlying cause of change is considered to be that chemicals mimic the action of true hormones at the receptor site and thus trigger the reactions that would normally be caused by the natural hormone. The problem is caused by the fact that these unnatural 'hormones' trigger reactions in the wrong sex.

Another series of problems can be caused when chemicals block the hormone receptor site. In this case, the normal action of the hormone is inhibited as it cannot react with the receptor. [An example of this mechanism is the action of tetrachlorodibenzodioxin (TCDD)]. The means of assessing this compound and other dioxins and related chemicals is considered in Chapter 12.

7.4.8 REACTIONS WITH PROTEIN SULPHYDRYL (SH) GROUPS

The ions Hg^{2+} and Cd^{2+} are toxic to many animals. The main reason for this appears to be their ability to combine with sulphydryl (thiol) groups, thereby preventing normal function. Sulphydryl groups on enzymes and other proteins have important functional roles, e.g. the formation of disulphide bridges and consequent conformational changes in the proteins. With this kind of toxic interaction, it may be difficult to establish which sulphydryl groups on which proteins represent the sites of toxic action.

Organomercury is more lipophilic than inorganic mercury and has a different distribution in the body. It tends to move into fatty tissues and to cross membranes, including those of the blood–brain barrier. As a consequence, the toxicity of organomercurial compounds tends to be expressed in the brain, whereas that of inorganic mercury is expressed in peripheral tissues. Other organometallic compounds, for example tetra-alkyl lead, also tend to have their toxic action in the brain.

7.4.9 PHOTOSYSTEMS OF PLANTS

A number of herbicides which show little toxicity to vertebrates or insects act as poisons of the photosynthetic systems of plants. Substituted ureas and triazines are examples. By mechanisms that are not yet clear, they interrupt the flow of electrons through the photosystems that are responsible for the light-dependent reaction of photosynthesis, i.e. the splitting of water to release molecular oxygen.

7.4.10 PLANT GROWTH REGULATOR HERBICIDES

A group of herbicides, sometimes called the phenoxyalkanoic herbicides, have growth-regulating properties. MCPA, 2,4-D, CMPP and 2,4-DB are well-known examples. The molecular mechanism by which they affect growth has never been clearly established, and the site at which they act has not been defined. However, it is interesting that exposure of plants to them causes the production of ethylene, itself a growth-regulating compound.

These herbicides have very low toxicity to vertebrates and insects which, it may be presumed, do not have receptors for growth regulation similar to those of plants.

7.5 *Summary*

Pollutants cause a wide variety of biochemical effects in organisms. Broadly, these can be divided into protective and non-protective responses. Protective responses include the induction of enzymes which have a detoxifying function and the induction of proteins which can bind heavy metals. Microsomal monooxygenases are examples of inducible enzymes which usually have a detoxifying function. However, the situation is complicated by the fact that in a relatively small (yet critically important) number of cases monooxygenases cause activation rather than detoxication. It should also be emphasized that protective responses such as these bear an energy cost which may have a detrimental effect on the organism (see Chapter 8).

Many responses, such as inhibition of cholinesterase, electrophysiological changes caused by organochlorine insecticides, antagonism of vitamin K, endocrine disruption and formation of DNA adducts, are non-protective and in many instances are harmful to the organism. It is very important to elucidate the mechanisms by which chemicals express toxicity. Understanding the mechanisms of toxicity can provide the basis for biomarker assays and can lead to the development of antidotes.

7.6 *Further reading*

HASSALL, K. A. (1990) *The Biochemistry and Uses of Pesticides*, 2nd edn. Describes the mode of action of many pesticides.

HODGSON, E and KUHR, R. J. (1990) *Safer Insecticides – Development and Use*. Detailed account of the mode of action of certain insecticides.

HODGSON, E. and LEVI, P. (1994) *Introduction to Biochemical Toxicology*, 2nd edn. Many examples given of mechanism of action of toxicants, including most of those mentioned in this chapter.

TIMBRELL, J. A. (1999) *Principles of Biochemical Toxicology*, 3rd edn. A long chapter is devoted to biochemical mechanisms of toxicity of importance in medical toxicology.

CHAPTER 8

Physiological effects of pollutants

8.1 *Introduction*

Pollutants may damage organisms with lethal consequences (as described in Chapters 6 and 7). The effect on the population is then an increase in the mortality rate of at least one age class. Alternatively, there may be damage to, or effects on, the machinery of reproduction, resource acquisition and uptake. These effects are described in detail in this chapter. Where resource uptake is reduced, there are consequent reductions of birth rate and/or somatic growth rate (here, referred to jointly as 'production rate'), and these depress population growth rate.

Detoxication mechanisms generally use resources, including energy, and these resources are consequently not available for production. Thus, production is likely to be reduced when detoxication occurs. The overall effects of detoxication mechanisms on production and mortality are considered at the end of this chapter. Note at this stage, however, that either damage or detoxication may result in reduced production in a polluted environment.

With these points in mind, the effects of pollutants will be considered at several different levels of organization from organelles to whole organisms (figure 8.1). In this chapter, we are concerned mainly with physiological effects above the biochemical level (covered in Chapter 7). Physiology is defined as the branch of science concerned with the functioning of organisms and the processes and functions of all or part of an organism. Thus, in ecotoxicology, we are concerned with describing the disruptive effects of pollutants on 'normal' physiology, normal referring to the state of the organism when not exposed to pollution although subject to other forms of stress. Abnormal stresses may be completely novel and occur in response to man-made chemicals that have appeared in the environment recently (on the geological time-scale), or they may simply be an increase in a response to a substance to which

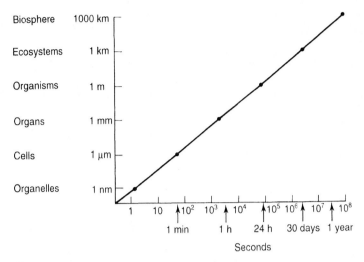

FIGURE 8.1 *Schematic graph of the relationship between complexity and size of natural systems and 'compartments' (= 'black boxes') and typical response times to pollutant 'insults'. After various authors.*

the organism has evolved natural protection mechanisms (e.g. metals).

8.2 *Effects of pollutants at the cellular level*

When a pollutant enters a cell it may trigger certain biochemical responses which have evolved either to break the chemical down (Chapter 7) or to store it in such a way that it is 'hidden' within a compartment, preventing interference with essential biochemical reactions within the cell. For example, in invertebrates there are clearly defined pathways for metal detoxication (Hopkin, 1989, 1990; Dallinger, 1993). Perhaps the simplest case is the epithelium of the digestive system of terrestrial invertebrates which is usually only one cell in thickness and acts as a barrier between the internal environment of the animal (i.e. the blood bathing the organs) and the food in the lumen. Therefore, storage mechanisms and/or methods of exclusion have to be extremely efficient because terrestrial invertebrates (unlike aquatic

organisms) are not able to excrete xenobiotics from the blood into the external medium across the respiratory surfaces if they are taken up to excess. In land mammals, lipophilic organics are converted to water-soluble metabolites and conjugates which are then excreted in the bile and urine (see Chapter 5).

Three main detoxication pathways have evolved for the binding of metals which enter the epithelial cells (figures 8.2 and 8.3). Although the chemistry of binding appears to be similar in all terrestrial invertebrates so far examined, the subsequent fate of the waste material is controlled by the digestive processes of the animals in question. An example of these differences is provided by the contrasting ways in which the metals are bound in the hepatopancreas of both the terrestrial isopod *Porcellio scaber* and a major predator of isopods, the woodlouse-eating spider *Dysdera crocata* (figure 8.4, see box 8.1).

Osmoregulation has been found to be affected in a wide variety of organisms ranging from crabs to birds and by a wide variety of pollutants (heavy metals, organochlorines and organophosphorous compounds). The basic

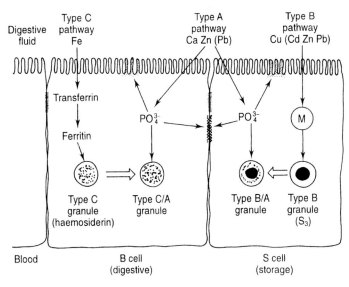

FIGURE 8.2 *Schematic diagram showing the three pathways of detoxication of metals from the digestive fluids by the B and S cells of the hepatopancreas of the woodlouse,* Porcellio scaber. *Reproduced from Hopkin (1990) with permission from Blackwell Science Ltd.*

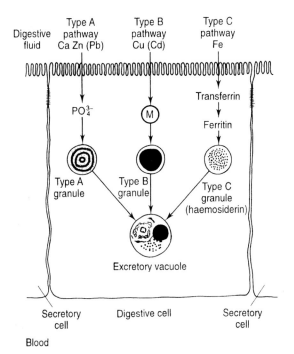

FIGURE 8.3 *Schematic diagram showing the three pathways of detoxication of metals from the digestive fluids by the digestive cells in the hepatopancreas of the woodlouse-eating spider,* Dysdera crocata. *Reproduced from Hopkin (1990) with permission from Blackwell Science Ltd.*

mechanisms for water and salt transport in a salt-water fish are given in figure 8.6. Kinter *et al.* (1972) found that the sodium space was increased and the activity of **ATPase** was significantly reduced by both DDT and PCBs and the sodium concentration of the serum was increased. The avian salt gland is ATPase dependent, and this enables seabirds to maintain their water balance in the marine environment by ingesting sea water and excreting a highly concentrated salt solution. However, in this case, experiments on seabirds did not indicate that organochlorines had a significant effect on osmoregulation. Another ATPase-dependent gland is the avian oviduct. The inhibition of **Ca-ATPase** is considered to be involved in **DDE**-induced eggshell thinning which caused marked declines in many raptorial species. This phenomenon is covered in Chapter 15.

8.3 *Effects at the organ level*

When pollutants are ingested by organisms, or

FIGURE 8.4 *(A)* Dysdera crocata *attacking a specimen of* Porcellio scaber *of 1 cm in length. (B) Scanning transmission electron micrograph (bright-field image) of an unstained resin-embedded section (0.5 μm in thickness) through two type A calcium phosphate granules in a digestive cell of the hepatopancreas of* D. crocata. *Diameter of each granule section = 1 μm. (C–F) Electron-generated X-ray maps for phosphorus (C), calcium (D), zinc (E) and iron (F) of the type A granules shown in (B). Philips CM12 STEM, EDAX 9900 X-ray analyser, screen resolution 256–200 pixels, dwell time on each pixel 50 ms, spot diameter 20 nm. Reproduced from Hopkin (1990) with permission from Blackwell Science Ltd.*

pass into the blood across respiratory epithelia or the external surface of the body, they may be 'compartmentalized' in particular organs within the body. This may be accidental or deliberate. For example, radioactive isotopes of essential elements will travel to the same locations in tissues as their non-radioactive counterparts. Radioactive iodine accumulates in the thyroid gland of mammals and may cause thyroid cancer. Cadmium is accumulated in the liver and kidneys of mammals. The symptoms of cadmium poisoning are proteinuria (excretion of proteins in the urine) as a result of the breakdown of the kidney cells when cadmium levels exceed a 'critical concentration'. This concept of a critical target organ concentration is

BOX 8.1 *Detoxication pathways for metals.*

The **type A pathway** is involved in the intracellular precipitation of calcium and magnesium as phosphates. In the hepatopancreas of the isopod *Porcellio scaber* (figure 8.2), zinc and lead may be present in this type A phosphate-rich material which is deposited on the cytoplasmic side of the cell membranes and around existing metal-containing granules. However, in the spider *Dysdera crocata* (figure 8.3), the type A material forms discrete granules with a characteristic concentric arrangement of layers in thin section (figures 8.4B and 8.5). Zinc (figure 8.4E) and lead are also found associated with these granules.

The **type B pathway** is followed by metals such as copper and cadmium which have an affinity for sulphur-bearing ligands. The sulphur-rich type B granules probably contain breakdown products of metallothionein, a cysteine-rich protein involved in the intracellular binding of zinc, copper, cadmium and mercury, and have their origin in the lysosomal system (Dallinger, 1993). Lead may also occur in type B granules. In isopods (but not in spiders), some type B granules may be surrounded by a layer of type A material (type B/A granules, following the convention that the type of material first accumulated in the granule is given precedence), but granules with a type A core surrounded by type B material have not yet been discovered.

The **type C pathway** is exclusively for the accumulation of waste iron in isopods and spiders. Type C granules are probably composed of haemosiderin, a breakdown product of ferritin. In isopods, type A material may be mixed with the iron-rich type C material to form type C/A granules (figure 8.2). In *Dysdera crocata*, iron is not found in the type A granules (figure 8.4F).

Once the type A, B and C material has been precipitated, there is no evidence that it is remobilized. Indeed, the only route by which the granules can be excreted is by voiding of the contents of the cell into the lumen of the digestive system for subsequent excretion in the faeces. The granules therefore represent a storage detoxication system.

In isopods, type A material occurs in both cell types of the hepatopancreas, but type B and C material is restricted to the S and B cells respectively. Large numbers of B cells break down during each 24-h digestive cycle, but S cells are permanent and never void their contents into the lumen of the hepatopancreas. Thus, material deposited in the S cells remains there until the isopod dies. In contrast, the spider stores all three types of material in a single cell type, the digestive cell (figure 8.3). Large numbers of these cells break down at the end of each digestive cycle and void their contents. This waste contains large numbers of type A, B and C granules which are excreted subsequently in the faeces (Hopkin, 1989).

Thus, terrestrial isopods accumulate metals in the S cells throughout their lives, so that in contaminated sites concentrations in the hepatopancreas reach very high levels. In contrast, the spider regulates the concentrations in the hepatopancreas by excreting metals assimilated in the digestive cells. Consequently, concentrations of zinc, cadmium, lead and copper do not deviate from normal over the long term in the hepatopancreas of *Dysdera crocata*, even when the spiders are fed on heavily contaminated woodlice for an extended period.

very important in ecotoxicology. With small organisms, analysis of concentrations of pollutants in whole individuals may be misleading if the contaminant is localized strongly within a specific part of the body. For example, over 90% of the cadmium, copper, lead and zinc in woodlice from metal-contaminated sites is contained within the hepatopancreas, an organ which con-

FIGURE 8.5 *Concentrically structured type A granule from the hepatopancreas of the woodlouse-eating spider* Dysdera crocata *(diameter 2 μm). Reproduced from Hopkin (1989) with permission from Elsevier Applied Science.*

stitutes less than 10% of the weight of the whole animal (Hopkin, 1989). In regions contaminated heavily with metals, concentrations of zinc can exceed 2% of the dry weight of the hepatopancreas of woodlice. In these circumstances, the detoxification capacity of the organ is eventually exceeded, the cells begin to break down and the woodlouse becomes moribund because of zinc poisoning (figure 8.7).

Specific methods to measure **cardiovascular** and **respiratory** responses have been developed. Methods are now available so that cardiovascular monitoring can be carried out non-invasively. Depledge and Andersen (1990) developed a method by which recorded heart rate in reflected light associated with the heart beat could be detected by a transducer. The only requirement is that the heart of the test organism is close enough to the body surface to allow detection of the reflected optical signal. This method has been used successfully with bivalves, crabs and crayfish (Bloxham *et al.*, 1999).

Respiratory responses have been measured in fish and invertebrates. Respiratory responses are rapid and are therefore useful for identifying short pollution events. Several parameters

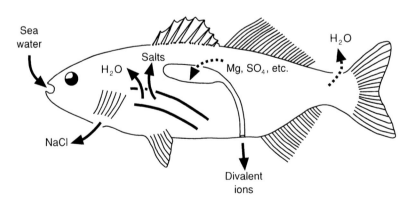

FIGURE 8.6 *Salt and water transport in a salt-water fish. Reproduced after Kinter* et al. *(1972) with permission from the Society of Environmental Technology and Chemistry.*

may be measured, including ventilatory frequency and volume and the exchange of respiratory gases. Oxygen consumption rates show a clear dose–response relationship in many organisms exposed to chemicals and can be used as a surrogate for metabolic rate. Portable, self-contained respirometry systems have been devised (Handy and Depledge, 2000).

8.4 Effects at the whole organism level

The effects of pollutants on the whole organism are considered under three main headings, namely neurophysiological, behavioural and reproductive effects. However, these effects can

FIGURE 8.7 *Light micrographs of B cells (B) and S cells (S) within the hepatopancreas of two specimens of the woodlouse,* Oniscus asellus, *6 weeks after release from the same brood pouch of a female from an uncontaminated woodland. The S cells of the juvenile reared on leaf litter contaminated with metal pollutants from a woodland near to a smelting works (B) contain far more metal-containing granules (g) than the S cells of the juvenile reared on uncontaminated litter (A). h, haemocoel; lip, lipid; lum, lumen of hepatopancreas tubule. Scale bars: 20 μm. Reproduced from Hopkin and Martin (1984) with permission from the Zoological Society of London.*

often be inter-related, neurological changes can affect behaviour, changes in behaviour can affect reproduction and so on.

8.4.1 NEUROPHYSIOLOGICAL EFFECTS

All animals apart from protozoa and sponges have nervous systems, and these are most highly developed in vertebrates. Among the invertebrates, certain groups, for example arthropods, have well-developed nervous systems. Nerves provide communication between receptors (e.g. retinal cells, taste receptors) and effectors (e.g. muscle and gland) (figure 8.8). Information is passed through the nervous system by the transmission of electrical impulses travelling along the axons of the neurones (nerve cells). Information is passed from one neurone to another across a synapse (nerve junction) by chemical messengers known as neurotransmitters. Acetylcholine, noradrenaline and serotonin are examples of neurotransmitters, being released from nerve endings, crossing the synaptic cleft and interacting with receptors on the post-synaptic membrane of the adjacent neurone (figure 8.8). More advanced nervous systems (e.g. of vertebrates) are differentiated into two separate but interrelated parts: a central nervous system in which information is integrated and a peripheral nervous system through which impulses are transmitted.

Many toxic chemicals, both naturally occurring and man-made, act upon the nervous system. With a wide range of potential sites of action both on the axonal membrane and at the synapse, this is not altogether surprising, considering the vital role that it has for regulating the function of the organism. It is noteworthy that the four main groups of insecticides – organochlorine, organophosphorous, carbamate and pyrethroid – all act as neurotoxins. For a description of their modes of action, see Eldefrawi and Eldefrawi (1990).

Chemicals can have effects on the nervous system at different types of receptor and at different locations. Broadly speaking, effects may be upon the central nervous system, upon the peripheral nervous system or upon both. Central effects can be monitored using an electroencephalogram (EEG), whereas peripheral effects are detectable with an electromyogram (EMG). Both techniques can detect changes in the passage of nervous impulses. The actual sites of action may be at the synapses or on the axonal membrane.

Anticholinesterases such as **organophosphorous** and **carbamate** insecticides are examples of poisons which act at synapses. They disturb synaptic transmission at cholinergic junctions by inhibiting the enzyme acetylcholinesterase. Acetylcholinesterase has the function of destroying the neurotransmitter acetylcholine, so that inhibition of the enzyme has the effect of lengthening the residence time of acetylcholine and prolonging the stimulation of cholinergic receptors on the post-synaptic membrane (see section 7.4.2). Thus, chemical messages are not rapidly terminated as they should be, and normal transmission of impulses across the synapse is disrupted. If the disruption is sufficiently prolonged, the system for relaying impulses across the post-synaptic membrane and beyond will be run down, leading to synaptic block, i.e. no synaptic transmission at all. Synaptic blockage causes tetanus (rigid fixation of muscle) and in vertebrates death is likely to follow quickly as a result of consequent paralysis of the diaphragm muscles and respiratory failure. Anticholinesterases act upon cholinergic junctions throughout the vertebrate nervous system, both central and peripheral, the pattern depending on the compound in question and the species affected. Typical symptoms include tremors, convulsions, respiratory and circulatory failure, coma, dizziness and depression. The organochlorine insecticide **DDT** acts upon the **sodium channels**

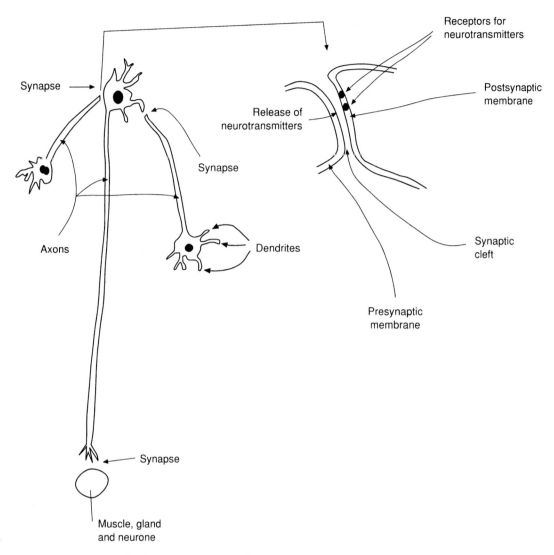

Synapse

Postsynaptic
membrane

Release of
neurotransmitters

Synapse

Synaptic
cleft

Axons

Dendrites

Presynaptic
membrane

Synapse

Muscle, gland
and neurone

FIGURE 8.8 *Schematic diagram for neurones and synapses.*

of the axonal membrane (see section 7.4.2). This has the effect of disturbing the action potential of the nerve and thus the transmission of impulses along the nerve. DDT can have both central and peripheral effects upon the nervous system, ranging from tremors and twitches due to peripheral effects to more co-ordinated disturbances caused by central effects. **Dieldrin**, on the other hand, acts upon **GABA receptors**, which are located predominantly upon pre- and post-synaptic membranes adjoining synapses.

In insects, the effects of dieldrin are on inhibitory synapses of the central nervous system. Symptoms of poisoning are the consequence of effects upon the central nervous system, e.g. tonic convulsions.

Neurotoxic compounds can have diverse and wide-ranging effects upon the whole animal. These effects can include endocrine and behavioural disturbances. Organophosphorous insecticides, for example, have behavioural effects upon birds, including reduction of movements and of singing. Endocrine function can be affected by disturbances in the nervous system and vice versa.

8.4.2 EFFECTS ON BEHAVIOUR

The effects of pollution on behaviour have been reviewed with primary reference to aquatic animals by Atchison *et al.* (1996), which should be consulted for further information. Although all behaviours are potentially vulnerable to alteration by pollutants, most work has been carried out on foraging, with some attention to vigilance. Impaired foraging results in reduced resource acquisition and so in reduced production. Impaired vigilance results in increased vulnerability to predators and so to increased mortality rate. In these ways, the effects of pollution on behaviour may result in lowered production and increased mortality rate.

The components of foraging behaviour are illustrated diagrammatically in figure 8.9. All these components may be adversely affected by pollution. Thus, although little work has been undertaken on the effects on appetite, a common consequence of chemical stressors is the cessation of feeding. Prey encounter rates depend on many factors, including search strategy, learning and sensory systems. All can be affected by chemical stressors, reducing the efficiency of searching for prey. Little work has been carried out on prey choice, but a few studies show that it, too, can be affected. Most importantly, capture rates are known to be affected by toxicants in some species, at least when larger or more evasive prey are being hunted. The time from capture to ingestion ('handling time') has often been shown to be increased by toxicants, e.g. by copper in bluegill feeding on *Daphnia*, by zinc and lead in zebrafish (*Brachydanio rerio*) and by alkyl benzene sulphonate detergent in flagfish (*Jordanella floridae*) (Atchison *et al.*, 1996). However, these increases in handling

times seem to be due to repetitive rejection and recapture, perhaps caused by blockage of the gustatory senses by the contaminant. These predators use vision to identify, pursue, capture and start processing their prey. Rejection may result from lack of gustatory confirmation that the captured item is edible.

Foraging is not the only goal, however, because as mentioned above an animal must also avoid being eaten itself. 'Few failures in life are as unforgiving as the failure to avoid predation' (Lima and Dill, 1990). Most studies of vulnerability to predation have been simple experiments in which prey were first exposed to a pollutant and then to a predator. The predators were generally not exposed to the pollutant. Such studies have shown, for example, that ionizing radiation and mercury both increased

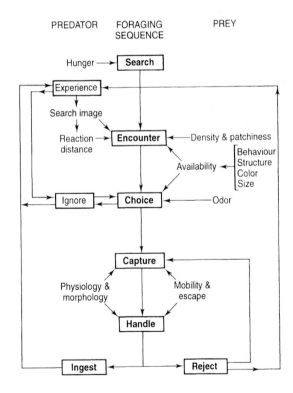

FIGURE 8.9 *Schematic representation of the components of foraging behaviour. reproduced from Atchison* et al. *(1996) with permission from Lewis Publishers, and imprint of CRC Press.*

the risk to mosquitofish (*Gambusia affinis*) of predation by largemouth bass (*Mircopterus salmoides*) (Atchison *et al.*, 1996). In other species, predation risk has been shown to increase with exposure of prey to thermal stress, insecticides, pentachlorophenol, fluorine and cadmium. Beitinger (1990) found that prey vulnerability to predation was increased in 23 of 29 experiments that he reviewed.

The relationship between depression of **acetylcholinesterase (AChE)** activity by OPs and carbamates and behaviour has been studied in many species ranging from invertebrates to mammals. The mechanism whereby these compounds prevent the normal functioning of the nervous system has been considered in section 7.3.

Impaired burrowing activity in **earthworms (*Pheretima posthuma*)** related to AChE inhibition caused by the carbamate carbaryl was studied by Gupta and Sundararaman (1991). They found that the ability of the earthworm to burrow in soil decreased in a dose–response fashion over the entire range of concentrations of carbaryl that were studied. The activity of AChE and locomotory behaviour in the **carabid beetle (*Pterostichus cupreus*)** exposed to dimethoate was studied by Jensen *et al.* (1997). In the case of males, all locomotory indices (path length, time during which the organism is active, average velocity, stops per walked metre and turning rate) were affected at the lowest dosage. Females were less affected, despite the fact that the degree of inhibition of AChE was similar.

Behavioural responses of mammals and birds after depression of AChE activity has been reviewed by Grue *et al.* 1991 (Grue and Heinz, in Dell'Omo, 2000). One difficulty of these behavioural studies is to follow the changes in AChE activity non-destructively, i.e. in blood. Unfortunately for the investigator, the recovery of activity of AChE in the blood is rapid and the coefficient of variation is three times

greater than the activity in the brain (Holmes and Boag, 1990). Thus, plasma measurements can be used only over a much shorter time (up to 12 h) and with considerably less precision than brain measurement. In practice, most investigators have sacrificed subsamples during the experiment and relied on measurements of brain AChE activity.

Detailed quantitative analyses at low levels of toxication have only recently been attempted. Hart (1993) investigated the relationship between behaviour and AChE activity in the starling (*Sturnus vulgaris*) and found that, although most behavioural effects occurred when brain levels fall below 50%, some subtle effects on behaviour, such as effects on posture (time spent standing on one leg while resting), were found at relatively high levels of AChE activity (88% of normal), and may reflect impaired balance or co-ordination.

Another example of a quantitative study is shown in figure 8.10. House sparrows (*Passer domesticus*) were dosed with an organophosphate, chlorfenvinphos, and their feeding behaviour was then recorded during four successive 1.5-h periods. The percentage seeds dropped is plotted against dose (assayed at the end of the day) in figure 8.10. Birds whose brain AChE activity was most reduced initially dropped some 30% more seeds than on control days (figure 8.10A). However, the effect had worn off 3 h after dosing (figure 8.10C and D). Figure 8.11 shows that the birds that dropped most seeds lost most weight.

Extrapolating from the laboratory to what happens in the wild is notoriously difficult (Hart, 1993). Animals may be able to detect and avoid toxicants in their diet. If they do ingest them, they may be able to compensate for any disabilities by retreating to safe places. Alternatively, disabilities may make them more vulnerable to predation, or less able to maintain their territories or to care for offspring. In general, some field studies have succeeded but

FIGURE 8.11 *Body weight loss in relation to percentage food items dropped for the birds of Figure 8.10. Reproduced from Fryday et al. (1994) with permission from Springer-Verlag.*

FIGURE 8.10 *Differences in percentage seeds dropped between dosing and control days plotted against brain acetylcholinesterase activity in house sparrows. A–D show results in successive 1.5-h time bins after dosing. Doses are given in the key. Reproduced from Fryday et al. (1994) with permission from Springer-Verlag.*

others have failed to find effects on behaviour when brain acetylcholinesterase activity was depressed by 40–60% (Hart, 1993). This could be consistent with the existence of a general threshold for behavioural effects at about that level.

A good example of a field study is that of Busby *et al.* (1990), who examined the effects of the organophosphorous insecticide fenitrothion on white-throated sparrows (*Zonotrichia albicollis*) in spruce fir forest in New Brunswick in Canada (see box 8.2).

The extrapolation to nature has also been tackled by carrying out similar studies both in the laboratory and in the field. In one study, the effects of the OP dimethoate on behavioural parameters of the **wood mouse (*Apodemus sylaticus*)** were studied in the laboratory. The parameters included frequency and duration of sniffing, rearing, grooming and general activ-

ity. Similar responses were demonstrated with radiotagged mice in the field (Dell'Omo and Shore, 1996a,b).

8.4.3 REPRODUCTIVE EFFECTS

Some of the best documented cases of effects of pollutants on reproduction are considered elsewhere in this volume. These are the effect of tributyl tin (TBT) on molluscs (section 15.3), DDE-induced eggshell thinning in raptorial and fish-eating birds (section 15.1) and reproductive failure of fish and birds in the Great Lakes caused by a mixture of pollutants (section 15.2). The mechanisms involved in these reproductive failures are diverse. TBT causes females to develop male characteristics (imposex), which can lead to the female becoming sterile. DDE causes the eggshell to be thinner than normal, leading to breakage of the egg. In the case of the Great Lakes (apart from eggshell thinning in some species), the mechanisms were less clearly defined.

Disruptors of the endocrine system have received considerable attention in recent years. Some of the basic biochemistry has already been covered in the previous chapter, here we con-

Box 8.2 *Forest spraying in Canada.*

The forest has been sprayed aerially each year since 1952 to control spruce budworm (*Choristoneura fumiferana*). Fenitrothion is the preferred insecticide. In the study of Busby *et al.* (1990), fenitrothion was applied aerially twice: once at 420 g ha^{-1} and once at 210 g ha^{-1} 8 days later. An earlier study had shown that spraying at these levels would have resulted in brain acetylcholinesterase inhibition of 42% and 30% respectively. A team of experienced field workers ringed and were able to identify individually 13 breeding pairs in the sprayed area, and seven in a nearby control area. Each pair was followed until it abandoned its territory or its offspring fledged. The results showed that the adult population of white-throated sparrows in the sprayed area was reduced by one-third, primarily as a consequence of mortality and territory abandonment after the first spray. Other behavioural responses of breeding birds included inability to defend their territories, disruption of normal incubation patterns and clutch desertion. Pairs which did manage to hatch at least one chick only produced one-third of the normal number of fledglings. Overall, the reproductive success of the pairs in the sprayed area was only one-quarter of that of the pairs in the control area.

sider some of the evidence for effects on reproduction in wildlife. The mechanisms whereby endocrine disruptors can affect the reproductive health and survival of wildlife are shown diagrammatically in figure 8.12.

The decline of populations of **ringed seals (*Phoca hispida*)** and **grey seals (*Halichoerus grypus*)** in the Baltic even after hunting pressures were reduced sparked an investigation. The main reason was that females were unable to reproduce; there was an increase in miscarriages and many females were sterile because of damage to the uterine wall. In addition to the uterine problems, there were indications of metabolic disorder, immunosuppression and hormonal imbalance, a condition known as hyperadrenocorticism. Although it is generally considered that the persistent organochlorines and their metabolites are the cause, 'it is not possible to make conclusive statements about the relation between disease and poor reproduction among Baltic seals and the concentration of organohalogen compounds. Even though PCB and DDT together with metabolites are the dominating xenobiotics in the seals, synergistic and additive effects of other compounds cannot be excluded' (Ollson *et al.*,

1992). Jaw bone damage was also found and this symptom had the advantage that it was possible to assemble a historical record from museum specimens. It was found that the increase in the lesions (from less than 10% to over 50%) started to occur in the late 1950s and the 1960s, which correlates with the increase of organochlorine pollution (Bergman *et al.*, 1989).

Endocrine disruptors have been shown to cause effects on fish in the UK (Allen *et al.*, 1999; Harries *et al.*, 1999). Vitellogenin is a yolk precursor, the synthesis of which is triggered by gonadotropin hormones in females. The induction of **vitellogenin** synthesis by endocrine disruptors in male fish is considered in Chapter 10. Although the abnormal induction of vitellogenin in male fish is well documented, it has not, as yet, been linked to altered reproduction.

Studies linking damage to DNA with reproduction have been carried out on **mosquito fish (*Gambusia affinis*)** in streams contaminated with radionuclides (Theodorakis *et al.*, 1997). These workers were able to relate the number of strand breaks with the number of embryonic abnormalities and fecundity. There was an

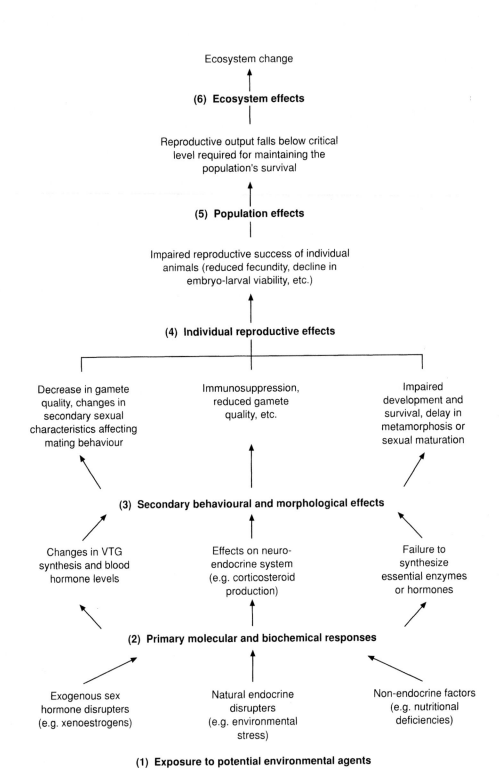

Ecosystem change

(6) Ecosystem effects

Reproductive output falls below critical
level required for maintaining the
population's survival

(5) Population effects

Impaired reproductive success of individual
animals (reduced fecundity, decline in
embryo-larval viability, etc.)

(4) Individual reproductive effects

Decrease in gamete
quality, changes in
secondary sexual
characteristics affecting
mating behaviour

Immunosuppression,
reduced gamete
quality, etc.

Impaired
development and
survival, delay in
metamorphosis or
sexual maturation

(3) Secondary behavioural and morphological effects

Changes in VTG
synthesis and blood
hormone levels

Effects on neuro-
endocrine system
(e.g. corticosteroid
production)

Failure to
synthesize
essential enzymes
or hormones

(2) Primary molecular and biochemical responses

Exogenous sex
hormone disrupters
(e.g. xenoestrogens)

Natural endocrine
disrupters
(e.g. environmental
stress)

Non-endocrine factors
(e.g. nutritional
deficiencies)

(1) Exposure to potential environmental agents

FIGURE 8.12 *Mechanisms by which endocrine disruptors affect the reproduction and survival of wildlife. Reproduced after Campbell and Hutchinson (1998) with permission from the Society of Environmental Toxicology and Chemistry.*

interesting seasonal variation in the findings. In the spring, the percentage of broods with at least one abnormality was 14% in one contaminated pond and 57% in another, whereas none were found in the control ponds. In summer, although the percentages were still significantly higher in the contaminated ponds, the percentage of abnormal broods in the control ponds was 20% and 40%. The relationship of fecundity (brood size/length of female) to strand breaks is shown in figure 8.13.

8.5 *Energy costs of physiological change*

In this chapter, we have seen cases in which the machinery of resource acquisition or uptake is damaged. In other cases damage is avoided, but the organism may still be affected because detoxication generally consumes energy and other resources which are therefore denied to production. Thus, either damage or detoxication is likely to lead to a loss of production.

The effect of pollution on production is usually measured by its effect on 'scope for growth' (SFG), defined as the difference between energy intake and total metabolic losses (Warren and Davis, 1967; Widdows and Donkin, 1992; figure 8.14). An example showing the effects of tributyl tin (TBT) concentration on SFG in the mussel *Mytilus edulis* is given in figure 8.15. Note that above a threshold of 2 μg g^{-1}, SFG declines as TBT concentration increases, indicating a loss of production. In the field, this decline could translate into a lower abundance of animals (see Chapter 15). There is here no effect of TBT on SFG at low levels of TBT. However, in the case of essential nutrients (e.g. some metals), there is a decrease in SFG at very low levels of nutrient. SFG has been particularly useful in assessing the effects of pollution on aquatic animals (e.g. fish, Crossland, 1988; *Gammarus*, Maltby *et al.*, 1990a). Field and microcosm studies have confirmed that the long-term consequences to growth and survival of individuals can be predicted from measured effects on energy balance observed at the individual level (Widdows and Donkin, 1991).

Widdows and Donkin (1991) described how reductions in SFG in *Mytilus edulis* in contaminated sites can be apportioned between specific pollutants. In a study in Bermuda, Widdows *et al.* (1990) showed that the overall reduction in SFG of *Mytilus edulis* could be pro-

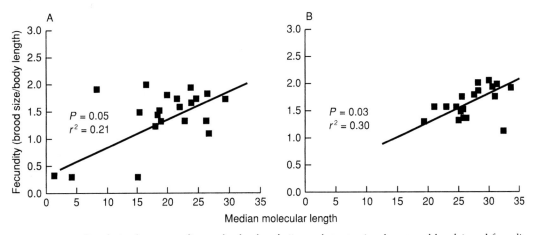

FIGURE 8.13 *Correlation between median molecular length (inversely proportional to strand breaks) and fecundity of mosquito fish collected from a radionuclide contaminated pond. (A) Total strand breaks (liver). (B) Double-strand breaks (liver). From Theodorakis* et al. *(1997).*

FIGURE 8.14 *Energy/nutrient allocation diagram illustrating the definition of 'scope for growth'.*

FIGURE 8.15 *Effects of TBT (tributyl tin) on scope for growth (SFG) in the mussel* Mytilus edulis. *Reproduced from Widdows and Donkin (1992) with permission from Elsevier Science Bv.*

portional such that, at the most contaminated sites, tributyl tin accounted for 21% and hydrocarbons for 74% of the observed effects.

A decline in SFG in the freshwater amphipod *Gammarus pulex* is due to increased stress; a decline in feeding rates leads to decreased offspring weight and increased numbers of abortions, with important consequences for the long-term viability of affected populations (Maltby and Naylor, 1990; Maltby *et al.*, 1990a,b).

Quantifying SFG relies on measuring parameters in organisms that have been exposed to pollutants and comparing these with unexposed individuals. Non-sedentary organisms can be caged in micro- and mesocosms to prevent them from migrating away from the pollutants. How-

ever, advances in microelectronics may enable some indicators of environmental stress to be measured remotely in the not-too-distant future.

Turning now to the effects of chemicals on growth, there is a vast literature for a wide variety of species. Virtually every class of pollutants – OCs, OPs, heavy metals, PAHs – have been shown to retard growth. Two examples are quoted, as follows. (i) During the investigation of the effects of high levels of boron and selenium in the Central Valley of California, the growth rate of ducklings was found to be reduced (Stanley *et al.*, 1996), although duckling survival was not affected. (ii) On the Great Lakes, the growth rate of young **ospreys** (***Pandion halieatus***) was inversely correlated with the concentration of dioxin in eggs (Woodford *et al.*, 1998), although overall breeding success was not affected. It appears that growth can be a sensitive, but non-specific, indicator of pollution.

In many cases, the organism can compensate for the effects of the pollutant on its physiology, but only at a price. Besides the direct damage, there is the energetic cost of detoxification mechanisms. In this case, the resources that the organism invests in detoxication reduce its chances of death, but at a cost in terms of lost production. In other words, the organism trades off a loss of production for a reduction of mortality rate.

The concept of a 'trade-off' is important in modern evolutionary ecology, and is described in detail in Chapter 13. Here, we note only that the genetic possibilities for species are in general limited by trade-offs. In this section, we are concerned with the trade-off between production rate and mortality rate. This can also be thought of as a trade-off between production and defence because mechanisms which reduce mortality rate serve to defend the organism.

A trade-off between production and defence

could come about in many ways (shells, spines, vigilance, etc.; Sibly and Calow, 1989), but here we are especially concerned with defences against toxins. Possible methods of defence include relatively impermeable exterior membranes (e.g. Oppenoorth, 1985; Little *et al.*, 1989), more frequent moults (e.g. in *Collembola*, in which metals stored in the gut cells are voided during ecdysis; Bengtsson *et al.*, 1985), a more comprehensive immune system and detoxification enzymes (Terriere, 1984; Oppenoorth, 1985). Many examples are given in this book. Although it is clear that such defences generally have energy costs (Sibly and Calow, 1989; Hoffman and Parsons, 1991), these could be small, e.g. in the case of inducible enzyme responses. Even here, however, there must be a cost because amino acids are required at every stage of the genetic response and because all genetic mechanisms involve overheads (molecular checking, DNA turnover and disposal of waste). However, in environments which are naturally highly stressful (wide fluctuation in

temperature, humidity, etc.), the costs of coping with pollution may only be a tiny proportion of the total stress that the organism has to cope with, and may consequently be difficult to quantify.

To illustrate what can be achieved in this area, we now consider two studies in more detail. In a classic study of insecticide detoxication in the aphid *Myzus persicae*, Devonshire and Sawicki (1979) used strains that differed in their insecticide resistance. The strains were genetically different, and the mortality and growth rates of each strain were examined in a standard environment. More resistant strains contained more detoxifying enzyme (phosphatase E_4, up to 1% of total protein) because they contained more duplications of a structural gene (figure 8.16A). LD_{50}s were measured for three of these strains, from which mortality rates could be obtained (figure 8.16B). Note that strains containing more copies of the structural gene, which presumably spent more on detoxication, achieved a reduction in mortality rate.

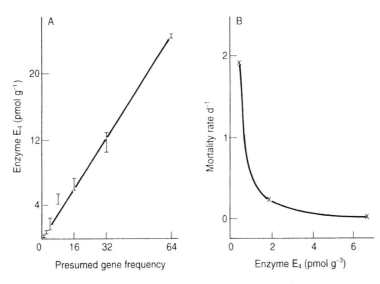

FIGURE 8.16 *(A) Concentration of the detoxifying enzyme E_4 in seven strains of the aphid* Myzus persicae *in relation to the number of copies of a structural gene hypothetically present in each strain. (B) Mortality rate of three of the strains in relation to E_4 concentration. Aphids were placed on potato leaves that had been dipped in an organophosphorus insecticide (Demeton-S-methyl). Data from Sawicki and Rice (1978) and Devonshire and Sawicki (1979). Reproduced from Sibly and Calow (1989) with permission from the Linnean Society of London.*

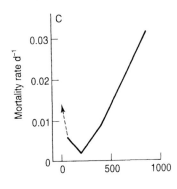

FIGURE 8.17 *(A) Initial peak levels and final steady-state levels of metals (Cu and Pb) in the bodies of Springtails in relation to levels in the nutrient broth. (B) Growth rate (reciprocal of time to first reproduction) and mortality rate (calculated from survivorship over the first 10 weeks of life). (C) Mortality rate in relation to the final steady-state levels of metals in the body. Mortality must increase at very low levels as the animals then suffer from copper deficiency. All data are from Bengtsson et al. (1983, 1985). Reproduced after Sibly and Calow (1989) with permission from the Linnean Society of London.*

Bengtsson *et al.* (1983, 1985) studied detoxication of metals by the springtail *Onychiurus armatus*. The springtails were fed on fungi grown on a nutrient broth contaminated with 0, 30, 90 or 300 μg g^{-1} of copper and lead in equal proportions. Concentrations of these metals within the animal reached high levels initially, but were then reduced by detoxication processes and reached steady state after a few weeks (figure 8.17A). Detoxication was achieved by more frequent moulting of the lining of the gut (where most metals are stored),

and this reduced the growth rate, as shown in figure 8.17B. Detoxication was not complete, however, so that body metal levels were elevated in more contaminated environments even at steady state, as shown in figure 8.17A. Mortality rate was higher in the contaminated environments (figure 8.17B), possibly as a result of the increased metal levels in the body (figure 8.17C). Note, however, that a certain amount of copper is needed physiologically, so that in a completely deficient environment growth rate must be reduced and mortality rate increased (figure 8.17B). In this study, growth rate was reduced at higher levels of pollution, presumably because the animal moulted more frequently, but metals were not completely eliminated from the body, so mortality rates were still increased.

Where there is a trade-off between production and defence, figure 8.18 illustrates the likely form of the relationship. The trade-off curve has the following features: (i) even when growth rate is maximal (zero allocation to defence), growth rate is finite – equal to the limiting production rate in the study environment; (2) even if all resources are allocated to defence, mortality rate is still greater than zero, equivalent to the 'extrinsic' mortality rate in that environment. The simplest curve having these general features is convex viewed from below, and has the general form of the curve shown in figure 8.18. The evolutionary implications of this trade-off are considered in section 13.3.

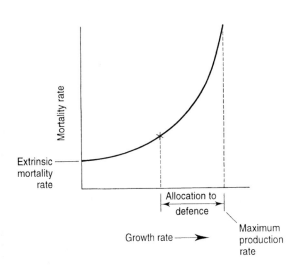

FIGURE 8.18 *The likely form of the trade-off between production rate and mortality rate that may constrain the operation of detoxication mechanisms. Allocation of resources to defence reduces mortality rate but simultaneously cuts growth rate. Reproduced after Sibly and Calow (1989) with permission from the Linnean Society of London.*

8.6 *Effects on plants*

The physiological effects of chemicals are much better understood in vertebrate animals than they are in plants, reflecting the much larger investment in human toxicology than in ecotoxicology. Some important effects in plants will now be briefly reviewed.

Photosynthesis is a complex process taking place in the chloroplasts of green plants, which can be inhibited by a number of **herbicides** (section 7.4.9). It involves the synthesis of organic compounds by the reduction of carbon dioxide using light energy absorbed by chlorophyll, as described by the following empirical equation.

$$6\,CO_2 + 6\,H_2O \rightarrow C_6H_{12}O_6 + 6O_2$$

glucose

The trapping of solar energy, and its utilization to biosynthesize organic compounds, is essential to the life of plants and to the animals that feed upon them. Also, the generation of oxygen by photosynthesis ensures the maintenance of an aerobic atmosphere and the operation of aerobic processes that are essential to most life forms on Earth. The evolution of photosynthesis provided the essential basis for the evolution of aerobic processes to follow the anaerobic ones which operated in the earliest life forms.

Herbicides such as **substituted ureas**, triazines and paraquat inhibit the process of photosynthesis, but there is some uncertainty about their exact mode of action. Two forms of chlorophyll and a number of carrier proteins are involved in the complex electron carrier systems known as photosystems I and II; a detailed discussion of how and where these compounds may act lies outside of the scope of the present text. The critical point is that the effects of herbicides, or other environmental chemicals, on the overall process of photosynthesis can be measured in living plants or in isolated chloroplasts.

Respiration, which is simply the reverse of the empirical equation shown above for photosynthesis, can also be inhibited by environmental pollutants, including nitrophenolic herbicides such as dinoseb which act as uncouplers of oxidative phosphorylation by mitochondria. The action of respiratory poisons can be determined by measuring changes in the rate of oxygen consumption in whole plants or in isolated mitochondria.

The *rate of growth* of plants is relatively easy to measure and often provides an index of the toxic effects of pollutants. A reduction in the rate of growth can lead to stunted development and dwarfism, a response that can be observed in the leaves of trees exposed to sulphur dioxide and other aerial pollutants. There is also the question of the action of chemicals which can affect growth regulation. The so-called 'plant growth regulator' herbicides (see sections 1.2.10 and 7.4.10) can cause characteristic growth disturbances in plants. Included here are the widely used phenoxyalkanoic acids MCPA, 2,4-D, CMPP and 2,4-DB. They cause distorted growth, malformed leaves and severe epinasty (downward curvature) of stems and petioles. They cause unequal growth of young rapidly developing tissues near the meristem. Effects of this kind will be familiar to gardeners who have used them as selective herbicides to control broad-leaved weeds (dicots) in lawns.

8.7 *Summary*

In this chapter, the effects of pollutants on physiological processes at the cellular, organ and whole organism level have been examined. The range of effects are very diverse, ranging from effects on the metabolism of cells to the behaviour of the individual. In this chapter, we have, largely, examined the effects of single pollutants, in the next chapter the interactive effects of pollutants are examined.

Pollutants may damage organisms directly by increasing their mortality rates or interfering with the processes of resource acquisition and uptake and so reducing production rates. These effects on individuals can result in slower population growth or even population decline. These effects are discussed in Chapter 12.

Alternatively, organisms may avoid or restrict damage by the use of detoxication mechanisms (e.g. induction of monooxygenases) or repair (e.g. DNA repair mechanisms). However, all of these use energy and resources which are therefore not available for production. It follows that the population effect can again be detrimental. The evolutionary implications of this trade-off are discussed in Chapter 13.

8.8 *Further reading*

ATCHISON, G. J. *et al.* (1996) A review of the effects of pollution on the behaviour of aquatic animals.

DALLINGER, R. and RAINBOW, P. S. (eds) (1993) *Ecotoxicology of Metals in Invertebrates*. Includes a large number of relevant papers.

DELL'OMO, G. (2000) *Behaviour in Ecotoxicology*. A new text on an important area of the subject.

DONKER, M. H. *et al.* (eds) (1994) *Ecotoxicology of Soil Organisms*. Includes a large number of relevant papers.

GRUE, C. E. *et al.* (1991) Covers the effects of acetylcholinesterase on the behaviour of mammals and birds.

HOPKIN, S. P. (1989) *Ecophysiology of Metals in Terrestrial Invertebrates*. Includes a comprehensive review of the distribution and effects of metals in terrestrial invertebrates.

LANGSTON, W. J. and BEBIANNO, M. J. (1998) *Metal Metabolism in Aquatic Environments*. A useful collection of papers describing current ideas on the role of metals in aquatic organisms.

WIDDOWS, J. and DONKIN, P. (1992) Reviews the effects of pollution on scope for growth in *Mytilus edulis*.

CHAPTER 9

Interactive effects
of pollutants

In the natural environment, organisms are frequently exposed to complex mixtures of pollutants and it is relatively uncommon to find any one pollutant dominant over all others. Yet, because of limitations of time and resources, nearly all regulatory toxicity testing is carried out using single compounds. It is not feasible to test the toxicity of more than a very small proportion of the chemical combinations that exist in terrestrial, marine or freshwater ecosystems or that may arise because of the release of new chemicals into the environment. The complexity of the situation is illustrated in figure 9.1, which gives analytical data for residues of PCBs in tissues of organisms from a polluted area. A number of different PCB congeners are found in both species, with a wider selection in the case of the mollusc than in the harbour seal. When effluents or contaminated environmental samples are subjected to testing, toxicity is often due to more than one chemical component of a mixture, and questions arise concerning possible potentiation. This issue has already been discussed in section 6.2.

9.1 Introduction

When regulatory authorities consider the toxicity of mixtures, it is usually assumed (unless there is definite evidence to the contrary) that the toxicity of combinations of chemical will be approximately additive. In other words, the toxicity of a mixture of compounds will approximate to the summation of the toxicities of its individual components. This is usually a correct assumption. However, in a relatively small yet very important number of cases, toxicity is substantially greater than additive. That is to say when organisms are exposed to a combination of two or more chemicals there is **potentiation** of toxicity. Sometimes, the term **synergism** is also used to describe this phenomenon. However, many scientists restrict the term 'synergism' to situations in which only one of two components is present at a level that can cause a toxic effect, whereas the other ('synergist') would have no effect if applied alone. This practice will be followed here. Some examples of synergism are given in table 9.1.

Gas chromatograms of extracts from *M. balthica* and *P. vitulina* (SE-54 capillary column).
The CB congeners are identified with their IUPAC numbers.

FIGURE 9.1 *PCB congeners in the tissues of marine organisms from the Dutch Wadden Sea. The organisms represented are a mussel* (Macoma baltica) *and the harbour seal* (Phoca vitulina). *The compounds were separated, identified and quantified by capillary gas chromatography. Each of the numbered peaks represents a PCB congener. HCB, hexachlorobenzene, used as an internal standard in the analysis. Reproduced from Boon et al. (1989) with permission from Elsevier Science Ltd.*

The effectiveness of a synergist is usually measured by a synergistic ratio (SR), which is:

$$\frac{\text{MLD (or conc.) for chemical alone}}{\text{MLD (or conc.) for chemical + synergist}}$$

Where there is synergism, the SR will be greater than 1; in effect, the synergist will lower the median lethal dose (MLD) or concentration of the chemical. In the following account, additive toxicity will be discussed before moving on to consider the questions of potentiation and synergism.

9.2 *Additive effects*

It is often the case that the toxicity of a mix-ture is roughly equal to the summation of the toxicity values of its individual components. In other words, each chemical expresses roughly the same toxicity in a mixture as it would when tested alone. Where there is no evidence of potentiation or antagonism, estimates of the toxicity of a mixture can be made by adding together the expected contributions from each of its components.

The toxicity of each individual component in a mixture depends on its concentration and can be estimated from a dose–response curve, e.g. a percentage mortality from an LD_{50} plot. Thus, a mixture containing three components at concentrations that would, if tested individually, cause 5%, 10% and 15% mortality, respectively, would be expected to cause 30% mortality overall if toxicity were simply additive. Toxicity data working to end-points other

TABLE 9.1 *Examples of synergism**

Organisms	Pesticide	Detoxifying enzyme system	Inhibitor (synergist)	Increase in toxicity
Strains of insect resistant to pyrethroids	Cypermethrin	Monooxygenase	Piperonyl butoxide	< 40x
Insects	Carbaryl	Monooxygenase	Piperonyl butoxide	< 200x
Mammals and some resistant insects	Malathion	Carboxy (B) esterase	Various organo-phosphorous compounds	< 200x

*For further details, see Chapter 7 of Hodgson and Levi (1994).

than mortality can be treated in a similar way, e.g. reduction of photosynthetic rate, expressed as a percentage. Clearly, where there is potentiation or antagonism, the estimated toxicity would differ markedly from the measured toxicity.

Where a group of compounds share a common mechanism of action and interact with the same site of action, it is probable that they will show additive toxicity when present as mixtures, unless the picture is complicated by toxicokinetic factors (see sections 9.3–9.5). Within such a group of compounds, there are bound to be differences between compounds in their affinity for the site of action (receptor), and thus corresponding differences in the relationship between concentration (or dose), and toxic effect. However, toxicity is often simply related to the percentage of receptor sites to which the toxic molecules bind. Thus, if the concentrations of individual compounds are corrected by affinity factors, then all toxicity data may be fitted to a single dose–response curve. In practice, this has been carried out by calculating 'toxic equivalency factors' (TEFs) in relation to the most toxic component of the group. The calculation of dioxin equivalents for mixtures of polychlorinated aromatic compounds gives an example of this approach (Safe,

1990; Ahlborg *et al.*, 1994). A standard toxic response is defined (e.g. 50% mortality), and a comparison is then made between the concentrations required to produce this effect by (i) the most toxic compound of the group and (ii) the concentration required by another chemical of the group. The ratio of (i):(ii) is then the toxic equivalency factor. The contribution of each compound to the overall toxicity of the mixture [the **toxic equivalent (TEQ)** of the compound] is then determined by multiplying its concentration in an environmental sample (e.g. water or tissue) by its toxic equivalency factor. The TEQs for individual compounds can then be added up to give the toxic equivalent for the whole mixture, i.e. equivalence relative to a reference compound. An important example of this approach is the calculation of **dioxin equivalents** for PCDDs and co-planar PCBs, in which TCDD ('**dioxin**') is the reference compound. See section 15.2 for further details. Recently, cellular systems have been developed which provide a measure of dioxin equivalents present in environmental samples. They depend on the principle that toxicity is the consequence of these compounds binding to the Ah receptor ('**Ah**'-**receptor-mediated toxicity**). The receptor is present in, for example, hepatocytes, and the extent to which polychlorinated compounds bind is indicated by a characteristic response such as light emission (see also section 6.7). Other examples in which members of a group of compounds share a common mode of action will now be briefly mentioned.

Warfarin and related anticoagulant rodenticides share a common binding site in hepatic microsomes, where they inhibit the operation of the **vitamin K cycle** (sections 1.2.1.1 and 7.4.4). Present evidence suggests that they express additive toxicity when present as mixtures. Several metabolites of PCBs act as **thyroxine antagonists**, competing with it for binding sites on the protein transthyretin (section 7.4.5). Finally, oestrogen and androgen binding sites represent the binding sites of certain environmental **endocrine disruptors** (section 7.4.7).

9.3 *Potentiation of toxicity*

The phenomenon of potentiation of toxicity requires further explanation. The general picture is illustrated in figure 9.2, where two compounds A and B are under consideration. The maximum dose of either of them will give the same degree of toxic response (X). (X could be a percentage reduction in respiration or mortality.) Between the two maximum doses are different mixtures of the two compounds. Doses for either compound run from 0% to 100% of the maximum dose. The summation of the contributions of the two components of the mixture will always be 100%, e.g. 30% of maximal dose of A + 70% of the maximal dose of B. If toxicities are simply additive, all of these combinations should give the same toxic response (X) as the maximal dose of A or B. If, however, there is potentiation or synergism, the toxic effect should be greater than expected. If there is antagonism, the toxic effect should be less.

Care is needed when deciding whether toxic effects of combinations of chemicals are truly greater than additive. In the first place, because of errors in measurement, interest is confined to situations where toxicity is *substantially* greater than additive (e.g. where SRs exceed 2). Smaller differences usually reflect no more than the compounding of errors. Also, account should be taken of the relationship between dose and toxic effect for each of the individual components of a mixture (figure 9.3). In particular, it is important to know whether there is a straight line relationship between dose (or log dose) and toxic response (figure 9.3). If this is the case, then increases in toxicity of combinations of chemicals which are substantially greater than additive are to be regarded as examples of potentiation. If, on the other hand, they are not linear, for example where the increase in toxicity of an individual compound is proportionately greater than the corresponding increase in dose, this conclusion does not necessarily follow (figure 9.3). An enhancement of toxicity above that which is simply additive may merely reflect what happens when the dose

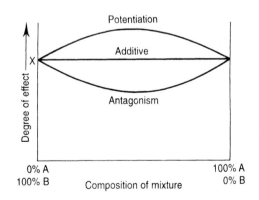

FIGURE 9.2 *Potentiation of toxicity. The vertical axis indicates the degree of toxic effect compound, and the horizontal axis represents composition of the mixture. The maximum dose of compounds A and B both give the same degree of toxic response X. Potentiation is seen when the toxicity of a mixture of two compounds exceeds the summation of toxicities of the individual components. After Moriarty (1999).*

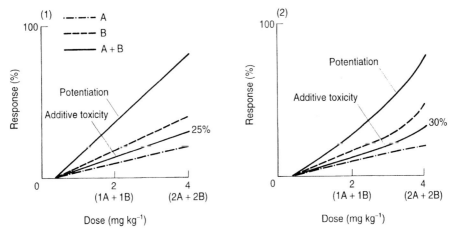

FIGURE 9.3 *Additive toxicity and potentiation. In (1), both compounds A and B produce a linear response over the dose range 0–4. If toxicity is simply additive, the response to 1 mg kg^{-1} A + 1 mg kg^{-1} B is intermediate between the responses to 2 mg kg^{-1} A and 2 mg kg^{-1} B. If potentiation occurs then the response to the combination will be higher than is shown for the additive response. In (2), the response to A is linear, but the response to B in non-linear. In this situation, the additive response to 2 mg kg^{-1} A + 2 mg kg^{-1} B is greater than in (1). This illustrates the point that linearity cannot be assumed for individual response curves. If, in this case, dose-response curves were known up to 2 mg kg^{-1} for A and B and the response to 2 mg $^{k-1}$ A + 2 mg kg^{-1} B were as shown, it might be wrongly assumed that potentiation had occurred. To establish that potentiation has occurred, the dose–response curves for the compounds A and B need to be determined above the doses used in combination.*

of an individual chemical (or chemicals) is increased, and may not therefore represent potentiation due to interaction between chemicals. Having said this, it should be re-emphasized that this chapter is mainly concerned with cases in which potentiation represents considerable enhancement of toxicity – frequently of at least one order of magnitude (i.e. 10-fold or more).

The question of interactive effects is also complicated by the question of which end-point is used for measuring toxicity. For example, in tests with the springtail *Folsomia candida*, Van Gestel and Hensbergen (1997) found that mixtures of **cadmium** and **zinc** were antagonistic to growth but were additive to reproduction compared with effects produced by the metals administered on their own.

The identification of combinations of pollutants which give rise to problems of potentiation might seem an impossible task. However, there are guidelines which aid the recognition of such combinations. In particular, recent rapid advances in biochemical toxicology have given more insight into the potentiation of toxicity due to interactions at the toxicokinetic level (Chapter 5). When one compound (A) causes a change in the metabolism of another (B), two types of interaction are recognized, as follows.

1. Compound A inhibits an enzyme system that detoxifies compound B. Thus, the rate of detoxication of B is slowed down because of the action of A.
2. Compound A induces an enzyme system which activates compound B. Thus, the rate of activation of B is speeded up because of the action of A.

These two phenomena will now be considered separately.

9.4 *Potentiation due to inhibition of detoxication*

In terrestrial vertebrates and invertebrates, the effective elimination of lipophilic xenobiotics depends upon their enzymic conversion to water-soluble products which are readily excreted (Chapter 5).

The inhibition of enzymes concerned with detoxication can lead to an increase in the toxicity of the compounds that they metabolize. This general phenomenon is well illustrated by the effect of synergists upon insecticide toxicity (table 9.1). Some of the most striking examples of synergism involve inhibition of monooxygenases by piperonyl butoxide and other methylenedioxyphenyl compounds. Synergists of this type can increase the toxicity

of pyrethroid and carbamate insecticides by as much as 60-fold and 200-fold respectively. (The synergistic ratios provide measures of increases in toxicity.)

9.5 *Potentiation due to increased activation*

Metabolism of lipophilic xenobiotics brings a reduction of toxicity in the great majority of cases. However, there are some very important exceptions to this rule. Oxidation by the monooxygenase systems sometimes generates highly reactive metabolites which can cause cellular damage. In principle, therefore, the induction of MO by a non-toxic dose of one compound can increase the toxicity of a second

FIGURE 9.4 *Metabolism of permethrin. Permethrin is detoxified by two different systems: monooxygenase, which attacks both acid and the alcohol moieties, and esterase ('B' esterase), which breaks the carboxyester bond. Inhibitors of monooxygenase can increase the toxicity of this and other pyrethroids.*

BOX 9.1 *Two examples of potentiation due to inhibition of detoxication.*

Pyrethroid insecticides, for example cyhalothrin, have considerable toxicity to bees. Often, however, they do not cause much damage when they are properly used in the field. Bees are repelled by pyrethroids, perhaps as the consequence of low-level sublethal effects. However, pyrethroids can become very much more toxic in the presence of certain ergosterol biosynthesis inhibitor (EBI) fungicides. Synergistic ratios of 5–20 have been reported for bees when EBI fungicides are added to pyrethroid insecticides. This potentiation of toxicity has been attributed to the inhibition of detoxication of pyrethroids by the monooxygenase system. There have been several reports from France and Germany of field deaths of hive bees that have been attributed to this type of potentiation (figure 9.4).

In a second example, organophosphorus insecticides which contain the thion (P=S) group can inhibit microsomal monooxygenases of vertebrates. This happens when the enzyme converts them to oxons (P=O) (figure 9.5). The process is termed 'oxidative desulphuration' and leads to the binding of sulphur to cytochrome P_{450} with consequent loss of monooxygenase (MO) activity. Exposure to relatively low levels of organophosphorus compounds can make birds more sensitive to the toxicity of carbamate insecticides; carbamates are detoxified by MOs, so a reduction of MO activity can cause an increase in toxicity (figure 9.5).

compound which is subject to oxidative activation. This said, it does not follow that the induction of monooxygenase will automatically lead to increases in the rate of activation (or in the degree of consequent cellular damage). For one thing, induction may also lead to the induction of other enzymes which have a detoxifying function and can compensate for increased activation.

9.6 *The detection of potentiation in the field*

Although potentiation is a well-recognized phenomenon in the laboratory, the extent to which it occurs in the field is virtually unknown. There are good reasons for suspecting that it may occur in heavily polluted areas. In some marine areas, for example, a wide range of

FIGURE 9.5 *Metabolism of malathion. Malathion is detoxified by the action of a carboxyesterase but activated by monooxygenase. Toxicity depends on the relative importance of these two competing enzymes. Chemicals that induce the monooxygenase system can make malathion more toxic.*

BOX 9.2 *Two examples in which potentiation results from increased activation.*

Benzo(a)pyrene and certain other carcinogenic polycyclic aromatic hydrocarbons (PAHs) are activated by an inducible form of cytochrome P_{450} known as P_{450} IAI (Chapter 5). A range of planar organic compounds, not in themselves carcinogens or mutagens, can cause the induction of P_{450} IAI. Examples include certain PAHs, co-planar PCBs and 1,2,7,8-tetrachlorodibenzodioxin (TCDD) (see Chapter 7). Such compounds can act as promoters, which potentiate the carcinogenic action of other compounds. By increasing the rate of activation of carcinogens, they can also increase the rate of formation of DNA adducts; this may lead to an increased rate of chemically induced mutation. In the marine environment, there is much evidence that (i) fish, birds and mammals sometimes have elevated levels of P_{450} IAI and (ii) that the levels of P_{450} IAI are related to the degree of exposure to pollutants such as co-planar PCBs. It is suspected, but not yet proven, that individuals with elevated IAI will experience higher levels of DNA damage caused by environmental carcinogens and mutagens.

In the second example, OP insecticides, which contain the P=S group, are activated by P_{450} forms of the monooxygenase system (figure 9.4). Thus, the induction of monooxygenase by a non-toxic dose of xenobiotic can lead to enhanced activation and consequently to increased toxicity of an OP. Partridges exposed to EBI fungicides can show increased levels of hepatic monooxygenase due to induction. While in the induced state, they show increased susceptibility to OP insecticides such as malathion and dimethoate (Walker et al., 1993; Johnston, 1995). The time dependence of this potentiation needs to be emphasized. After exposure to the EBI, several hours will elapse before the rate of activation of the OP is increased. Furthermore, the monooxygenase activity will return to its normal level after a few days if exposure to the EBI does not occur again. Thus, birds may become more susceptible to certain OPs during the period between 6 h and 5 days after exposure to an EBI. Potentiation of this kind has, so far, been demonstrated only in laboratory studies. It has not yet been shown to occur in the field with normal approved use of pesticides (but see further discussion in section 9.6).

organochlorine compounds (PCBs, TCDDs and others) (figure 9.1) are found (Malins and Collier, 1981). Elevated levels of P_{450} 1A1 have been found in fish and birds from such areas, and relatively high rates of activation of mutagenic PAHs have been suspected.

In intensive agriculture and horticulture, animals, birds and non-target invertebrates are exposed to a variety of insecticides, herbicides and fungicides. In particular, combinations of pesticides are sometimes used as seed dressings and as sprays (e.g. tank mixes). Questions are now asked about the possibility of potentiation of toxicity when animals are exposed to mixtures such as these. Also, mobile species (e.g. birds, flying insects) can be exposed sequentially to different compounds when they move from field to field. As explained earlier, there is good evidence that bees can be poisoned by pyrethroid sprays when EBIs are mixed with them; EBIs can cause potentiation by the inhibition of detoxication (Pilling et al., 1995). There are also other combinations of pesticides which give cause for concern with regard to possible potentiation of toxicity under field conditions. However, these issues remain the subject of speculation because field studies which could resolve them have yet to be performed.

The difficulty of identifying harmful effects caused by pollutants in the field is a recurring theme of this book. The potential of biomarker strategies to aid the resolution of this problem was emphasized in the previous chapter and is discussed further in Chapter 15. This point can be illustrated by considering a hypothetical example in which the interaction of two pesticides is investigated in a field trial. One or more biomarkers of toxic effect would be chosen to measure responses to a pollutant in an appropriate indicator species. These responses would then be determined in three different field situations:

1. where only compound A was applied;
2. where only compound B was applied;
3. where both A and B were applied.

If potentiation occurred, then the responses measured in 3 would be substantially greater than the summation of responses measured in 1 and 2. This approach has the advantage that it can measure potentiation at the sublethal level, i.e. it can give early warning of enhancement of toxicity before lethal effects are produced.

There are advantages in using non-destructive biomarkers in situations of this kind. In particular, they make it possible to conduct serial sampling in individual animals or birds; any changes caused by chemicals can then be measured in relation to *internal* controls. In other words, biochemical or physiological parameters can be measured in individuals before and after exposure to the pesticide, thereby overcoming the serious difficulty of interindividual variation. Serial sampling is possible, for example, in the case of nestling birds in nest boxes, or in animals or birds which are restricted to limited areas (e.g. enclosures) from which they can readily be recaptured and sampled.

9.7 *Summary*

In the field, living organisms are exposed to mixtures of pollutants, and questions arise about possible interactions between the components of mixtures. When chemicals are tested during the course of environmental risk assessment, it is usually carried out compound by compound. Very rarely are mixtures tested. In the absence of evidence to the contrary, it is normally assumed that the toxicity of mixtures of compounds will be additive, i.e. the toxicity of the mixture will approximate to the summation of the toxicities of its individual components. This is often the case, and there are already good examples of environmental chemicals that share a common mode of action behaving in this way (e.g. Ah-receptor-mediated toxicity caused by mixtures of co-planar PCBs and dioxins). However, there are a small but significant number of cases in which toxicity is potentiated when organisms are exposed to mixtures; toxicity is substantially greater than additive. The best known cases of this involve interactions at the toxicokinetic level, where either one compound inhibits the detoxication of another or one compound increases the rate of activation of another. With increasing knowledge of the biochemical toxicology of pollutants, it becomes increasingly possible to predict such interactions and to test for them.

9.10 *Further reading*

Boon, J. P. *et al.* (1989) Gives examples of complex pollution patterns caused by PCBs.

Malins, D. C. and Collier, T. K. (1981) Discusses the problem of effects of mixtures upon marine organisms.

Moriarty, F. (1999) *Ecotoxicology*, 3rd edn. Discusses the problem of interpretation of data purporting to show potentiation.

Walker, C. H. *et al.* (1993) Discusses laboratory evidence for potentiation in birds.

Wilkinson, C. F. (1976) Deals with potentiation of toxicity of insecticides.

Biomarkers

The term 'biomarker' has been gaining acceptance in recent years, albeit with some inconsistency in definition. Here, we define biomarkers as 'any biological response to an environmental chemical at the individual level or below demonstrating a departure from the normal status'. Thus, biochemical, physiological, histological, morphological and behavioural measurements are to be considered as biomarkers. Some examples of biomarkers at different organization levels are given in table 10.1.

Biological responses at higher organizational levels – population, community and ecosystem – are considered as bioindicators. Despite the importance of any changes at these higher levels (considered in Chapters 12 and 15), these changes are too general to be considered as specific biomarkers. The relationship between biomarkers and bioindicators regarding their specificity and ecological relevance is shown diagrammatically in figure 10.1. In general, it can be said that it is difficult to relate biochemical changes to ecological changes (although eggshell thinning caused by p,p'-DDE and imposex in dog whelks caused by TBT, discussed in Chapter 15, shows that a physiologi-

cal change can be related to a massive population change). It is also difficult to relate ecological changes to specific chemical causes.

10.1 Classification of biomarkers

A number of classifications of biomarkers have been proposed. The most widely used is division into biomarkers of exposure and biomarkers of effect. Biomarkers of exposure are those that indicate exposure of the organism to chemicals, but do not give information of the degree of adverse effect that this change causes. Biomarkers of 'effect', or more correctly 'toxic effect' (because all biomarkers by definition show an effect), are those which demonstrate an adverse effect on the organism.

A biomarker approach based on changes in physiological parameters is shown in figure 10.2. The change in the health status of an individual with increasing exposure to a chemical is shown by a smooth curve running from healthy through reversible to irreversible changes leading to death. The important transition points along the way are: (i) when the

TABLE 10.1 *Biomarkers at different organizational levels*

Organizational level	Example of biomarker
Binding to a receptor	TCDD binding to Ah receptor Nonylphenols binding to oestrogen receptor
Biochemical response	Induction of monooxygenases Vitellogenin formation
Physiological alterations	Eggshell thinning Feminization of embryos
Effect on individual	Behavioural changes Scope for growth

organism is first stressed (*h*), i.e. when physiology is no longer normal, then moving into the area where the organism, although stressed, is able to compensate for this stress; (ii) when the organism is no longer able to compensate (*c*), but the changes are still reversible and removal of the stress enables the organism to recover; and (iii) point *r* beyond which the changes are irreversible and death ensues. The second part of figure 10.2 shows the responses of five different biomarkers which are used to measure the health status of the individual (Depledge, 1994).

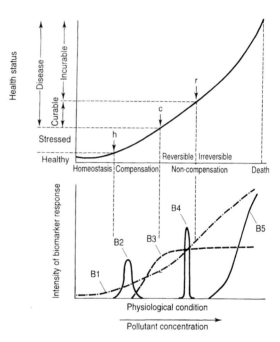

FIGURE 10.2 *Relationship between exposure to pollutant, health status and biomarker responses. Upper curve shows the progression of the health status of an individual as exposure to pollutant increases. h, the point at which departure from the normal homeostatic response range is initiated; c, the limit at which compensatory responses can prevent development of overt disease; r, the limit beyond which the pathological damage is irreversible by repair mechanisms. The lower graph shows the response of five different hypothetical biomarkers used to assess the health of the individual. Reproduced from Depledge et al. (1993) with permission from Springer-Verlag.*

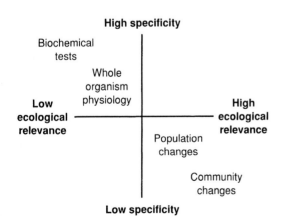

FIGURE 10.1 *Specificity and ecological relevance of biochemical effects measurements From Addison (1996). Reprinted with permission from* Environmental Reviews.

10.2 *Specificity of biomarkers*

Biomarkers range from those that are highly specific – an enzyme of the haem pathway aminolaevulinic acid dehydratase (ALAD) is inhibited only by lead – to those that are non-specific, whereas effects on the immune system can be caused by a wide variety of pollutants. A listing of biomarkers in order of their degree of specificity is given in table 10.2.

Both highly specific and highly non-specific biomarkers are of value in hazard assessment. When blood samples of waterfowl can be taken and the activity of ALAD determined then it is possible, without further measurements, to determine the percentage of the waterfowl that are at risk from lead poisoning. However, determination of ALAD does not give any information on the (many) other pollutants that may be present.

Inhibition of AChE can be used to provide legal proof of death by organophosphorous and carbamate pesticides (Hill and Fleming, 1982) and such inhibition is considered specific to these classes of chemicals. However, in recent years, evidence has emerged that the inhibition of AChE is not caused solely by OPs and carbamates. Payne *et al.* (1996) found depression of AChE activity as high as 50% in fish in Newfoundland remote from pesticide usage and considered that a complex mixture of pollutants might be involved. Experimental studies by Guilhermino *et al.* (1998) have shown that both detergents and metals can inhibit the activity of AChE and suggest that use of this enzyme as a biomarker could be extended.

The induction of monooxygenase is caused by a wide variety of chemicals and is often a sensitive indicator of the presence of pollutants (Besselink *et al.*, 1997). Thus, it is a useful indication that organisms are affected by pollutants, although it gives little information on the specific cause. In this case, the biomarker is valuable to target additional studies because

pollutants are present at a high enough level to cause a detectable effect.

10.3 *Relationship of biomarkers to adverse effects*

It is clearly useful to be able to relate the degree of change of the biological response measured to the harm that it causes so that when remedial action is proposed the cost of it can be defended. A list of the same biomarkers already given in table 10.2 are listed in table 10.3 in the order of our knowledge of the adverse effects that they measure. A quick examination of the two tables will reveal that they differ in ranking order.

At the top of table 10.3 is eggshell thinning. In this case, it is possible to define the critical degree of eggshell thinning; it has been found for a variety of species that eggshell thinning in excess of 16–18% is associated with population declines. This phenomenon is discussed in more detail in Chapter 15.

The fact that the relationship between the biomarker response and an adverse effect is not clear cut does not invalidate the use of that biomarker. First, it demonstrates that the organism has been sufficiently exposed to a pollutant or pollutants to cause a physiological change. In some cases, such as the induction of metallothionein, the change is a protective mechanism (Chapter 7); here, a knowledge of how much of the possible protective mechanism has already been induced is valuable to assess the risk to the individual. Second, in the case of vital systems, it is an indication that further investigations should be undertaken. For example, few would take damage to the integrity of DNA lightly, even though in many cases the damage is repaired and no adverse effects occur (see Chapter 7). In other cases, such as changes in porphyrin levels, it is clear that the

TABLE 10.2 *Some biomarkers listed in order of decreasing specificity to pollutants*

Biomarker	Pollutant	Comments and references to analytical procedures
Inhibition of ALAD	Lead	Sufficiently reliable to replace chemical analysis (Wigfield *et al.*, 1986)
Induction of metallothionein	Cadmium	More difficult to measure than cadmium levels (Hamer, 1986)
Eggshell thinning	DDT, DDE, Dicofol	Degree of eggshell thinning is easily measured (Ratcliffe, 1967)
Inhibition of AChE	OPs, carbamates	Easier and more reliable than chemical analysis (Fairbrother *et al.*, 1991)
Anticoagulant clotting proteins	Rodenticides	Measurements similar in complexity to chemical analysis (Huckle *et al.*, 1989)
Induction of monooxygenases	OCs, PAHs	Dioxin equivalent more easily measured than using chemical analysis (Murk *et al.*, 1997)
Porphyrin profiles	Several OCs	Separation by high-performance liquid chromatography is well developed (Kennedy and James, 1993)
Retinol profiles	OCs	Provides means of showing exposure to specific chemicals (Shugart, 1994). Considerable natural variations, ratios more reliable than absolute values (Spear *et al.*, 1986)
DNA and haemoglobin adducts	Largely PAHs	Several tests available but complicated by repair mechanisms
Induction of vitellogenin	Oestrogenic chemicals	Induction of this yolk protein in a male fish is a sensitive indicator of the presence of oestrogenic chemicals (Harries *et al.*, 1997)
Other serum enzymes	Metals, OCs, PAHs	Several different enzyme systems have been studied (Fairbrother, 1994)
Stress proteins	Metals, OCs	Wide range of stress proteins have been studied (Sanders, 1993)
Immune responses	Metals, OCs, PAHs	Many different tests are available (Wong *et al.*, 1992)

levels are much lower than has been shown to cause harm. Nevertheless, these biomarkers can be used to demonstrate exposure. The use of these various types of biomarkers in hazard assessment is considered later in this chapter.

TABLE 10.3 *Some biomarkers listed in order of decreasing specificity of adverse effect*

Biomarker	Organizational level	Comments and references
Eggshell thinning	Intact animal – population	Wide species variation in sensitivity. Related to reproductive success (Peakall, 1993)
Inhibition of AChE	Organ – intact animal	Degree of inhibition has been related to mortality and sublethal effects (Grue *et al.*, 1991)
Inhibition of ALAD	Organ – intact animal	Degree of inhibition has been related to mortality (Scheuhammer, 1989)
Anticoagulant clotting proteins	Intact animal – population	Has been related to mortality; risk assessed from blood protein levels (Hegdal and Blaskiewicz, 1984)
Induction of monooxygenases	Organ population	Analysis of dioxin equivalents has been related to reproductive success. Induction of P_{450} enzymes related to specific chemicals (Bosveld and van den Berg, 1994)
Depression of plasma retinol and thyroxine	Organ	Binding to specific protein has been shown. Relation to adverse effects tenuous (Brouwer and van den Berg, 1986)
DNA integrity	Organ	DNA damage is a serious indication of harm, but relationship to effects often tenuous (Everaarts *et al.*, 1998)
Immune responses	Organ	Proper functioning is critical to health, but system has considerable reserve (Richter *et al.*, 1994)
DNA and haemoglobin adducts	Organ	Good monitor of exposure, especially for PAHs; relation to effects is tenuous (Varanasi *et al.*, 1989)
Other enzymes	Organ	Relationship to effects are not clear (Fairbrother, 1994)
Porphyrin profiles	Organ	Levels in environmental samples are well below those causing adverse effects (Fox *et al.*, 1988)
Induction of vitellogenin	Organ – intact animal	Clear link between induction of vitellogenin and presence of oestrogenic chemicals, but biological significance is speculative (Archand-Hoy and Benson, 1998)
Induction of metallothionein	Organ	Protective mechanism, not related to mechanism of toxicity (Hamer, 1986)
Stress proteins chemicals	Organ	Difficult to separate effects from non-chemical stresses (Pyza *et al.*, 1997)

10.4 *Discussion of specific biomarkers*

Some of the biomarkers listed in tables 10.2 and 10.3 have been covered elsewhere in this book. Specifically, eggshell thinning is discussed in Chapter 15, and some information on the inhibition of AChE and induction of monooxygenases has been given in Chapter 7. The sections that follow discuss some, but by no means all, of the available biomarkers. More complete coverage is given in the books on biomarkers given in the reading list at the end of this chapter.

10.4.1 INHIBITION OF ESTERASES

From the point of view of ecotoxicology, AChE is particularly useful as it represents the site of action and its degree of inhibition is related to toxic effects. Butyrylcholinesterase (BuChE) is sometimes studied in parallel with AChE, but its physiological role is unknown and its degree of inhibition is not simply related to toxic effect. The study of neuropathy target esterase, the interaction of which with organophosphorous compounds (OPs) can lead to organophosphorous compound-induced delayed neurotoxicity, has been confined to laboratory studies.

The mode of action is well established and has already been considered in some detail in Chapter 7. Two classes of compounds, the OPs and carbamate, inhibit AChE, causing an accumulation of acetylcholine at the nerve synapses and disruption of nerve function. Disruption of nerve function has obvious effects: tremors, motor dysfunction and death. The assay for AChE is more straightforward, quicker and cheaper than chemical analysis for OPs or carbamates. The degree of inhibition of AChE has been related to the symptoms observed.

With vertebrates, the inhibition of brain AChE has often been used to establish that death has been caused by OP or carbamate pesticides (Mineau, 1991). Under ideal conditions, inhibition in the range of 50–80% can be taken as proof of mortality from the pesticide (Hill and Fleming, 1982). In practice, the degree of denaturation is often unknown and adequate controls are often difficult to obtain. Also, inhibition by carbamates is readily reversible (cf. OPs) and can be quickly lost after death. However, the measurements of OPs and carbamates themselves are even more difficult because they are rapidly metabolized and eliminated; in fact, chemical analysis for residue levels has not been widely used in the diagnosis of poisoning by these compounds. The use of AChE inhibition is discussed in the diagnosis of damage caused by pesticides in Chapter 15 as a consequence of forest spraying in eastern Canada.

The usual organ studied in the case of wild vertebrate samples is the brain, which is the principal site of action of OPs and carbamates. Although using inhibition of blood AChE or BuChE would be more acceptable, the relationship to inhibition of brain AChE is complex. Studies have shown that variability of esterase activity is much greater with plasma than with brain and that recovery of plasma AChE activity is much more rapid than in the case of brain AChE. Also, plasma AChE does not represent the site of action of OPs and carbamates and there is no simple relationship between degree of inhibition and toxic effect. Thus, diagnosis based on plasma AChE activity is difficult.

10.4.2 THE MONOOXYGENASES

The haem-containing enzymes known as cytochromes P_{450} are major components of the defences of organisms against toxic chemicals in their environment. Originally evolved, perhaps as long as 2000 million years ago, to handle naturally occurring toxic compounds (Nebert and Gonzalez, 1987), they now play an important role in the detoxication of manmade chemicals.

The monooxygenase system is a coupled electron-transport system composed of two enzymes – a cytochrome and a flavoprotein (NADPH-cytochrome reductase). The system occurs in the endoplasmic reticulum of most organs, but the activity is much higher in the liver than in most other tissues (Chapter 5). Initially, these two enzymes were divided into two major classes, the enzymes P_{450} and the enzymes P_{448} (the numbers refer to the wavelengths at which the cytochrome could be detected). Recent work has shown the complexity of the system. A recent listing gave more than 750 isoforms belonging to 74 different gene families (Nelson, 1998). Cytochrome P_{450}I-dependent reactions include N-oxidation and S-oxidation; widely studied enzyme activities include ethoxyresorufin O-deethylase (EROD), benzo(a)pyrene hydroxylase (BaPH) and aryl hydrocarbon hydroxylase (AAH). Cytochrome P_{450}II-dependent oxidations include aromatic hydroxylation, acyclic hydroxylation, dealkylation and deamination. The most widely studied enzyme activities are benzphetamine N-demethylase and aldrin epoxidase.

The first demonstration of the effect of an organochlorine pesticide on hepatic microsomal metabolism (see Chapter 5) was made over 30 years ago. This important finding was the fortuitous result of an experiment to examine the effects of food deprivation on drug metabolism. These workers noted an unexpected decrease in sleeping time of rats given phenobarbital after their cages were sprayed with chlordane. They discovered that this was due to stimulation of hepatic microsomal drug metabolizing enzymes (Hart *et al.*, 1963), and this finding was soon extended to many other organochlorines. This includes not only all the

organochlorine pesticides but also the PCBs, PCDFs and PCDDs. The assessment of complex families of chemicals by means of the calculation of dioxin equivalents is considered in Chapter 15.

Activity of monooxygenases is affected by a wide variety of compounds. Classes of compounds of environmental interest besides the organochlorines include the organophosphorous compounds, pyrethroids and polycyclic aromatic hydrocarbons (PAHs).

The concept of using induction of monooxygenase activity in fish as a monitor of pollution of the marine environment by oil was put forward by several workers in the mid 1970s. Since then, a wide variety of studies have been published, ranging from the induction of AHH in fish off the Los Angeles sewage outlets to EROD induction by pulp mill effluent in Sweden. Induction of monooxygenases by paper mill effluent has proved to be one of the most sensitive biomarkers for tracking this particular type of pollution.

Monooxygenase activity is shown by a wide range of species (see Chapters 5 and 7) and some studies on fish-eating birds in the Great Lakes are detailed in Chapter 15. The usefulness of monooxygenases for biological monitoring has been clearly demonstrated in the case of hydrocarbon pollution in fish and aquatic invertebrates and for both PAHs and OC contamination in a wide range of organisms. As the response is caused by a very wide variety of chemicals, it means that the system is capable of detecting exposures sufficiently high to cause a biological response to many xenobiotics. Conversely, it tells little about the causative agent(s), but can be used to delimit an area for which it is worth the time and expense of more detailed investigations.

From a practical viewpoint, the considerable variation within a specific population means that the sample size usually has to be fairly large. The fact that the system is induced by a large variety of natural compounds as well as xenobiotics and is affected by a wide variety of other parameters – temperature, diet, etc. – means that great care must be taken to ensure that there are reliable control levels.

10.4.3 STUDIES ON GENETIC MATERIAL

The fundamental role of DNA in the reproductive process is well known and will not be discussed further here. The end-points used to assess the damage caused to DNA by environmental pollutants are specific genotoxic effects, especially the increase of carcinogenesis, rather than effects on the reproductive process.

There is a sequence of events between the first interaction of a xenobiotic with DNA and consequent mutation, which may be divided into four broad categories. The first stage is the formation of adducts. At the next stage, there may be secondary modifications of DNA, such as strand breakage or an increase in the rate of DNA repair. The third stage is reached when the structural perturbations to the DNA become fixed. At this stage, affected cells often show altered function. One of the most widely used assays to measure chromosomal aberrations is sister chromatid exchange. Finally, when cells divide, damage caused by toxic chemicals can lead to the creation of mutant DNA and consequent alterations in gene function.

The covalent binding of reactive metabolites of environmental pollutants to DNA – adduct formation – is a clear demonstration of exposure to these agents and an indication of possible adverse effects. The relationships between environmental levels, the degree of adduct formation and the ultimate effect are complex. For example, although a direct relationship between the extent of cigarette smoking and the number of DNA–BaP adducts has been clearly shown, the relationship between DNA–BaP adducts

and the occurrence of lung cancer is less well defined. In the field of wildlife toxicology, the establishment of the sequence of events from the initial DNA lesion to harm is even more difficult. Nevertheless, it is reasonable to conclude that reaction of chemicals with DNA can have harmful consequences, such as tumour formation.

Two different approaches have been used to study the formation of DNA adducts after exposure to pollutants:

1. radioactive post-labelling (usually with ^{32}P), leading to the separation of a range of adducts by two-dimensional thin-layer chromatography;
2. techniques to identify specific adducts, including fluorescence spectrometry, chromatographic techniques and enzyme-linked immunosorbent assays (ELISA) – techniques which are sensitive, if properly used, and can detect one adduct among 10^8 normal nucleotides.

The two techniques give different information. The former gives an index of the degree of total covalent binding, whereas the latter gives information on the actual degree of binding for a few specific compounds.

Monitoring of adduct formation provides one of the best means to detect exposure to polycyclic aromatic hydrocarbons (PAHs). The stability of DNA and haemoglobin adducts formed by this class of compounds means that evidence of exposure to them remains after they have been cleared from the body. The fact that adduct formation by haemoglobin can be studied means that non-destructive testing is possible.

In studies at Puget Sound, Washington, fish and sediments were sampled. The levels of PAHs in sediment and gut were determined and the extent of DNA–xenobiotic adducts in the liver measured by ^{32}P labelling. Additionally, concentrations of PCBs and the degree of induction of MFOs were determined. The various indices enabled these workers to discriminate between sites which exhibited a considerable range of differences in chemical contamination by both PCBs and PAHs (Stein *et al.*,1992).

DNA fingerprinting using the polymerase chain reaction has also been used as a biomarker for the detection of genotoxic effects of environmental chemicals (Savva, 1998; Atienzar *et al.*, 1999). The differences between control and experimental fingerprints are thought to be due to presence of DNA adducts. In studies of bivalves in the Lagoon of Venice, good correlation was found between DNA fingerprinting and the degree of strand breakage, as measured by the alkaline unwinding assay (Castellini *et al.*, 1996).

Breakage in chromosomes can be examined directly under the microscope or by the alkaline unwinding assay (Peakall, 1992). This technique is based on the fact than DNA strand separation takes place where there are breaks. The amount of double-stranded DNA remaining after alkaline unwinding is inversely proportional to the number of strand breaks, provided that renaturation of the DNA is prevented.

Chromosome breaks in the gills of the mud minnow have been used to study pollution of the Rhine. Also, increased chromosomal aberrations were found in rodents collected from areas close to a petrochemical waste disposal site. Although damage to chromosomes can lead to serious effects, it must be remembered that repair mechanisms are capable of preventing this from happening.

Sister chromatid exchange (SCE) is the reciprocal interchange of DNA during the replication of chromosomal DNA. Chromosomes of cells that have gone through one DNA replication in the presence of labelled thymidine or nucleic acid analogue 5-bromodeoxyuridine and then replicated again in the absence of the la-

bel are generally labelled in only one of the chromatids. The label is observed to be exchanged from one chromatid to the other. The sister chromatid exchanges can be easily visualized using the light microscope or by differential staining.

Chromosomes that have undergone SCE should not be regarded as damaged in the conventional sense because they are morphologically intact. Nevertheless, SCE occurs at sites of mutational events, including chromatid breakage. Good correlations have been observed between the number of induced SCEs per cell against the dosage of X-rays and the concentration of a number of chemicals known to cause chromosomal aberrations.

A relationship between SCE level (Nayak and Petras, 1985) and distance from an industrial complex was demonstrated in wild mice in Ontario, Canada, and a variety of chromosomal aberrations were found in cotton rats living close to hazardous waste dumps in the USA. Fish exposed to water from the Rhine showed a marked increase of SCE levels (van der Gaag *et al.*, 1993).

Flow cytometry can detect chromosomal aberrations in a large number of cells rapidly and accurately. It has been shown to detect mutagenic and clastogenic effects in a wide variety of species (Bickham *et al.*, 1998; Whittier and McBee, 1999).

As has already been mentioned, under specific assays a number of changes in the genetic material can be used to monitor for pollution. Monitoring has been carried out on the incidence of tumours in fish. Fish are frequently sampled in considerable numbers and, more importantly, many of their tumours are visible externally. Such data could not be readily collected from other classes of species, even when large sample sizes are available such as muskrat from trappers or duck from hunters, because of the cost of dissection.

In the North American Great Lakes, surveys

to determine the instance of tumours have been carried out as part of the surveillance programme. The levels of occurrence of tumours in brown bullheads (*Ictalurus nebulosis*) and white suckers (*Catostomus commersoni*) are greatest in the most polluted areas. Given the large number of contaminants in the Great Lakes, it is virtually impossible to link carcinogenesis to a specific chemical but circumstantial evidence for a chemical origin is strong in many cases, although it should be cautioned that viral agents and parasites can cause neoplasms. The occurrence of liver neoplasms in brown bullhead and the levels of PAHs in sediment has been monitored for 20 years on the Black River, Ohio (Baumann and Harshbarger, 1998). In the early 1980s, the prevalence of liver cancer was high (22–39% of fish over 3 years old). The coke factory was closed in 1983, and by 1987 levels of PAHs in sediment had decreased by two orders of magnitude and cancer rates had fallen to one-quarter of the previous figure. Dredging of the most contaminated sediments was carried out in 1990, following which there was a marked rise in the cancer rates followed by a decrease over the next few years.

Overall, the studies involving DNA have reached an interesting stage. A great deal of medical research is being carried out and there are strong indications that this information can be used in assessing the impact of pollutants, especially PAHs, on wildlife.

10.4.4 PORPHYRINS AND HAEM SYNTHSESIS

Porphyrins are produced by the haem biosynthetic pathway, which is a vital system for most of the animal kingdom. Two major disruptions of haem biosynthesis by environmentally important agents have been studied. These are the formation of excess amounts of porphyrins after exposure to some organochlorines (OCs) and the inhibition by lead on the enzyme aminolaevulinic acid dehydratase (ALAD).

Haem biosynthesis is normally closely regulated, and levels of porphyrins are ordinarily very low. Hepatic porphyria is characterized by massive liver accumulation and urinary excretion of uroporphyrin and heptacarboxylic acid porphyrin. Although the mechanism of OC-induced porphyria has not been completely elucidated, it is considered by several workers that inhibition of the enzyme uroporphyrinogen decarboxylase is the proximal cause. The two OCs that are most involved in inducing porphyria are hexachlorobenzene (HCB) and the PCBs. Although HCB has been shown in both mammals and birds to induce porphyria, the dosages required are high compared with environmental levels. PCBs have also been shown to be potent inducers, although the various congeners act quite differently.

Studies on the Rhine showed that the patterns of hepatic porphyrins were markedly different and that the total porphyrin levels were much higher in pike collected than in those from the cleaner River Lahn. The levels of organochlorines were up to 40-fold higher in the fish from the Rhine than in those from the Lahn.

The variation in the mean of the hepatic levels of highly carboxylated porphyrins (HCPs) in seven species of birds (covering five orders) was only twofold and the total range was 4–22 pmol g^{-1} (Fox *et al.*, 1988). No similar study appears to have been carried out on any other class of organism. These baseline data, collected from areas of low contamination, show only small variation, but in view of variability of response to OCs in experimental studies variability in areas of high contamination is a problem. Nevertheless, the levels of hepatic HCPs were markedly elevated in herring gulls (*Larus argentatus*) collected from the North American Great Lakes when compared with those from the Atlantic coast (Fox *et al.*, 1988).

Aminolaevulinic acid dehydratase (ALAD) is an enzyme in the haem biosynthetic pathway. Inhibition of ALAD was first studied over 30 years ago as a means of detecting environmental lead exposure in humans and has since become the standard bioassay for this purpose; it has subsequently been used in wildlife investigations. The assay is highly specific for lead, with other metals being 10 000 times less active in causing inhibition. ALAD inhibition is rapid, but the effect is only slowly reversed, with ALAD values returning to normal values only after about 4 months.

Inhibition of ALAD has been used as an indicator of lead exposure for general problems, such as in urban areas and along highways, and also specifically to study the 'lead-shot problem' in waterfowl. A threefold difference in the blood ALAD activity was found between rats in a rural and rats in an urban site in Michigan (Mouw *et al.*, 1975). The main physiological indications of lead toxicity in the urban rats were an increase in kidney weight and the incidence of intranuclear inclusions. Both effects could be correlated with lead levels. Similarly, marked differences in the ALAD activity were found among feral pigeons (*Columbia livia*) from rural, outer urban, suburban and central London areas (Hutton, 1980).

The lead levels, ALAD activity and reproductive success of barn swallows (*Hirundo rustica*) and starlings (*Sturnus vulgaris*) along highways with different traffic densities have been studied in North America (Grue *et al.*, 1986). It was found that there was a significant increase of the lead levels in the feathers and carcasses of both adults and nestlings and a 30–40% decrease in plasma ALAD activity. However, the number of eggs laid, number of young fledged and prefledgling body weights were not affected, indicating that lead from automotive emissions does not pose a serious hazard to birds nesting close to motorways.

Mortality of ducks and other waterfowl due to the ingestion of lead shot has been of serious concern for many years, the issue first being raised in North America over 70 years ago. The problem is caused by ducks and geese ingesting spent lead shot during the course of their feeding. A nationwide survey found that 12% of the gizzard samples examined contained at least one lead shot and considered that 2–3% of all waterfowl in North America died from lead poisoning. Secondary poisoning of bald eagles feeding on waterfowl has also been of concern. National surveys of eagles found dead in the USA showed that about 5% died from lead poisoning. The inhibition of ALAD has been shown to be sensitive enough to detect the effect of a single pellet. A strong negative correlation between blood lead concentration and log ALAD activity has been found by many workers. The ALAD assay is a simple one, which can be carried out without expensive equipment or lengthy training.

ALAD inhibition represents one end of the biomarker spectrum. It is a sensitive dose-dependent measurement that is specific for a single environmental pollutant, lead.

10.4.5 INDUCTION OF VITELLOGENIN

The discovery of hermaphrodite roach in stretches of rivers close to sewage outlets was the trigger for research into the effect of oestrogenic disruptors in fish because natural hermaphroditism is assumed to be low. Examination of these fish showed that males had high levels of vitellogenin, which is an egg yolk protein usually produced only by females. Thus, increased levels of vitellogenin in male fish provided the basis for a biomarker assay for studies into endocrine disruptors.

BOX 10.1 *Induction of vitellogenin in fish.*

A survey of five rivers in England was conducted in the summer of 1994 (Harries *et al.*, 1997). For this survey, caged male rainbow trout were deployed at five sites in each river, one upstream from the suspected source (waste treatment plants), one at the point of effluent discharge and the other three at different distances downstream. In four cases, the fish placed in the neat effluent showed very marked and rapid increases in vitellogenin levels. In two of these cases, none of the downstream sites showed any oestrogenic activity; in another, activity was detected 1.5 km downstream. The fourth river was quite different, the effluent was extremely oestrogenic and so was water sampled at all other sites (maximum distance 5 km). The effluent in these cases contained much higher levels of **alkylphenolic compounds**. On the fifth river, even the neat effluent did not cause increased levels of vitellogenin. In this case, the waste treatment plant was a small one that did not receive any industrial waste. A disturbing finding is that some UK estuaries show a high degree of oestrogenic contamination. **Flounder** (***Platichthys flesus***) in the Tyne and Mersey estuaries had vitellogenin levels four to six orders of magnitude higher than controls (Scott *et al.*, 1999). Elevation of vitellogenin was less marked in the Crouch and the Thames.

The implications of these findings at the population level are not known and would be difficult to determine. Sewage discharges may well contain other compounds that cause reproductive toxicity by other means. The effect on the lowland coarse fish may be different from that upon trout, which usually does not occur in these waters. In the case of estuaries, it would be important to examine for effects on fish that reproduce in these areas rather than the flounder which breeds at sea. Besides these ecotoxicological considerations, there are a myriad of other factors that affect wild populations of fish.

10.4.6 BEHAVIOURAL BIOMARKERS

Behavioural changes represent a higher organizational level of biomarker than any considered so far. One of the early proponents of the value of behavioural toxicology stated that 'the behaviour of an organism represents the final integrated result of a diversity of biochemical and physiological processes. Thus, a single behavioural parameter is generally more comprehensive than a physiological or biochemical parameter'. Although much interesting work has been carried out in the years since this statement was made, behavioural biomarkers have still not reached the stage where they are accepted as part of formal testing procedures.

There are two fundamental difficulties facing the use of behavioural tests in wildlife toxicology: first that the best studied and most easily performed and quantified are those that have the least environmental relevance, and second that the most relevant behaviours are the most strongly conserved against change.

Operant behaviour, such as conditioning to respond to a coloured key to obtain food, is too remote from real life to be capable of being related to survival. It can merely be presumed that a decrease in learning ability is an unfavourable response. Avoidance behaviour is more directly related to survival, although the relationship has not been quantified. The ability to capture food is clearly important to predatory species, but is difficult to measure under field conditions.

Field observations are usually difficult to quantify. It was suggested that behavioural changes were a possible cause in the decline of the peregrine falcon. However, observations by time-lapse photography at the eyries of highly contaminated peregrines revealed little in the way of abnormal behaviour. This study was based on seven peregrine eyries in Alaska, using battery-powered time-lapse motion pictures cameras that took pictures about every 3 min and the film cartridges needed to be replaced every 6–7 days. Even so, it was a full-time job to replace the film cartridges as the eyries were widely separated and the terrain was difficult. In two of the nests, the eggs broke, but no evidence of abnormal behaviour was observed. The other five nests were successful. In all, some 70 000 pictures covering 4200 h were obtained. One of the drawbacks of this type of experiment is the time taken to analyse the data obtained. In another study (Nelson, 1976), observations from a hide, totalling over 300 h, were made on 12 peregrine eyries. Four clutches lost single eggs, probably by breakage, but no abnormal behaviour was observed. Although 300 h is a lot of time to spend sitting in a hide, it is only 25 h per clutch out of 400 h of daylight during the incubation period.

These two studies illustrate the difficulties of making field observations on behavioural changes. An additional problem is the fact that even if behavioural changes are documented it is difficult to relate them to a specific chemical or chemicals.

The best documented studies that can be extended to real life situations are those involving the organophosphorous pesticides and the subsequent inhibition of AChE. These have already been covered in section 8.4 and their use under field situation is further considered in section 15.4.

Studies of avoidance responses of fish to toxicants dates back over 80 years. From the first simple studies on acetic acid, the field has grown enormously and the equipment used has become highly complex. We now have sophisticated fish avoidance chambers with video monitors and computer-interfaced recording systems.

Recent studies include many on the effects of heavy metals. If one compares the lowest observed effect concentration (LOEC) obtained from behavioural studies (avoidance, attractance, fish ventilation and cough rates) with chronic toxicity studies, one finds that some of the behaviour tests are more sensitive than life cycle or early life stage tests. Other tests involving predator avoidance, feeding behaviour, learning, social interactions and a variety of locomotor behaviours have been insufficiently studied to enable a judgement of their sensitivity or utility. At the moment, behavioural tests have not replaced conventional toxicity tests. However, behavioural tests may provide ecological realism, e.g. the effects of pollutants on predator–prey relationships. Such tests must be capable of field validation.

Behavioural tests that are most advanced are those involving fish. The fish avoidance test is well established in the laboratory as a means of showing effects well below the lethal range and highly automated procedures are available. Nevertheless, a note of caution should be injected. It has been found that pre-exposure to effluent reduced the avoidance behaviour and pre-exposed fish were observed more often in contaminated than in clean water (Hartwell *et al.*, 1987). This desensitization caused by pre-exposure makes it likely that laboratory experiments will overestimate the responsiveness of fish to metal pollution in the wild.

These difficulties do not imply that behavioural effects caused by pollution are unimportant. Studies such as those examining the predation pressure on fiddler crabs (*Uca pugnax*) have shown that operational levels of pesticide use can cause population effects through behavioural changes. Substandard prey are more readily captured by predators, but field studies of the impact of chemicals on behaviour are difficult.

The overall conclusion is that behavioural parameters are not especially sensitive to exposure to pollutants and that biochemical and physiological changes are usually at least as sensitive. Further, the variability of biochemical data are generally less and the dose–response relationship clearer than those obtained from behavioural studies. In general, physiological and biochemical changes are more readily measured and quantified.

10.4.7 BIOMARKERS IN PLANTS

Plants have been widely used as biomonitors to localize emission sources or to analyse the impact of pollutants, especially gaseous air pollutants, on plant performance, one of the earliest ones being by Angus Smith, who coined the term 'acid rain' after examining the damage done to plants around Manchester in the middle of the last century.

However, biomarkers should go beyond the visible parameters of sentinel species. They should establish such processes and products of plants, which enable a recognition of environmental stress earlier than visible damage. Biomarkers must therefore be able to predict the environmental outcome and consequential damage. Ideally, biomarkers should be selected from the events of biochemical or physiological pathways, but at the present time we have not yet reached this stage with plant biomarkers (Ernst and Peterson, 1994).

Specific biomarkers have been identified in sensitive plants. In a few cases, it is known that excess of a specific chemical will give rise to the production of a metabolite which is different between tolerant and sensitive plants.

In the presence of an excess of selenium, Se-sensitive plants fail to differentiate between S and Se. They incorporate Se in sulphur amino acids, leading to the synthesis of enzymes of lower activity, which can lead to plant death. In contrast, Se-tolerant plants biosynthesize and accumulate non-protein seleno-amino acids

such as selenocystathionein and Se-methyl-selenocysteine which do not cause metabolic problems for the plant. Thus, the occurrence of selenoproteins in plants are excellent biomarkers for Se stress, although their use in the field has not been widely reported.

Another example is that after an exposure to an excess of fluor (a mineral containing fluorine), plants synthesize fluoroacetyl-coenzyme A, and then convert it, via the tricarboxylic acid cycle, to fluorocitrate. The latter compound blocks the metabolic pathway by inhibiting the enzyme aconitase. As a result of this process, fluorocitrate accumulates and is a very reliable biomarker for fluor poisoning.

There are some plant biomarkers for free heavy metals. Phytochelatins are synthesized during exposure to a number of heavy metals and anions as SeO_4^{2-}, SeO_3^{2-} and AsO_4^{3-}. Dose- and time-dependent relationships have been established under laboratory conditions for cadmium, copper and zinc. For monitoring purposes, research is needed into the phytochelatin production of plants in the field.

General plant biomarkers, which respond to a variety of environmental stresses, may be useful to indicate that something in the environment is a hazard to plant life. For example, the activity of the enzyme peroxidase has been used to establish the exposure of plants to air pollution, especially SO_2.

Changes of enzyme systems during the development stage of the plant, as well as the effects of seasonal and climatic processes, are not yet sufficiently well known for the activity of enzymes to be reliable as monitoring devices. At present, plant biomarkers are not as well advanced as animal biomarkers. However, this situation is likely to change. The fact that plants are stationary aids greatly in measuring exposure to pollutants; monitoring surveys for measuring levels of heavy metals in lichens and organochlorines in pine needles are in place and would be valuable if these measurements could be linked with biological changes within the plant.

10.5 Role of biomarkers in environmental risk assessment

The most compelling reason for using biomarkers in environmental risk assessment is that they can give information on the effects of pollutants. Thus, the use of biomarkers in biomonitoring is complementary to the more usual monitoring involving the determination or prediction of residue levels (see section 6.5 for discussion of risk assessment). The first point in any assessment process is to decide what is being assessed. This may sound self-evident, but it is surprising how seldom precise objectives are defined. This is the case with many monitoring programmes to determine the levels of environmental chemicals. Take, for example, the International Mussel Watch Programme which measures heavy metals in mussels in many parts of the world. The justification of such surveys is that we should know what pollutants are, where and at what concentration. However, what is going to be done with the information? Only if it is known what concentrations are hazardous and if effective remedial action can be taken will this information be of practical use. Similar considerations apply to the use of biomarkers. Again, action levels have to be decided if the monitoring programme is to be effective. An advantage of the biomarker approach is that it may show that the physiology of the organisms is within normal limits, indicating that no action is necessary. By contrast, zero levels are rarely found in analytical determinations. Ideally, both approaches (i.e. residues and biomarkers) should be used together in an integrated manner (Chapter 11).

Hazard is a function of exposure and toxicity (Chapter 6). In the simplest terms, if there is no exposure there is no hazard, and if there is no toxicity there is no hazard. Legislation, such as EEC 6th and 7th Amendments, refer, in general terms, to protecting man and the environment. This legislation has specific requirements for data that must be provided at various stages of production. There is a basic set – called the minimum premarket data – that must be provided for all new chemicals before any production occurs. Increasing amounts of data must be produced as the amount of the chemical produced increases. The use to which these data are put in hazard assessment is not spelled out in the EEC legislation. Rather, this is left to individual governments. A comparison of the various approaches that have been used in prioritization and standardization of hazard assessment are considered by Hedgecott (1994).

Before legislation on limiting hazard can be enforced, two fundamental questions must be answered: (i) how much damage are we prepared to tolerate? and (ii) how much proof is enough? At one extreme, there will not be much concern if a few aquatic invertebrates die within a few metres of the end of an outlet pipe. At the other end of the scale, an event such as the destruction of most of the biota of the Rhine (Deininger, 1987) resulted in a worldwide reaction. Our concerns are also species dependent and tend to increase as we move from algae to mammals. Even within a class of animals, there are widespread differences: it is easy to arouse concern over pandas and whales, but difficult in the case of rats. Further, some types of damage are considered more serious than others; alteration of the genetic material, which may be passed onto future generations, being considered one of the most serious effects.

The question 'how much damage are we prepared to tolerate?' is a question for society to answer. Once it is answered, it is possible to design protocols to meet the standards required. Unanswered, it leaves the scientists with the problem of trying to set up regulations in an impossible situation.

The second question 'how much proof is enough?' is largely a scientific one, although it requires the first question to be answered before it can be tackled. There is considerable inertia in decision making, but decisions must be made. 'No decision' is a decision.

Why use biomarkers in hazard assessment? One important reason lies in the limitations of classic hazard assessment. The basic approach of classic hazard assessment is to measure the amount of the chemical present and then relate that, via animal experimental data, to the adverse effects caused by this amount of chemical. The limitation of this approach is that only for a very few compounds has it been possible to define the levels of a chemical that are critical to an organism. Under real life situations, a wide variety of organisms are exposed to complex and changing levels of mixtures of pollutants. Further chemical monitoring only works if the material is persistent. Chemicals such as the PAHs and many pesticides have very short biological half-lives in most species but may, nevertheless, have long-term effects. Biological and chemical monitoring systems should be complementary to each other. It is important to know both what is there and what it does.

The first question that biomarkers can be used to answer is 'are environmental pollutants present at a sufficiently high concentration to cause an effect?'. If the answer is positive, further investigation to assess the nature and degree of damage and the causal agent or agents is justified. If negative, it means that additional resources do not have to be invested. In this way, biomarkers can act as an important 'early warning' system.

The role of biomarkers in environmental assessment is envisaged as determining whether or not, in a specific environment, organisms are physiologically normal. The approach has similarities to the use of clinical biochemistry in human medicine. A suite of tests can be carried out to see whether the individual is healthy. It is necessary to select both the tests and the species to be tested. The selection of indicator species is going to be, at least to some extent, site specific. However, there is merit in having as much commonality of species as is feasible between studies. It is important to see that the main trophic levels are covered and not to rely completely on organisms at the top of the food chain. Although these have been the species most affected by some pollutants such as the persistent organochlorines, there is no reason to believe that this will be a universal truth.

In the selection of tests, the specificity of the test to pollutants and the degree to which the change can be related to harm need to be considered. Both specific and non-specific biomarkers are valuable in environmental assessment. In an ideal world, we would have biomarkers to indicate the exposure to, and to assess the hazard of, all major classes of pollutants and non-specific biomarkers that assess accurately and completely the health of the organism and its ecosystem.

From the above, it is clear that there are two main aspects to a definition of harm, one scientific and the other social. Scientifically, it is important to demonstrate unequivocally that changes have occurred as a result of pollution. But whether or not that change is sufficiently serious to make it essential to bear the cost of the remedial action is something for society to decide.

10.6 *Summary*

The term biomarker is defined as 'any biological response to an environmental chemical at the individual level or below demonstrating a departure from the normal status'. The specificity of biomarkers to pollutants varies greatly. Highly specific biomarkers are valuable in detecting exposure to and possible effects of specific chemicals but give no information on other pollutants. In contrast, non-specific biomarkers give information that exposure to pollutants has occurred without identifying the specific pollutant responsible. A number of specific biomarkers are considered in some detail. Biomarker assays are particularly useful when they relate to toxic effect and not just exposure. The most important reason for using biomarkers in environmental risk assessment is that they can give information on the effects of pollutants. Thus, the use of biomarkers is complementary to biomonitoring which involves the determination of levels of environmental chemicals.

10.7 *Further reading*

Fossi, M. C. and Leonzio, C. (eds) (1994) *Nondestructive Biomarkers in Vertebrates*. Proceedings of a workshop devoted to the use of blood and other tissues that can be collected non-destructively in the measurements of biomarkers.

Huggett, R. J. *et al.* (1992) *Biomarkers. Biochemical, Physiological and Histological Markers of Anthropogenic Stress*. In addition to those techniques listed in the title, there is extensive coverage of DNA alterations and immunological biomarkers.

McCarthy, J. F. and Shugart, L. R. (1990) A compilation of a large number of papers presented by both American and European scientists at an American Chemical Society meeting.

Peakall, D. B. (1992) *Animal Biomarkers as Pollution Indicators*. A single-author work looking at the use of biomarkers in higher animals in environmental assessment.

Peakall, D. B. *et al.* (1999) *Biomarkers: a Pragmatic Basis for Remediation of Severe Pollution in Eastern Europe*. NATO workshop.

CHAPTER 11

In situ *biological monitoring*

11.1 *Introduction*

It is difficult to predict the effects of pollutants on organisms to an acceptable degree of accuracy by simply measuring concentrations of a chemical in the abiotic environment (figure 11.1). Factors which affect bioavailability of chemicals to organisms include temperature fluctuations, interactions with other pollutants, soil and sediment type, rainfall, pH and salinity. Even using biotic monitoring of chemical residue levels (see section 11.3), there are considerable difficulties in knowing the effects of this level of chemical on the organism. This process is made more difficult by the presence of mixtures (see Chapter 10) and the considerable interspecies differences in response. *In situ* biological monitoring attempts to get around these problems by analysing various parameters of natural populations which reflect the situation in the field rather than the standardized conditions of laboratory experiments.

There are four main approaches to *in situ*

biological monitoring of pollution (Hopkin, 1993a). Each will be examined in this chapter and will be illustrated with examples of recent studies on a range of animals and plants from terrestrial and aquatic ecosystems. The four involve:

1. monitoring the effects of pollution on the presence or absence of species from a site or changes in species composition, otherwise known as 'community effects' (see also Chapter 14);

2. measuring concentrations of pollutants in indicator or 'sentinel' species (see also Chapters 4 and 5);

3. assessing the effects of pollutants on organisms and relating them to concentrations in those organisms and other biotic and abiotic indicators (see also Chapters 7 and 8);

4. detecting genetically different strains of species which have evolved resistance in response to a pollutant (see also Chapter 13).

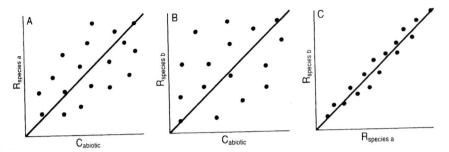

FIGURE 11.1 *Schematic graphs to illustrate a principle of* in situ *biological monitoring of pollution. In this hypothetical example, the responses, R (as measured by concentrations or effects of the pollutant on the y axis), of species 'a' (A) and 'b' (B) in sites with different levels of contamination are not closely related to concentrations of the pollutant (x axis) in abiotic samples (soil, air, water sediment) from the same sites. Because the relationship between species in the same sites is much closer (C), the responses of species 'a' to the pollutant can be used to predict the responses of species 'b' more accurately than similar predictions from abiotic samples (see figures 11.5 and 11.6 for data that support this hypothesis). Reproduced from Hopkin (1993a) with permission from Blackwell Scientific.*

11.2 *Community effects (type 1 biomonitoring)*

The most frequent response of a community to pollution is that some species increase in abundance, others (usually the majority) decrease in abundance and populations of others remain stable. The patterns of the species abundances reflect effects integrated over time and are used widely to monitor effects of pollutants on communities. Effects of pollution on communities and ecosystems are covered in greater detail in Chapter 14.

11.2.1 TERRESTRIAL ECOSYSTEMS

To be able to recognize an unusual assemblage of species, it is necessary to monitor changes over time or to have sufficient background knowledge of the 'normal' ecology of similar but unpolluted sites. In the former approach, sites must be monitored for several years to distinguish effects of pollutants from natural fluctuations. Perkins and Millar (1987) showed that emissions from an aluminium works on

Anglesey, North Wales, were responsible for the almost complete elimination of lichens within 1 km of the factory soon after its opening in 1970. Some recovery has taken place since 1978, when new emission controls were introduced, but the lichen flora is still very impoverished in comparison with pre-1970 populations (figure 11.2). In the latter 'background knowledge' approach, one or more

FIGURE 11.2 *Cover (as percentage of type maximum) of foliose (—), fruticose (— — —) and crustose (— · —) corticolous lichens in eight permanent quadrats set up on broad-leaved trees within 1 km of the aluminium works in Anglesey, north Wales, during 1970–85. Reproduced from Perkins and Millar (1987) with permission from Elsevier Applied Science.*

sites are examined in the contaminated area and are compared with at least one 'reference' site (e.g. figure 11.3). However, if the differences between polluted and reference sites are subtle, the investigator is left with the difficult problem of deciding when a change indicates a toxic effect or natural between-site variations.

11.2.2 FRESHWATER ECOSYSTEMS

In Britain, three main approaches have been adopted to assess the effects of pollution on communities of freshwater organisms (British Ecological Society, 1990, on which this section is based). These are:

1. the *biotic* approach, based on the differential sensitivities of species to pollutants;
2. the *diversity* approach, based on changes in community diversity;
3. *river invertebrate prediction and classification* (**RIVPACS**), which combines an assessment in terms of both the types of species present and the relative abundances of families.

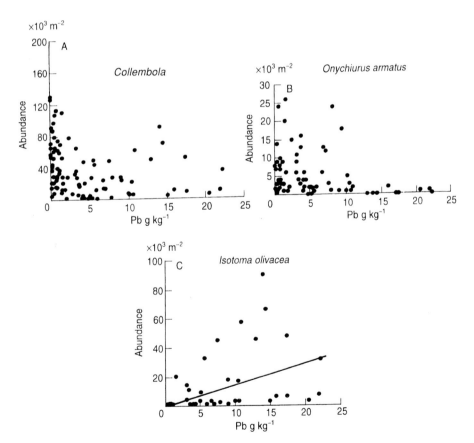

FIGURE 11.3 *Abundance of (A) total* Collembola, *(B)* Onychiurus armatus (Collembola) *and (C)* Isotoma olivacea (Collembola) *in the 0- to 3-cm layer in lead-contaminated soils in the vicinity of a natural metalliferous outcrop in a Norwegian spruce forest. The concentrations of lead represent metal extracted from soil over 18 h in 0.1 M buffered acetic acid. Note that* O. armatus *is sensitive to lead pollution whereas* I. olivacea *reached higher population densities in contaminated soils. Reproduced from Hågvar and Abrahamsen (1990) by permission from the Entomological Society of America.*

The most frequently used biotic indices have been the Trent Biotic Index (TBI), Chandler Biotic Score (CBS) and Biological Monitoring Working Party (BMWP). All three are based largely on relative tolerances of macroinvertebrates to organic pollution. TBI and CBS require identification of species, whereas BMWP requires only family level identification (see box 11.1). Only the CBS takes abundance into account. However, these scores are generally assumed rather than experimentally derived. Furthermore, sensitivity rankings are assumed to apply across a range of toxicants, even though laboratory experiments have shown that this is not necessarily the case.

The most frequently used diversity indices have concentrated on species richness and the distribution of individuals among species. The Shannon–Weiner diversity index is the most commonly used. However, there are three main problems with diversity indices. First, many factors influence community structure and as diversity indices do not take account of the species present their usefulness as a measure of water quality can be questioned. Second, there is an unresolved debate as to which index to use to measure diversity and as to which taxonomic group and level should be considered.

Third, it is not clear how diversity responds to pollution. For example, diversity of plankton reduces continuously with organic enrichment, but for benthic invertebrates the response is 'bell-shaped' with the greatest diversity at intermediate pollution levels (British Ecological Society, 1990).

The RIVPACS approach is used to predict the fauna of a site using environmental variables (Wright *et al.*, 1993). Hypothetical target communities are provided against which the combined effects of physical and chemical stresses can be assessed. Comparison of observed values with these predictions provides environmental quality indices. Although RIVPACS is the most widely used of the techniques described, it is still a 'broad brush' approach more useful for highlighting rivers and streams in need of more detailed study than giving a final verdict on the level of environmental damage.

11.2.3 MARINE ECOSYSTEMS

One of the simplest methods of detecting a pollution-induced change in communities of marine benthic organisms is to analyse the log normal distribution of individuals per species

BOX 11.1   *Use of the BMWP score to assess the health of fresh waters.*

A system for rapid appraisal of the 'health' of freshwater ecosystems was developed in the late 1970s by the **Biological Monitoring Working Party** (**BMWP**). Scores ranging from 1 to 10 are given to families of invertebrates depending on their sensitivity to organic pollution. Families with a low tolerance to pollution are given a high score (e.g. the mayfly family *Heptageniidae* is allocated a score of 10), whereas those tolerant of severe pollution are given the lowest score of only 1 (e.g. tubificid oligochaete worms). At a particular site, scores for all families are added up to give the total BMWP score. The **average score per taxon** (APST = total BMWP score divided by the number of taxa) is a particularly valuable index because it is less sensitive to sampling effort and can be predicted with greater reliability than the number of taxa or the BMWP score.

Sites can be graded into four biological classes on the basis of these scores (see table 11.1). Thus, the highest quality of site gets a grade of A, whereas the grades B, C and D are indicative of progressive loss of water quality.

TABLE 11.1 *Biological banding of average score per taxon (ASPT), number of taxa and biological monitoring working party score based on sampling in three seasons (from Wright* et al.*, 1993)*

Biological class	Observed/ expected ASPT	Observed/expected number of taxa	Observed/expected BMWP score
A	≥ 0.89	≥ 0.79	≥ 0.75
B	0.77–0.88	0.58–0.78	0.50–0.74
C	0.66–0.76	0.37–0.57	0.25–0.49
D	< 0.66	< 0.37	< 0.25

in sediment samples (Gray, 1981). In many samples of benthic communities, the most abundant class is not that represented by one individual per species but often lies between classes with three and those with six individuals per species. Thus, the curve relating numbers of individuals per species (x axis) to number of species (y axis) is often strongly skewed. This curve can be 'brought back' to a normal shape by plotting the number of individuals per species on a geometric scale (usually × 2). Plotting the geometric classes on the x axis (class I = 1, class II = 2–3, class III = 4–7, class IV = 8–15, and so on) against the cumulative percentage of species on the y axis invariably gives a straight line. In polluted sites, there is often a break in the line indicating a departure from an equilibrium community. If this persists over several sampling occasions, it is indicative of pollution-induced disturbance.

The responses of marine-fouling communities to pollution stress, in terms of changes in species composition, can be monitored *in situ* by reciprocal transplants. Climax communities are allowed to develop on submerged surfaces in a clean and a polluted site and are then moved between the sites. An experiment in Australia at Woolongong Harbour (uncontaminated) and Port Kembla Harbour (polluted by discharges from heavy industry) demonstrated rapid changes in the community structure in response to pollution (Moran and Grant, 1991). Indeed

within 2 months, those communities on submerged plates that had been transferred from Woolongong to Port Kemblar were similar in structure to those that had developed entirely in Port Kemblar. Most changes occurred in short time periods when sensitive species were killed by periodic discharges (an effect difficult to predict by measuring levels of pollutants in the water). Space previously occupied by these species was quickly colonized by opportunists more tolerant to the pollutants, thus leading to changes in community structure.

11.3 *Bioconcentration of pollutants (type 2 biomonitoring)*

Destructive measurement of levels of pollutants in organisms provides an indication of how much is present at a particular moment in time and may enable effects on predators to be assessed (see Chapters 12 and 15). Take, for example, a species of wading bird that feeds primarily on estuarine bivalve molluscs. 'Critical' (safe?) concentrations for the birds could be set based on levels in bivalves rather than sediment or water or indeed their own tissues. The bivalves provide a critical pathway from the abiotic environment to the waders, the importance of which can be monitored biologically by analysing the molluscs.

11.3.1 TERRESTRIAL ECOSYSTEMS

Contamination of plants has been monitored either by collecting samples directly from the field or by exposing material (usually bags of *Sphagnum* moss) for a specified period and returning it to the laboratory for analysis. The recent decline in concentrations of lead in air in the UK following the reduction of permitted levels of lead in petrol was mirrored by a decline in lead concentrations in plants (Jones *et al.*, 1991).

Biological monitoring of radioactive fall-out in Italy derived from the Chernobyl disaster in 1986 has shown that levels of ^{137}Cs in mushrooms have been increasing since the accident (Borio *et al.*, 1991). Basidiomycete hyphae in the soil accumulated the radioisotopes but it was not until they produced sporophores that the caesium became available to above-ground fungivores. No correlation was found between the level of ^{137}Cs in the mushrooms and the soil in this study, emphasizing the importance of

FIGURE 11.4 *Radiocaesium (^{134}Cs and ^{137}Cs) activity from 1987 to 1989 in goat milk during the grazing seasons from 15 June to 15 September in the Jotunheimen mountain range (mean ± SE, three to eight animals in 1987 and eight in 1988 and 1989). Reproduced from Hove et al. (1990) with permission from the Health Physics Society and Williams & Wilkins, Baltimore.*

biological monitoring of radioactive pollution. Traditional risk assessment does not make allowances for effects such as this.

In Norway, where similar results have been obtained, the level of ^{137}Cs in the milk of goats increased fivefold in 1988 after the abundant growth of mushrooms in grazing land (figure 11.4). The sporophores contained levels of radioactivity up to 100 times those in green vegetation. Thus, fungi provide an important critical pathway for the concentration and transport of radioactive isotopes along food chains.

As far as invertebrates are concerned, it is clear that some groups, such as woodlice, snails and earthworms, accumulate significant amounts of metals from their diet (Hopkin, 1989; Van Straalen, 1996), whereas most insects are able to regulate concentrations to relatively low levels (Hopkin, 1995). Thus, species in the former three groups provide a significant route for the transfer of metal pollutants to their predators (figures 11.5 and 11.6).

Monitoring concentrations of pollutants in vertebrates may pose some difficulties. They are usually difficult to catch, population densities are lower than in invertebrates and a licence may be required (or may even be unobtainable for some species). Ways around this are to take blood samples, or use a non-living product of the animal which indicates previous exposure to a pollutant. Analysis of eggs has been widely used for organochlorines and this is less destructive than the collection of adults. Feathers have been proposed as possible indicators. In Finland, feathers from nestlings of a range of birds provide a good indicator of mercury exposure of the adults (Solonen and Lodenius, 1990). Regurgitated pellets from owls can be used to monitor exposure to rodenticides.

11.3.2 FRESHWATER ECOSYSTEMS

Predicting bioaccumulation in aquatic systems

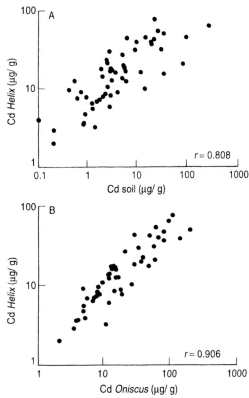

FIGURE 11.6 *Relationships between concentrations of cadmium (dry weight) in the snail* Helix aspersa *and soil (A) and* Oniscus asellus *(B) collected from the same region as in figure 11.5. Each point represents the mean of seven snails, 12 woodlice or six samples of soil from each site. Note that the concentrations of cadmium in* H. aspersa *can be predicted more accurately from the concentrations on* O. asellus *(B) than from levels in soil (A). Reproduced from Hopkin (1993a) with permission from Blackwell Scientific.*

FIGURE 11.5 *Relationships between concentrations of cadmium (dry weight) in the terrestrial isopods (woodlice)* Oniscus asellus *(A) and* Porcellio scaber *(B) and soil, and between the two species (C) collected from sites in Avon and Somerset, south-west England in 1998 and 1989. The region includes a primary zinc, cadmium and lead smelting works and disused zinc mining areas. Each point represents the mean of 12 isopods and six samples of soil from each site. Note that the concentrations of cadmium in* P. scaber *in this region can be predicted more accurately from the concentrations in* O. asellus *(C) than from levels in soil (B). Reproduced from Hopkin (1993a) with permission from Blackwell Scientific.*

is more difficult than in terrestrial ones because of the greater mobility of water and sediments in comparison with soils and because of the difficulty of knowing whether the main route of exposure is via water or food. Organisms can be collected directly from the field for analysis or caged in polluted and unpolluted sites to assess bioavailability. The freshwater amphipod *Gammarus pulex* has been used extensively for such work.

Monitoring concentrations of pollutants in fish is carried out all over the world. For example, a monitoring programme for mercury in Brazil showed that levels in edible parts of fish from gold mining areas (where mercury is used extensively in gold extraction and refining) were five times greater than the 'safe' level for human consumption (Pfeiffer *et al.*, 1989). The long-term studies of Schmitt and Brumbaugh (1990) detected a decrease in the concentrations of lead in fish in the USA between 1976 and 1984. This coincided with regulatory measures that reduced the influx of lead to the aquatic environment. A similar downward trend has been found also in the levels of PCBs in the eggs of herring gulls from the Great Lakes (figure 11.7).

11.3.3 MARINE ECOSYSTEMS

Marine mammals at the top of the food chain are particularly difficult to study as they are usually protected, therefore many studies are based on beached specimens. This procedure is likely to be biased in favour of high residue levels. Some progress has been made in obtaining skin samples non-destructively (Aguilar and Borrell, 1994), but systematic biomonitoring of marine mammals is still at an early stage of development.

Colonial nesting marine seabirds have been studied with eggs being used for determination of OC levels and feathers for the levels of heavy metals. In the latter case, it is possible to extend the time backwards by using museum

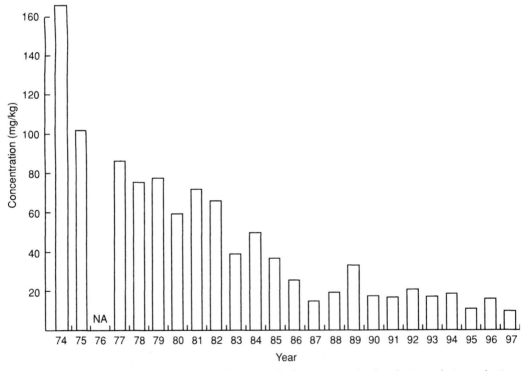

FIGURE 11.7 *Concentrations of PCB in eggs of herring gulls from Muggs Island/Leslie Spit colonies, Lake Ontario, 1974–97. Data supplied by Environment Canada.*

specimens. For example, Applequist *et al.* (1995) were able to examine mercury levels in guillemots (*Uria aalge*) over a 150-year period and found that levels increased from 1–2 ug g^{-1} during the period 1830–1900 to 3–6 ug g^{-1} by the 1970s.

By far the greatest research effort has been directed towards bivalve molluscs. The reasons for this are fourfold. First, many species are a source of food for predatory vertebrates, particularly birds. Second, they are widespread and common, are easily collected in large numbers and are sedentary. Third, because they are filter feeders, bivalves pass large volumes of water through their bodies, accumulate pollutants continuously and act as integrators of exposure over long periods. Fourth, there is good background knowledge of their basic biology to allow results to be put into an ecotoxicological context.

Mytilus edulis has been analysed most frequently as it is common and has a global distribution. Research on *Mytilus* has been so extensive that global 'mussel watch' schemes have been established. These have shown trends in pollutant levels which, in many cases, have declined. For example, Fischer (1989) was able to demonstrate that concentrations of cadmium in *Mytilus edulis* in Kieler Bucht in the western Baltic in 1984 had declined to about 30% of their level in 1975. Many of the schemes are ongoing and have been running for several years. This will ensure that long-term trends in the bioavailability of inorganic and organic pollutants to bivalves and their predators can be separated from natural fluctuations (for recent examples, see Cattani *et al.*, 1999; Gunther *et al.*, 1999).

Vertebrates are at the top of marine food chains and are vulnerable to poisoning from pollutants contained in their diets. This is particularly true of some organic pollutants which may accumulate in fatty tissues of marine mammals and birds. Residence times may be very long in organic pollutants which are highly lipophilic ($K_{ow} > 10^5$) or which are only slowly metabolized to water-soluble products that can be excreted (see Chapter 5). Thus, even after the complete removal of the source of pollution, contaminant levels in the tissues of vertebrates may take several years to decline to background levels. Levels of PCBs and DDT in open ocean dolphins did not decline between 1978 and 1986, despite a reduction in organochlorine contamination of the marine environment during this period (Loganathan *et al.*, 1990).

11.4 *Effects of pollutants (type 3 biomonitoring)*

The primary aim of ecotoxicologists should be to describe and predict *effects* of pollutants on organisms and ecosystems. The basis of such studies is that biochemical, cellular, physiological and morphological parameters can be used as screening tools or 'biomarkers' in environmental monitoring (see Chapter 9).

11.4.1 TERRESTRIAL ECOSYSTEMS

One of the simplest *in situ* indicators of air pollution is the Bel W3 variety of tobacco. The plants develop a mottling of the leaves in response to low levels of ozone pollution. The test is easy to perform and has been used in several surveys involving schoolchildren (Heggestad, 1991). Forest dieback is one of the clearest examples of the effects of pollution on ecosystems. Growth rates of trees can also be useful and can be retrospective if determined from the widths of annual growth rings.

Ecological and physiological effects due to pollution may have unexpected side-effects on behaviour. For example, air pollution from a copper smelter in Finland resulted in a decline in caterpillars in the vicinity of the factory. Great tits (*Parus major*) obtain the carotenoids that give them their yellow colour by feeding on these caterpillars. Birds near to the factory had paler plumage than did birds further away, and it is thought that this will reduce fitness in terms of choice of mate and survival (Eeva *et al.*, 1998). Similar results have been obtained for radioactively contaminated barn swallows (*Hirundo rustica*) near Chernobyl, in which there appears to be a trade-off between increased use of carotenoids for free radical scavenging and their role in sexual signalling (Camplani *et al.*, 1999).

In some situations, effects can be related directly to a local source of pollution. Walton (1986) obtained direct evidence for the effects of fluoride emissions from an aluminium plant on small mammals. Moles and shrews collected less than 1 km from the plant had extremely high levels of fluoride in their bones and teeth. Several manifested the symptoms of fluoride poisoning, including chipped and broken teeth and brittle bones.

An *in situ* bioassay using earthworms for assessing the toxicity of pesticide-contaminated soils was developed by Callahan *et al.* (1991). At each site, five *Lumbricus terrestris* were placed in enclosures distributed in transects throughout areas of high and low contamination at a 'superfund' site in Massachusetts formerly used for mixing pesticides. Mortality, morbidity (coiling, stiffening, swelling, lesions, etc) and whole body concentrations of a wide range of organic pollutants in worms were related to levels in the soils. This *in situ* method does not require removal of highly contaminated soils from the site. It provides an accurate dose–response relationship which can be used to predict the soil levels below which worms will return to the site and the 'safe' levels in soils at which worms will not accumulate sufficient concentrations of pollutants to be harmful to their predators.

11.4.2 FRESHWATER ECOSYSTEMS

At the organism level, most recent research has been directed towards developing *in situ* bioassays for detecting sublethal effects. The most widely used parameter is 'scope for growth' (SFG) (see section 8.5). SFG measures the difference between energy input to an organism from its food and the output from respiratory metabolism and, at least in principal, can be related to population and community processes. Animals which are 'stressed' (expending energy on detoxifying and excreting pollutants) have less energy available for somatic growth and reproduction. This is manifested as lower reproductive and growth rates compared with unstressed controls. Most measurements on SFG in fresh water have been on amphipod and isopod crustaceans. The SFG test using *Gammarus pulex* for measuring the effects of zinc and low pH is at least an order of magnitude more sensitive than acute 24-h LC_{50} tests (Naylor *et al.*, 1989).

Many rivers and streams are affected by acute episodic pollution rather than long-term chronic contamination. 'Bursts' of pollutant run-off can occur after thunderstorms, rapid snow melt (see figure 4.8), accidents at factories or deliberate release. These effects of intermittent increases in pollutant concentrations can be detected by continuous monitoring of caged organisms. Seager and Maltby (1989) described such a system which used rainbow trout (*Salmo gairdneri*). Fish were caged *in situ* in Pendle Water, a polluted urban river in Lancashire, England. The trout responded to sewer outflow discharges by increasing their breathing rate. This was monitored by measuring the small oscillating voltage produced by the muscles involved in gill ventilation.

11.4.3 MARINE ECOSYSTEMS

The effects of pollutants on marine ecosystems are difficult to study because of the vastness of the system and the difficulty of relating any effects seen to specific chemicals. The use of biological effects to monitor marine pollution has been reviewed by Addison (1996). He discusses the use of three biochemical responses (monooxygenase induction, metallothionein induction and AChE inhibition), measurements of energy partitioning in molluscs and analysis of benthic community structure to assess the impact of marine pollution.

One of best studied examples of effects of pollutants on marine organisms is that of imposex in barnacles and bivalve molluscs caused by TBT. This phenomenon is considered in more detail in Chapter 15. In the present chapter, only the biomonitoring aspects are discussed (see box 11.2 and figure 11.8).

Studies to determine the effects of pollutants on marine mammals are few. Detailed studies on the effects of PCBs on seals have been carried out in The Netherlands (Brouwer *et al.*, 1989, 1990), but this is an exception. Usually, extrapolation has been made from unrelated species when assessing risks to marine mammals.

Work on the beluga whale (*Delphinapterus leucas*) in the St Lawrence estuary in Canada illustrates the difficulties of field studies with marine mammals. This population has failed to increase despite the cessation of hunting, and toxic chemicals have been blamed for the problems of this isolated population. Here, collec-

BOX 11.2 *Imposex in the dog whelk.*

The mean size of the female penis relative to males in a population of dog whelks is called the level of 'imposex' or relative penis size index (RPSI). The RPSI is calculated by dividing the cube of the mean length of the female penis by the cube of the mean length of the male penis (both in mm) and multiplying by 100. Thus, if the RPSI is 100% then the mean size of the female penis is the same as that of the males. This index provides an indicator of the exposure of dog whelks to TBT at a site (figure 11.8).

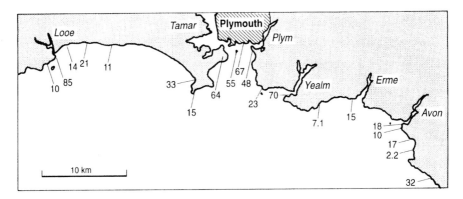

FIGURE 11.8 *Levels of relative penis size index (RPSI) (percentage size of female penis relative to males) in dog whelks* (Nucella lapillus) *collected from south-west England in 1984–5. Imposex develops in females in response to TBT leached from antifouling paints and is most prevalent in areas of high boating activity (e.g. the Looe and Yealm estuaries). Reproduced from Bryan* et al. *(1986) by permission from the Marine Biological Association of the United Kingdom.*

tion of specimens for scientific study cannot be made and thus material can only be obtained from animals found dead or by non-destructive sampling. Determination of the residue levels revealed high levels of PCBs but as the animals had been dead for some time measurements of the biomarkers studied by the Dutch workers were not possible. At the present time, there is no firm evidence that PCBs are causing effects in whales.

It should be pointed out that although both pinnipeds and cetaceans are termed 'marine mammals', they are not closely related. Studies on DNA adducts (sections 7.4.1 and 10.4.3) from stranded whales in the St Lawrence have revealed high levels of benzo(a)pyrene adducts which were not detected in the brains of belugas killed by native hunters in the Arctic (Martineau *et al.*, 1988). The levels of adducts have been correlated with high incidence of tumours in the St Lawrence beluga whales.

The possibility that the widespread mortality of seals in various parts of the world caused by viruses is linked to effects of chemicals (especially PCBs) on the immune system has been put forward, but the Scottish verdict of 'not proven' seems to be as far as one can go at the moment. Detailed immunological studies on beluga whales have been started based on non-destructive sampling of skin and it will be interesting to learn whether the high levels of PCBs in the St Lawrence population is affecting the immune system.

11.5 *Genetically based resistance to pollution (type 4 biomonitoring)*

Strains of plants which possess genetically based resistance to high concentrations of metals in soils have been recognized for many years

(Baker and Walker, 1989). Resistance is also well documented in insects. Such resistance is inheritable and should be distinguished from phenotypic tolerance which all members of a species may posses ('preadaptation'). The latter may consist of avoidance strategies, high excretion ability or possession of enzymes that break down organic pollutants. Phenotypic tolerance can be induced (e.g. increased synthesis of metal-binding proteins). However, genetically distinct pollution-resistant strains will evolve only if the selection pressure persists for several generations.

Terrestrial invertebrates which have been shown by breeding experiments to develop genetic resistance to high concentrations of metals include races of earthworms (Spurgeon and Hopkin, 2000), *Collembola*, terrestrial isopods (woodlice) and *Drosophila* (Posthuma and Van Straalen, 1993). However, the selective advantage that this conveys is quite small. Resistant animals typically survive concentrations of metals only some 30–50% higher than controls. Freshwater oligochaetes and marine polychaetes have also evolved resistance to metals. In some of these cases, the basis of the increased tolerance is an increase in the copy number or transcription rate of the gene coding for detoxifying proteins. In the case of organophosphate insecticides, up to a 256-fold amplification of the genes coding for non-specific esterases that break down the insecticide have been found in mosquitoes (see also figure 13.6). For metals such as copper and cadmium, the gene that codes for the metal-binding protein metallothionein may be duplicated up to four times. The metals are bound more rapidly by resistant animals after ingestion. Such amplification has been found in wild *Drosophila* and has probably evolved in response to the spraying of fruit trees with copper-containing fungicides. Metallothioneins are ancient proteins known from such diverse groups as *Collembola* (Hensbergen *et al.*, 1999) and snails (Dallinger *et al.*, 1997). Thus, the ability to develop resistance in sites naturally enriched with metals such as copper and zinc has been around for many millions of years.

11.6 *Conclusions*

In situ biomonitoring organisms should satisfy the '5Rs' if they are to be used successfully (Hopkin, 1993a). These are as follows.

1. *Relevant* – to be ecologically meaningful, ecotoxicological tests should use species that play an important role in the functioning of the ecosystem.
2. *Reliable* – species should preferably be widely distributed, common and easily collected to facilitate comparison between sites separated by large distances.
3. *Robust* – bioindicators should not be killed by very low levels of pollutants (with the exception of type 1 community structure monitoring where sensitivity is important) and should be robust enough to be caged in polluted field sites.
4. *Responsive* – the organisms should exhibit measurable responses to pollutant exposure by having greater concentrations of the contaminant(s) in the tissues (type 2 biomonitoring), by exhibiting effects such as reductions in scope for growth and fecundity, increased incidence of disease or induction of a biochemical response (type 3 biomonitoring) or by possession of genetically based resistance (type 4 biomonitoring).
5. *Reproducible* – the species chosen should produce similar responses to the same levels of pollutant exposure in different sites.

11.7 *Summary*

Numerous factors influence the bioavailability of pollutants. This makes it very difficult to predict accurately the extent to which chemical residues are assimilated by animals and plants and their biological effects in the field. Consequently, many researchers have used *in situ* biological monitoring to gather such information. Four main approaches have been adopted. First, the presence or absence of taxa from clean and contaminated sites can be assessed ('community effects'). Second, concentrations of pollutants are measured in organisms collected from the field; this approach also allows field deployment of 'sentinel' species to monitor pollutant availability (e.g. the mussel *Mytilus edulis* as part of the global 'mussel watch' programme). The third type of monitoring quantifies effects of chemicals on organisms and includes a wide range of biochemical and physiological parameters which come under the general heading of 'biomarkers'. One of the most successful biomarker assays is the measurement of 'imposex' in dog whelks. This phenomenon is caused by tributyl tin (TBT), which has been widely used as an antifouling coating on boats. TBT imposes male characteristics on females, leading to a reduction in reproductive performance and local extinction of dog whelk populations in the most contaminated sites. The fourth type of monitoring detects the evolution of genetic resistance. This is most obvious in insects where, for example, resistant strains of mosquitoes may tolerate concentrations of OP insecticides more than 200 times greater than those which will eliminate non-resistant populations. Natural selection has favoured those mutations able to produce greater quantities of non-specific esterases with a detoxifying function in response to the toxic action of the insecticides.

11.7 *Further reading*

BAKER, A. J. M. and WALKER, P. L. (1989) Concise review of metal tolerance in plants (see also Ernst *et al.*, 1992).

BRITISH ECOLOGICAL SOCIETY (1990) A useful summary of the background to monitoring water quality.

BRYAN, G. W. *et al.* (1986) A classic paper on the effects of TBT on dog whelks.

CALLAHAN, C. A. *et al.* (1991) Excellent paper describing *in situ* monitoring with earthworms at a 'superfund' site in the USA.

HOVE, K. *et al.* (1990) Interesting study on radiocaesium in goats which emphasizes the importance of long-term monitoring.

LOWE, V. P. W. (1991) Use of laboratory data to interpret 'unexplained' mass mortality of seabirds near the Sellafield nuclear reprocessing plant.

NAYLOR, C. *et al.* (1989) Study on scope for growth in *Gammarus pulex*.

POSTHUMA, L. and VAN STRAALEN, N. M. (1993) Comprehensive review of metal tolerance and resistance in terrestrial invertebrates.

WRIGHT, J. F. *et al.* (1993) A good summary of the RIVPACS technique by some of the workers who developed it.

PART 3

*Effects of pollutants
on populations
and communities*

CHAPTER

12

Changes in numbers: population dynamics

So far, this book has described the effects of pollutants on individual organisms, and Chapter 11 described methods of collecting data on populations in the field. In this chapter, we consider the uses to which population data may be put, how they should be analysed and the interpretations that may be made as to cause and effect. The first five sections of the chapter are an exposition of population ecology theory, and some readers will find this quite heavy going. An attractive route for those encountering this material for the first time may be to begin by reading the case studies in section 12.6.

When pollutants enter an ecosystem, the species within it may be affected in any one of the ways shown in figure 12.1. The numbers of some species will decline, perhaps to zero (figure 12.1, curve i), if the species becomes locally extinct. Alternatively, numbers may decline but level out lower than before (curve ii) and the population may persist at this level if the pollution endures *(chronic pollution)*. A third possibility is that population size initially increases (curve iii). If the pollution is chronic, resistance may evolve within the population, allowing

population numbers eventually to increase to a new equilibrium. The evolution of resistance is considered in Chapter 13. If the pollution is transient then the population may eventually recover, either rising from the level to which it was depressed by pollution (curve iv) or returning through immigration–recolonization if pollution had rendered the population extinct (curve v). In these last two cases, the population does not necessarily return to its original level. We shall see examples of some of these curves in the case studies described later in this

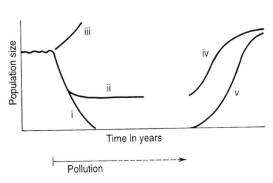

FIGURE 12.1 *Possisble responses of population size to pollution.*

chapter. To begin with, however, we need a little ecological theory with which to interpret and understand the processes that may operate to produce curves such as those in figure 12.1.

12.1 *Population growth rate*

A key feature of population growth or decline, as shown in figure 12.1, is the rate at which it occurs. The curve i population, for example, seems to be declining at a constant rate. The curve ii population initially declines, but then steadies and neither grows nor declines – its population growth rate is zero. The curve iii population initially increases in size – its population growth rate is positive. Thus, population growth rate is positive, zero or negative according to whether the population increases, is stationary or decreases in size.

Population growth rate is the most important characteristic of the population, and population ecologists spend much time and effort measuring it and establishing the factors that affect it. Some of these factors are described in section 12.2. Right at the outset, however, it is worth noting that there is one set of factors which is particularly important and particularly difficult to study in practice. These are density-dependent factors, i.e. factors which are affected by population density.

Typically, the effect of density dependence is that population growth rate reduces as population density increases. In the most straightforward case, the population density is then stabilized at a level characteristic of the environment in which the population exists. This level is referred to as the **carrying capacity of the environment**.

It will be immediately apparent that density dependence is an unwelcome complication for ecotoxicologists. It means that for a full understanding we need to know not only the way that pollution affects population growth rate but also the way that density dependence affects it.

Curve iii in figure 12.1 represents a population that increases in size when the environment becomes polluted. Let us call this species A. If species A was the only one adversely affected by pollutants (e.g. decreased reproductive success, increased mortality rate), its increase would constitute a paradox. The only way species A can increase is as a result of complex interactions with other species. Perhaps some other species, call it species B, exists which depresses species A. If pollution were to depress species B, the consequence for species A could be beneficial. This could come about in several ways. A common case is that species B is a predator of species A; thus, species A increases because the pollutant removes its predators. For instance, the red spider mite increased in numbers when its predators were killed by pesticides. Interactions between species are examined further in section 12.5.

12.2 *Population growth rate depends on the properties of individual organisms*

Growth rates of a population can be measured in different ways, as shown in box 12.1.

It is intuitively clear that population growth rate depends on individuals' birth and death rates and on the timing of their breeding attempts. Together, these characterize the individuals' life histories. In general, however, mortality, birth and growth rates may change with age. For a complete description of the life history, therefore, we need a record of *age-specific* growth, birth and death rates. Collectively, these are sometimes referred to as the organism's *vital rates*. At this point, the reader may

BOX 12.1 *Different ways of assessing population growth.*

The measure we favour is population growth rate, r. Population growth rate is defined as the population increase per unit time divided by the number of individuals in the population. The definition of population growth rate can be understood mathematically as follows. If population size $N(t)$ is plotted as a function of time t, as in figure 12.1, then dN/dt represents mathematically the population increase per unit time, in units of animals per unit time. Population growth rate puts this on a per capita (i.e. per animal) basis by dividing the number of individuals in the population. Mathematically, $r = 1/N \, dN/dt$. This is measured as animals per capita per unit time. Thus, if population size is 1000 and is increasing by one animal per year, the population growth rate is 0.001 per capita per year.

If r is constant, then $N(t) = N(0)e^{rt}$. This shows that exponential population increase occurs if population growth rate is constant.

Another measure that is sometimes used is *net reproductive rate*, usually given the symbol λ, defined as:

$$\lambda = e^r \tag{12.1}$$

Conversely,

$$r = \log_e \lambda \tag{12.2}$$

λ is the factor by which the population is multiplied each year. Thus, if the population doubles each year, then $\lambda = 2$. For example, if $\lambda = 2$ and initial population size is 10, then in subsequent years population size is 10, 20, 40, 80 ... In this case, $r = \log_e 2 = 0.693$. Further properties of λ are discussed in the Appendix.

wish to look at the real life example provided in case study 12.1 at the end of this section.

We will here consider two methods of recording a life history that differ in their level of detail. Both can be applied to organisms with discrete breeding events – for example, they might breed annually. We start with the simpler approach.

Suppose the organism breeds for the first time at age t_1, for the second time at age t_2, for the third time at age t_3 and so on, as shown in figure 12.2. Suppose the number of offspring produced by each female is n_1 at her first breeding attempt, n_2 at her second and so on. Last, suppose that the probability of a female surviving from birth to age t_i is l_i. The population growth rate, designated r, can be calculated from the formula:

$$1 = \tfrac{1}{2}n_1 l_1 e^{-rt_1} + \tfrac{1}{2}n_2 l_2 e^{-rt_2}$$
$$+ \tfrac{1}{2}n_3 l_3 e^{-rt_3} + \ldots \tag{12.3}$$

Equation 12.3 is known as the Euler–Lotka equation. In deriving the equation, it is assumed that the proportion of females in each age class is invariant (the assumption of 'stable age distribution'). However, estimates of population growth rate calculated from equation 12.3 are still useful in practice even if the stable age distribution assumption does not hold.

It is usual to use a computer to calculate r from measurements of the life history parameters (the values of n_i, l_i and t_i). The simplest method is to calculate the right-hand side of equation 12.3 for each of a number of trial values of r (e.g. try –0.5, –0.4, -0.3, ... 0.3, 0.4, 0.5). Just one value of r makes the right-hand side of equation 12.3 equal to 1. That value of *r* satisfies equation 12.3 and so measures the population's growth rate.

The life history may be simpler than that shown in figure 12.2. For example, adult mortality rate may be constant, or birth rate may be constant or breeding attempts may be at regular intervals. Such simplifications often allow equation 12.3 to be written in a simpler, more tractable form (Calow *et al.*, 1997).

One other very important way of describing life histories is to record the number of organisms surviving and the number of offspring produced at regular intervals, e.g. daily. These records are conveniently tabulated in a matrix, referred to as a population projection matrix. Powerful methods of matrix algebra have been developed and applied to the analysis of these matrices (Caswell, 1989). A brief introduction to their use is given in the Appendix.

In the following case study, life histories were recorded by daily counting of the numbers of animals surviving and of the numbers of offspring produced.

12.2.1 THE LIFE HISTORY AND POPULATION GROWTH RATE OF THE COASTAL COPEPOD *EURYTEMORA AFFINIS*

The life history and population growth rate of the coastal copepod *Eurytemora affinis* at different concentrations of dieldrin was studied by Daniels and Allan (1981). A population of *E. affinis* was obtained from Chesapeake Bay, MD, USA, where it undergoes annual population expansions between February and May, when water temperatures are increasing from 5 to 15–20°C. Animals were kept in the laboratory for 2 months (three or four generations) at 18°C before experimentation began.

The experiment consisted in recording the life histories of animals subjected to different

FIGURE 12.2 *A general life history. t_1, t_2, t_3, ... represent the ages at which the organism breeds. n_1, n_2, n_3 ... are the number of offspring then produced by each breeding female.*

concentrations of dieldrin. There were seven treatments, 0, 1, 2, 3, 4, 5 and 10 µg l^{-1} dieldrin, together with an acetone control because acetone was used as the 'carrier' of dieldrin. Sixty newly hatched larvae ('nauplii') were allocated to each treatment and maintained in groups of six in dishes containing 20 ml of bay water. The water was changed every other day, when the animals were fed a fixed number of algal cells. The numbers of survivors and births were counted daily.

The survivorship curves of animals undergoing the various treatments are shown in figure 12.3A. Survival to day 20 was worse at concentrations of 5–10 µg l^{-1} than at lower concentrations. Reproduction began around day 18, and birth rate was rather variable over time (two representative birth rate curves are shown in figure 12.3B).

The life history can be fully described by the three parameters mortality rate, birth rate and age at first reproduction if the simplifying assumptions are made that for each treatment mortality rate does not change with age and that birth rate does not change with age after first reproduction. Mortality rate, birth rate and age at first reproduction ('development period') are shown in relation to dieldrin concentration in figure 12.4A. As dieldrin concentration increased, mortality rate and development period increased and birth rate fell.

Any of these effects on its own would produce a reduction in population growth rate. The reductions in population growth rate that would

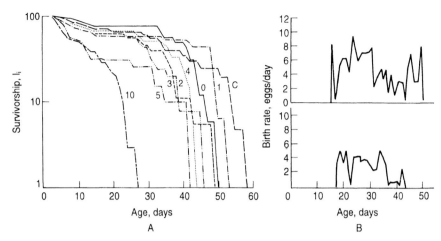

FIGURE 12.3 *Effect of dieldrin on the life history of* Eurytemora affinis. *(A) Survivorship curves. These show for each treatment the survivorship of animals to each age. Numbers indicate treatments, i.e. concentrations of dieldrin, in µg l⁻¹. C is the acetone control. (B) Birth rate in relation to age. Birth rate was measured at all concentrations, but only two representative concentrations, 2 and 4 µg l⁻¹, are shown here (upper and lower graphs respectively). Modified from Daniels and Allan (1981) with permission from the National Research Council of Canada.*

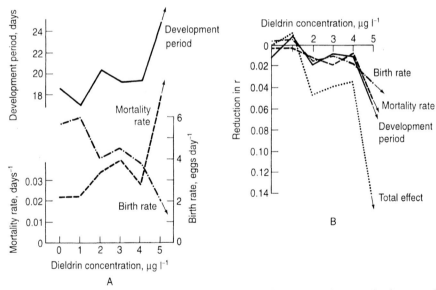

FIGURE 12.4 *Effects of dieldrin concentration in E. affinis. (A) Effects on mortality rate, birth rate and development period (age at first reproduction) estimated as described in the text. (B) Reductions in population growth rate, r, caused by the effects shown in (A). From Sibly (1996).*

be produced by each effect on its own are shown in figure 12.4B. The reduction due to birth rate depression is similar both to that due to increased mortality and to that due to increased development period. Thus, the reduction in population growth rate with increasing dieldrin concentration is due equally to birth rate, death rate and age at first reproduction effects (figure 12.4B).

This case study shows how population growth rate varied with dieldrin concentration (figure 12.4B). Population growth rate also depends on many physical aspects of the environment. For example, population growth rate of the rice weevil *Sitophilus oryzae* depends on the temperature and moisture content of the grain stores in which it lives. The dependence is shown in a contour plot in figure 12.5 (dashed lines). The axes represent the temperature and moisture content of the grain. The dashed lines are contours of equal population growth rate. Within the range of environments represented in figure 12.5, there are some in which the population flourishes (high values of *r*). In general, the higher the value of *r* the faster the population grows. Ecologists refer to those conditions for which $r \geq 0$ as the species' **ecological niche**.

In general, different species have different niches. The grain beetle *Rhizopertha dominica*,

for example, flourishes at higher temperatures than *S. oryzae* (figure 12.5).

In figure 12.5, population growth rates were not affected by interactions with other animals. When interactions within and between species are taken into account, population growth rates are reduced. In the long term, population growth rates do not exceed zero (i.e. long-term population explosions do not occur).

As we have seen, many factors affect population growth rate. If the long-term population growth rate is zero, however, some factors must operate more strongly when the population is large, but only weakly when the population is small (**density dependence**). We turn to these factors next.

12.3 *Density dependence*

As mentioned earlier, factors which vary in their effect with population density are said to be density dependent. Their net result is that

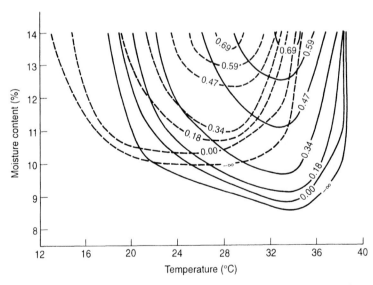

FIGURE 12.5 *Contour plots of population growth rate for two species of grain beetle,* Sitophilus oryzae *(– – –) and* Rhizopertha dominica *(——). Reproduced from Andrewartha and Birch (1954) with permission from the University of Chicago Press. (Contours are here labelled in terms of population growth rate; net reproductive rate was used in the original.)*

population growth rate is affected by population density. In the simplest case, population growth rate is a negative linear function of population density, as in figure 12.6. Note that when the population is small (left-hand side of figure 12.6), the population increases (population growth rate is positive). When the population is large (right-hand side of figure 12.6), the population decreases (population growth rate is negative). In between, there is an (equilibrium) population density for which population growth rate is zero. This population density is called the **carrying capacity of the environment** in which the population lives. Ecologists give it the symbol K. The effect of density dependence is here to push the population density towards the equilibrium density if other factors have increased or decreased it. The equilibrium is therefore a *stable equilibrium*.

Suppose population growth rate reduces with population density in a straight line relationship, as in figure 12.6. Let the equation of the straight line be:

$$\text{population growth rate} = r_0 - bN \quad (12.4)$$

where r_0 and b are constants. r_0 represents population growth rate at low population density.

We can write b in terms of r_0 and carrying capacity, K, as follows. When the population is at the carrying capacity of the environment, population density is K and population growth rate is zero. Substituting these values in equation 12.4, we obtain:

$$0 = r_0 - bK \quad (12.5)$$

Hence, $b = r_0/K$, and substituting this value of b into equation 12.4 we get:

$$\text{population growth rate} = r_0 - r_0N/K \quad (12.6)$$

As population growth rate $= 1/N \; dN/dt$ (from box 12.1), equation 12.6 can be written:

$$\frac{dN}{dt} = r_0 N \left(1 - \frac{N}{K}\right) \quad (12.7)$$

Ecologists call this the **logistic equation**. It produces a 'sigmoidal' pattern of population growth (figure 12.7). When small, the population grows exponentially (left-hand side of figure 12.7). As population density approaches carrying capacity, population growth rate declines, resulting in a slow approach to the final equilibrium value (right-hand side of figure 12.7).

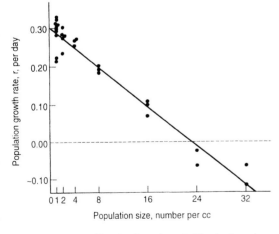

FIGURE 12.6 *Density dependence in* Daphnia pulex. *From Frank* et al.*(1957).*

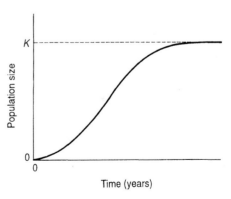

FIGURE 12.7 *The 'sigmoidal' growth curve that results from the logistic equation (12.7).* K *is the carrying capacity of the environment.*

12.4 *Identifying which factors are density dependent:* k-*value analysis*

Population growth rate depends on the life history traits of the individuals in the population, as described in section 12.2. In particular, population growth rate depends on individuals' age-specific birth and death rates and on the timing of their breeding attempts. Any or all of these may be density dependent. The analysis of which traits are density dependent is referred to as *k*-value analysis by population ecologists. This is because age-specific mortalities are known as *k*-values. Mortality at age (or stage) *i* is given the symbol k_i. In practice, it is usually mortalities that are analysed.

k-value analysis assesses how the *k*-values vary as population density varies. Usually, natural variation in population density is used. Population density is measured repeatedly, over a period of years, together with the *k*-values. Mortality at age *i*, k_i, is then plotted against population density. The most appropriate measure of population density is usually that of individuals of age *i*. An example of a *k*-value analysis is shown in figure 12.8. It was obtained from a 25-year study of sea trout by J. M. Elliott and collaborators (Elliott, 1993). By electric fishing at fixed times of year, population density was established at various points in the life history, as shown in figure 12.9.

k-values, which are measures of mortality, were calculated using the formula:

$$k_i = \log_e R_{i-1}/R_i \qquad (12.8)$$

where R_i represents the population density of stage *i*, as indicated in figure 12.9. These age-specific mortalities are plotted against population density for five phases of the sea trout life history in figure 12.8. In the first two phases (figure 12.8A and B), there are clear positive

relationships between population density and age-specific mortality, but there is no relationship in the later phases of the life history. The effect of a positive relationship as shown in figure 12.8A and B is to stabilize the population because higher mortality occurs at higher population density, making the population decrease when population density is high. Conversely, at low population density, mortality rate is relatively low, and this allows the population to increase.

Although population ecologists generally work with k_i-values, as in equation 12.8, it is sometimes better to use mortality *rates*. These can be calculated as:

$$\text{mortality rate} = k_i/t_i \qquad (12.9)$$

$$= 1/t_i \log_e R_{i-1}/R_i \qquad (12.10)$$

This example shows how population density affects mortality rate, and it may also affect somatic growth rate and birth rate. Moreover, as noted before, when considering the effects of pollutants, mortality, somatic growth and birth rates together determine population growth rate.

The effects of population density on mortality rate are central to population ecology and have been reviewed by Sinclair (1989). Considering the importance of the topic and the attention it has received over the years, it is perhaps disappointing to record Sinclair's conclusion that 'we still have a poor understanding of where density dependence occurs in the life cycle of almost every group of animals'.

12.5 *Interactions between species*

So far we have considered some of the factors which affect the numbers of a single population. If understanding a single population is

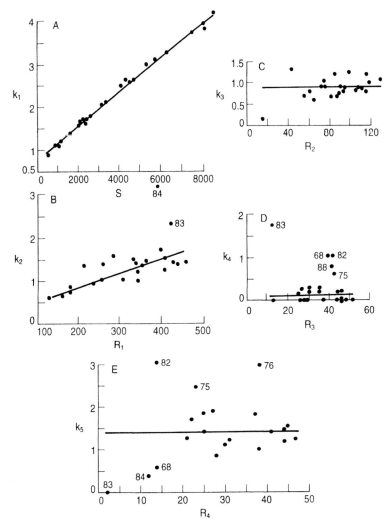

FIGURE 12.8 *Sea trout mortalities, k_1–k_5, in relation to population density in each of the five periods depicted in figure 12.9. Thus, (A) refers to alevin, (B) to young parr, and so on. Population densities S, R_1 ... R_4 are defined in figure 12.9; k_1–k_5 were calculated for the periods shown in figure 12.9 using equation 12.8. Each datum point refers to a single year. Reproduced from Elliott (1993) with permission from the National Research Council of Canada.*

difficult, untangling interactions between species is more than twice as hard. Here, population growth rate depends not only on the population's own density but also on the population density of the other species. To establish these dependencies in the field is generally prohibitively expensive. For this reason, detailed study has usually been restricted to simple laboratory systems or to mathematical models.

Despite these difficulties, a number of general points can be made.

Interactions between species can be logically classified as being of one of three types, as follows.

FIGURE 12.9 *The life history of the sea trout at a stream in north-west England. The eggs hatch after about 5 months. The young trout are known as alevin from hatching until they resorb their yolk sacs; after this, they are known as parr. Population density at each age was measured by electric fishing and was designation S, R$_1$, R$_2$ ... as shown.*

1. **Competition**, in which the population growth rate of each species decreases the more there are of the other species. Generally, species compete for common resources (e.g. food, space, breeding sites), so the more competitors there are, the lower the average success of each.
2. **Mutualism**, in which the population growth rate of each species increases the more there are of the other species. Thus, under mutualism, the species in effect help each other. Mutualism is the opposite of competition.
3. **Predator–prey**, in which the population growth rate of species A increases the more there are of species B, but the population growth rate of species B decreases the more there are of species A. These asymmetries are also evident in host–parasite, host–parasitoid and plant–herbivore interactions, and these can therefore be treated under the same heading as predator–prey.

Although, logically, all interactions must fit into one of the above categories because population growth rates are density dependent, the interaction of two species may not be in the same category at all population densities.

A species whose numbers vary substantially with time can persist in its environment only if it has a **refuge** that protects it and allows it to increase when its numbers have been reduced to low levels. The term 'refuge' is here used in a very broad sense. It may refer to a physical refuge. Examples of physical refuges include defendable holes or cracks in rocks, as used by whelks to escape predation by crabs, or areas accessible to only one species, such as the splash zones on upper shores where barnacles live but their whelk predators cannot follow them.

Environmental patchiness can also result in refuges for prey species. For instance, examples are known in which infested patches go extinct, but prey are able to escape to uninfested patches. If there is a sufficient time lag before the predators find the uninfested patches, this effectively creates a 'refuge' for the prey in which, for a time, they can increase.

Refuges can occur as a result of predators' foraging behaviour. There are many cases known in which foraging effort decreases as the density of prey decreases and this may result in a reduced mortality rate of prey at low prey densities, allowing prey populations to grow. Reduced foraging effort at low prey densities may result if predators switch to a different type of prey.

It follows from the definition of predator–prey interactions that removal of predators can lead to an increase in the population density of the prey species. Many cases are known in which pollutants are known to have had these two related effects (Dempster, 1975). For example, the red spider mite, *Panonychus ulmi*, appeared as a pest on outdoor fruit trees after the elimination of the slow-breeding predatory insects which previously controlled the mites. Fruit farmers used to apply pesticides to orchards in Britain as many as 20 times in a season, and this killed the mite's predators, so upsetting the natural balance which had kept the mite population under control (Mellanby, 1967). More recently, Inoue *et al.* (1986), investigating the effects of spraying Kanzawa spider mites, *Tetranychus kanzawai*, with six kinds of insecticide and three kinds of fungicide, showed that the population density of the Kanzawa spider mites increased with the application of certain insecticides and fungicides, probably because of their adverse effects on the natural predators (three species of predatory mite, three species of predatory insect and a spider).

12.6 Field studies: three case studies

Ecological studies of the effects of pollutants are generally based on circumstantial evidence, although experimental studies are possible (see below). Commonly, the time-course of pollution is monitored and is compared with observed changes in species numbers. Negative correlations between pollution and changes in species numbers do not, however, necessarily imply causal links, and there is always a worry that parallel changes could have been coincidental. This makes it attractive, as in all science, to carry out replicated experiments in which a treatment is applied to experimental areas, but not to control areas. The use of proper replication allows an assessment of whether the variation between treated and untreated areas is significantly greater than natural variation between control areas. The effect of the treat-

ment can then be measured as the difference between the two types of areas, and this can be assessed statistically provided that a suitable experimental design has been used.

12.6.1 THE DECLINE OF THE PARTRIDGE

The decline of the partridge was due in part to the *indirect* effects of pesticides killing the insect prey necessary for chick survival. Eighteen years of intensive study of the partridge (*Perdix perdix*) in Sussex, England, are summarized and wordwide trends are reviewed in a book by G. R. Potts (1986), from which the following case study is taken.

The decline in numbers in Britain and worldwide are shown in figure 12.10. In seeking the reasons for the decline, it is sensible to start by examining the trends in mortality rates ('k-values'). Table 12.1 summarizes data from 34 populations, individually studied for 2–29 years between 1771 and 1985. This suggests that the main reason for the population decline lies in the increase in chick mortality k_3. This raises the question as to what factors cause increases in chick mortality rate?

The causes of partridge chick mortality have long been a matter of controversy among gamekeepers and ecologists. Both weather and the availability of insect food could be important. The advocates of weather are struck by the fact that small partridge chicks can produce only about one-third of their own body heat, the rest comes from brooding parents. In cold weather, chicks therefore cannot afford to spend too long away from their parents, so feeding time is limited. This might lead to chick starvation. Arguing on this basis, there have been many attempts to correlate chick production with summer weather, but none of these have been particularly successful.

The other school of thought holds that in-

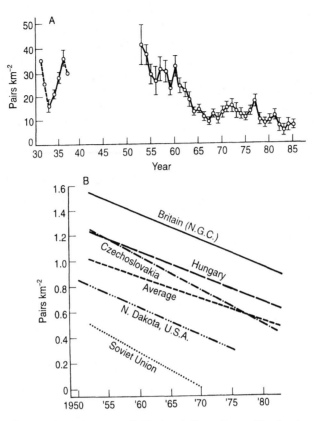

FIGURE 12.10 *(A) The UK Game Conservancy's National Game Census March pair counts for the partridge from 1933 to 1985, with estimated minimum densities for the early 1930s ± 2 standard errors. (B) The trend in density of breeding pairs km⁻² over the period 1952–85 in various regions of the world range. Reproduced from Potts (1986) with permission from HarperCollins.*

sects must be an important food for chicks because chicks go to a great deal of trouble to find insects, and they eat them in large quantities. Laboratory studies have shown that insects are nutritionally necessary for growing chicks; also in the Sussex study, there was a good relationship between chick mortality k_3 and the density of preferred insects (figure 12.11).

Distinction between the effects of insect availability and weather was achieved by multiple regression. This showed that 48% of the variation in chick mortality was explained by the density of preferred insects, and an additional 10% by average daily temperature in the critical period 10 June to 10 July. Both of these were statistically significant ($P < 0.001$), but

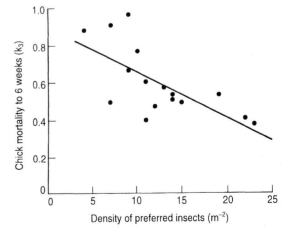

FIGURE 12.11 *Annual chick mortality in relation to the density of preferred insects in the third week of June. Data from the Sussex study 1969–85. reproduced from Potts (1986) with permission from HarperCollins.*

TABLE 12.1 *Comparison of mortality rates (k-values) in stable and declining populations ± standard errors*†*

		Populations		
		Stable (21)	Declining (13)	Significance
Nest loss	$(k_1 + k_2)$	0.26 ± 0.02	0.21 ± 0.03	ns
Chick mortality	k_3	0.29 ± 0.02	0.44 ± 0.02	$P < 0.01$
Shooting mortality	k_4	0.07 ± 0.01	0.08 ± 0.02	ns
Winter loss	k_5	0.38 ± 0.03	0.41 ± 0.05	ns
Total loss		1.00	1.14	

*The last column gives the results of statistical tests comparing stable and declining populations.
†From Potts (1986).
ns, not significant.

it seems that insect density was much more important than temperature in the Sussex study.

Given that reduced availability of insects is the key to the partridge population decline, it is natural to ask whether the insects declined as a result of pesticide usage. The increase in the use of herbicides is shown in figure 12.12 and chick mortality rates k_3 are shown in relation to pesticide usage in table 12.2. Note that chick mortality rates appear to be considerably increased by the use of herbicides, and increased again when some insecticides were also used.

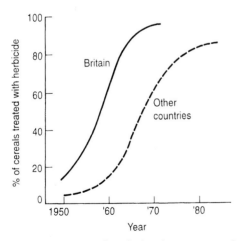

FIGURE 12.12 *Trend in herbicide use on cereals. Reproduced from Potts (1986) with permission from HarperCollins.*

The effect of herbicides was presumably indirect, removing the food necessary for the survival of the insects eaten by the partridge chicks. The effects of insecticides on insects appear less important. It seems reasonable to conclude that the decline of the partridge population occurred largely because of increased chick mortality, and that this in turn was the result of a decrease in the density of insects consequent on increased use of pesticides.

The above are not the only factors known to affect partridge populations. Nest losses are known to be density dependent, the form of the relationship depending on whether gamekeepers are present (figure 12.13). Dispersal and adult mortality due to shooting are also density dependent. These density-dependent effects would return populations to carrying capacity within a few years were it not for the high levels of chick mortality. Not surprisingly, equilibrium levels are considerably higher if gamekeepers are present.

Because fewer gamekeepers are employed in the UK now than formerly, it is reasonable to ask whether the population decline could be better ascribed to reduced gamekeepers than to increased pesticide use. The Sussex study attempted to answer this by entering the key known relationships into a simple model of partridge population dynamics. The model

TABLE 12.2 *Summary of estimates of partridge chick mortality rates grouped according to herbicide and insecticide use**

	Mean $k_3 \pm$ SE		
	Up to 1952 (no herbicide)	1953–61 (some herbicide)	1962–85 (herbicide + some insecticide)
National game census	0.33 ± 0.02	0.45 ± 0.05	0.51 ± 0.02
Damerham	0.32 ± 0.05	0.50 ± 0.06	0.50 ± 0.05
Lee Farm	–	0.44 ± 0.06	0.67 ± 0.07
North Farm	–	0.45 ± 0.05	0.63 ± 0.04
Sussex study	–	0.40 ± 0.05	0.61 ± 0.04
Mainland Europe	0.29 ± 0.03	0.37	0.45 ± 0.03
North America	0.25 ± 0.07	0.28	0.36 ± 0.04

*From Potts (1986).

comprised four basic equations showing how the four k values listed in table 12.1 were affected by population density and some environmental features. The model was used to see how the population would have reacted if gamekeepers had been employed and if pesticides had not been used. Figure 12.14 shows that, according to the model, use of gamekeepers would have reduced the rate of population decline, but would not have prevented it. Only when herbicides are not used, restoring chick

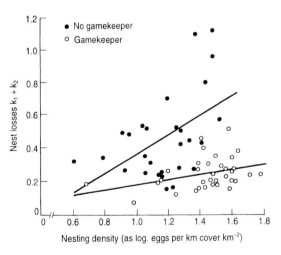

FIGURE 12.13 *Dependence of nest losses on nesting density with and without gamekeepers. Reproduced from Potts (1986) with permission from HarperCollins.*

mortality rates to their former levels, is the population decline prevented altogether.

The final message of Potts's (1986) book was that, as far as could then be seen, density dependence alone would not be sufficient to save partridge populations from extinction. Partridge preservation could only be achieved by increasing the supply of insects to chicks, so reducing the mortality rate of chicks. At the time of writing, little has changed except very locally where appropriate management has been introduced, and the partridge continues to decline (Potts, 2000). This huge loss of a valuable natural resource is arguably worth over £500 million annually in Europe alone.

Although the decline of the grey partridge is the best documented case, many other species of birds have declined markedly, especially on farmland, in the UK over the last 20–30 years. A listing of some of these species is given in table 12.3. For them, there is much less information upon which to base an analysis of causation. However, using an epidemiological approach (Peakall and Carter, 1997), intensive agriculture does appear to be the primary cause.

The data bases of the **British Trust for Ornithology (BTO)** (table 12.4) have proved valuable in examining the problem. The Com-

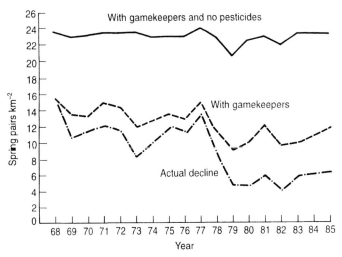

FIGURE 12.14 *Simulations of the Sussex partridge population showing the actual decline, the less severe decline that would have occurred if gamekeepers had been employed at 1968 levels and how no decline would have occurred if gamekeepers had been employed and pesticides had not been used. Reproduced from Potts (1986) with permission from HarperCollins.*

mon Bird Census of the BTO documents population changes without identifying their causes. However, from the bird ringing scheme, it is possible to calculate the annual survival rates of adults and first-year birds and from the nest record scheme it is possible to see whether clutch size, hatching and fledging success have been altered. These data bases are now used as one of the 13 indicators of sustainability by the UK government.

The basic problem appears to have been the onset of intensive farming practices, and in western Europe these changes have been fuelled by the European Union Common Agricultural Policy. This massive ecotoxicological experiment has been reviewed in detail by Pain and Pienkowski (1997). Although it is to be expected that several factors are involved, the present evidence suggests that indirect effects of herbicides and insecticides are probably the most important causes of the declines shown in table 12.3.

A recent concern given much prominence by the media has been over the development of

TABLE 12.3 *Population trends of selected species of birds in the UK*

Species	Year decline started	Population trends 1972–96 (%)	
		Farmland	Countrywide
Tree sparrow (*Passer montanus*)	1978	−87	−76
Turtle dove (*Streptopelia turtur*)	1979	−85	−62
Grey partridge (*Perdix perdix*)	1978	−82	−78
Spotted flycatcher (*Muscicapa striata*)	Before 1969	−78	−78
Skylark (*Alauda arvensis*)	1981	−75	−60
Song thrush (*Turdus philomelos*)	1975	−66	−52
Lapwing (*Vanellus vanellus*)	1985	−46	−42

Data from Crick *et al.* (1998).

TABLE 12.4 *Some of the surveys undertaken by the British Trust for Ornithology (BTO)*

Survey	Technique	Characteristics	Outcome
Bird ringing	Rings (or bands) are placed on leg of birds. Subsequent recoveries reported to BTO	Recoveries allow calculation of annual mortality besides giving information on migration	Started in 1909, 22 million birds ringed; 450 000 subsequently recovered
Constant effect sites	Standardized mist netting of birds on 12 visits	Monitors change in abundance and productivity	Started in 1983; 120–130 sites annually
Nest record scheme	Using special cards, the contents of nests are recorded at each visit	Provides data on clutch size and nesting success	Started in 1939; 30 000 records per year, over a million in total
Common birds census (CBC)	Detailed census of plot on 10 visits during the breeding season	Detailed data on the population changes of common species	Started in 1962; 220–230 sites covered per year.
Breeding bird survey	Randomly selected sites, transects carried out on three visits per year	Breeding population data, less detailed but larger sample size than CBC	Started in 1994; 2169 squares covered in 1997
Garden bird watch	Recording numbers of 10 common species, presence or absence of others	Population data from gardens. Lack of fixed protocol balanced by large sample size	Started in 1995; now has 10 000 observers

herbicide-resistant crops. There is the possibility that farmers could treat fields with high levels of herbicides, thereby effectively eliminating food essential to wildlife. The contrary view is farmers could let the weeds grow alongside crops for a longer period, secure in the knowledge that weeds can be eradicated later in the season. Crops engineered to be resistant to insect pests could cause a problem in that they are designed to produce a steady stream of natural insecticides that could harm beneficial predators. It is possible that this could be more harmful than the intermittent use of chemicals which could allow insect populations to recover. The recent case of the monarch butterfly (*Danaus plexippus*) affected by transgenic pollen (Losey *et al.*, 1999) has focused attention on this problem. Research is needed to see whether these effects are seen under realistic conditions of use.

12.6.2 POPULATION STUDIES OF PESTICIDES AND BIRDS OF PREY IN THE UK

The study of the effects of pesticides on birds of prey in the UK has a special place in ecotoxicology. This is partly because some pesticide effects were reported first in these species and partly because of the intensive nature of the studies, which have been conducted over 50 years. Despite this, our knowledge of some of the population processes is less complete than in the case of the partridge. It is therefore necessary to keep in mind that even where pesticides are known to affect individual vital rates (e.g. mortality rate or breeding success) this does not necessarily lead to population decline. Density dependence can compensate for the effects of the pesticides on particular age classes, and the net result could be that carrying capacity is unchanged.

The Peregrine falcon, *Falco peregrinus*, was the first bird of prey to be analysed in detail from the point of view of pesticide impact. Very similar, and in some instances fuller, data have been obtained for another bird of prey, the sparrowhawk, *Accipiter nisus*, by I. Newton. Some of these data will be referred to where appropriate. The following account is based mostly on books by Ratcliffe (1993) and Newton (1986). We begin by reviewing the properties of the pesticides involved and their known effects on mortality, breeding success and behaviour, and then go on to consider their effects on populations.

The pesticides that have been shown to affect the birds of prey are insecticides belonging to the organochlorine group (section 1.2.6). They include DDT, introduced into agricultural use in the late 1940s, and the cyclodiene insecticides aldrin, dieldrin and heptachlor, introduced in the mid-1950s. Both these types are extremely stable in their original form and/or as metabolites and so persist in the environment for many years. In addition, they are readily dispersed in wind and water or in the bodies of migratory birds and insects.

The organochlorine pesticides have properties which make them particularly hazardous to birds of prey and their predators. Because of their high fat solubility and resistance to metabolism, they have long biological half-lives in many species (see Chapter 5). Consequently, they tend to bioaccumulate as they move up food chains, reaching their highest concentrations in predators. It follows that top carnivores are especially useful indicators of the environmental effects of these compounds.

In the case of both sparrowhawks and peregrines, organochlorine insecticides have two distinct types of effect upon wild populations. First, the cyclodiene insecticides aldrin, dieldrin and heptachlor (used, for example, as seed-dressing chemicals) can cause lethal toxicity – and so may increase mortality rates. Some sparrowhawks and peregrines found dead in the field contained lethal concentrations of these insecticides in tissues such as liver, brain or muscle. Evidence from both laboratory and field studies indicates that tissue concentrations of dieldrin and heptachlor epoxide above 10 p.p.m. will cause death of predatory birds because of their direct neurotoxic action. The cyclodiene insecticides affect the central nervous system by inhibiting GABA receptors and it is known from laboratory studies that they can have sublethal effects on the function of the nervous system (e.g. irritability, disorientation and convulsions) before lethal effects are produced. A high level of skill and co-ordination is required in predators if they are to catch their prey, so disturbances of the nervous system can seriously affect their hunting skills (Chapter 8, section 8.4.2). There can be no doubt that serious sublethal effects as well as lethal ones were produced in the field during and after the 1950s, but there is no means in retrospect of quantifying this.

A second type of effect of organochlorine insecticides is eggshell thinning, caused by p,p'-DDE, a persistent metabolite of p,p'-DDT. The association between eggshell thinning in British peregrines and sparrowhawks and exposure of DDT residues was first noted by Ratcliffe (figure 12.15). Initially, it was not clear whether this was a causal relationship, but dosing predatory birds with p,p'-DDE in the laboratory has established that low levels of the chemical do cause eggshell thinning. It is now known that p,p'-DDE reduces transport of Ca^{2+} to the developing eggshell, probably because of inhibition of calcium ATPase in the shell gland

FIGURE 12.15 *The decline in Peregrine eggshell thickness which commenced in the UK in 1947. Shaded areas represent 90% confidence limits. Reproduced with permission from* Environmental Reviews. *The eggshell thickness index is defined as: index = weight of eggshell (mg)/length × breadth (mm). From Peakall (1993).*

(Chapter 15). Thus, there are good scientific grounds for regarding the close negative correlation between residue of p,p'-DDE and eggshell thickness as a causal relationship (figure 12.16A). Furthermore, because eggshell thinning is correlated with fledging success in wild sparrowhawks (figure 12.16B), it is probable that DDT affects hatching success directly.

Although it is well established that lethal toxicity and eggshell thinning have occurred in the field in the UK, difficulties arise in quantifying them and relating them to population change. Consider p,p'-DDE and shell thinning first. In both sparrowhawk and peregrine, substantial reductions in shell thickness occurred during 1946–7 in the UK, coincident with the large-scale introduction of DDT as an insecticide. There was, however, no evidence of a general decline in these species at this time, or for several years after (see figure 12.17 for peregrine). As sparrowhawk and peregrine breed at ages 1–2 and 2 years, respectively, and then have relatively short life expectancies (about 1.3 and 3.5 years respectively), the very sharp decline in populations in the late 1950s cannot

be directly attributed to eggshell thinning, which commenced in the period 1946–7. Thus, although eggshell thinning was occurring, and this was having some effect on fledging success (figure 12.16B), it did not bring any reduction in population size. Indeed, there was an increase in numbers of peregrines at this time. Presumably, population sizes at this stage were still at or close to carrying capacity, despite the reduction in hatching success brought about by DDT. Probably some density-dependent process compensated for the reduction in fledging success brought about by DDT. Perhaps the smaller number of fledglings were individually more successful in obtaining food because they had fewer competitors and so grew faster and were more successful than they otherwise would have been.

Whereas the introduction of DDT did not lead to population decrease, the populations of both species declined rapidly in the mid- to late 1950s (see figure 12.17 for peregrine, figure 12.18 for sparrowhawk). These declines coincided in both time and space with the introduction of aldrin, dieldrin and heptachlor as

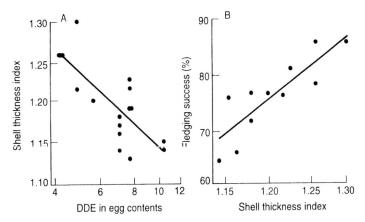

FIGURE 12.16 *In sparrowhawks, (A) DDE is negatively correlated with shell thickness and (B) eggshell thickness is positively correlated with the percentage of young raised per brood ('fledgling success'). Data from different areas, modified from Newton (1986) with permission from Academic Press. Data analogous to (A) for peregrine can be found in figure 15.2.*

FIGURE 12.17 *Peregrine population size in Britain (1930–9 = 100) showing the 1961 population decline and subsequent recovery, together with an outline of pesticide usage. Reproduced from Ratcliffe (1993) with permission from Academic Press.*

seed-dressing chemicals. As discussed above, these compounds caused deaths in the field and extensive sublethal effects were suspected but were not quantifiable.

A series of bans, placed between 1962 and 1975, led to the progressive removal of the cyclodienes and DDT from the UK (figure 12.17). These bans were followed by a decline in organochlorine residues and population recovery in both of these species. The pattern of recovery differed between areas. Particularly interesting was the late recovery of sparrow-

hawk populations in eastern England (figure 12.18). Sparrowhawk populations recovered when tissue concentration of dieldrin* fell below 1 p.p.m. (figures 12.19 and 12.20). Interestingly, the same was true of the kestrel during this period. Closer inspection of the data for

*In this account, the term dieldrin is used synonymously with the abbreviation HEOD, which is the chemical term for the active constituent of the commercial insecticide dieldrin.

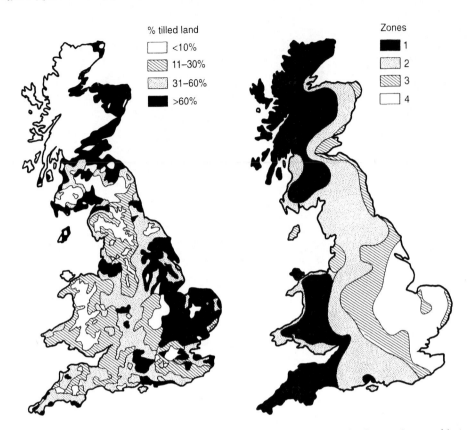

FIGURE 12.18 *Changes in the status of sparrowhawks in relation to agricultural land use and organochlorine use. The agricultural map (left) indicates the proportion of tilled land, where almost all pesticide is used. The sparrowhawk map (right) shows the status of the species in different regions and time periods. Zone 1, sparrowhawks survived in greatest numbers through the height of the 'organochlorine era' around 1960; population decline judged at less than 50% and recovery effectively complete before 1970. Zone 2, population decline more marked than in zone 1, but recovered to more than 50% by 1970. Zone 3, population decline more marked than in zone 2, but recovered to more than 50% by 1980. Zone 4, population almost extinct around 1960, and little or no recovery evident by 1980. In general, population decline was most marked, and recovery latest, in areas with the greatest proportion of tilled land (based on agricultural statistics for 1966). Reproduced from Newton and Haas (1984), reproduced in Newton (1986), with permission from Academic Press.*

both sparrowhawks and kestrels showed that there was a wide range of tissue dieldrin levels in individuals whose liver samples showed concentrations of 1 p.p.m. or more. Death could be attributed to direct dieldrin poisoning (on the grounds of high tissue residues ≥ 10 p.p.m. and symptoms of toxicity) in only a small proportion of the kestrels or sparrowhawks found dead during the period 1976–82. On the other hand, a substantial proportion had residues in

the range 3–9 p.p.m., increasing the suspicion that sublethal effects were more important than lethal ones (Sibly *et al.*, 2000; Walker and Newton, 1999). Population modelling for the sparrowhawk strengthened the case for suggesting that sublethal effects of dieldrin were important in causing population decline of the sparrowhawk in Britain over this period. There is, consequently, a strong suspicion that sublethal effects of dieldrin upon adults were wide-

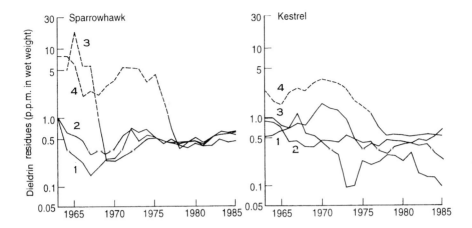

FIGURE 12.19 *Dieldrin (HEOD) levels in the livers of sparrowhawks found dead in the four zones shown in figure 12.18. HEOD is the active principle of the commercial insecticide dieldrin and accounts for some 80% of the technical product. Broken lines show periods when populations were depleted or decreasing, solid lines show periods when populations were normal or increasing. Population increase occurred when liver levels were less than about 1.0 p.p.m. wet weight. Reproduced from Newton (1988) with permission from Elsevier Science Ltd.*

spread and may have been important in causing population decline; it is also possible that embryotoxicity was a contributing factor.

The recovery of the sparrowhawk shown in figure 12.20 was rapid between 1980 and 1990, but levelled off in 1993 – a sigmoidal population increase of a type frequently encountered as populations approach carrying capacity (compare figure 12.7). Peregrines also have now recovered (figure 12.17).

It should be added that the situation was different in North America. Peregrines were less exposed to cyclodienes but more exposed p,p'-DDE (thus reflecting patterns of use of these insecticides; see also Chapter 15). Eggshell thinning was greater in North American peregrine populations than in British ones. In this case, there is strong evidence to suggest that p,p'-DDE was the principal cause of population decline, although the field data on populations is not so complete as that obtained in the UK.

The use of eggshell thinning as a biomarker is considered in Chapter 15.

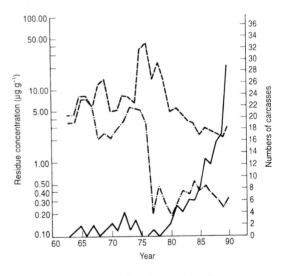

FIGURE 12.20 *Numbers of sparrowhawk carcasses (——) received in a region of eastern Britain, together with concentrations of DDE (– – –) and dieldrin (· – · – ·) found in their livers. Reproduced from Newton and Wyllie (1992) with permission from Blackwell Science Ltd.*

12.6.3 THE BOXWORTH PROJECT (AN EXPERIMENTAL ANALYSIS OF THE EFFECTS OF PESTICIDES ON FARMLAND)

The Boxworth project was designed to assess experimentally the effects of three contrasting regimes of pesticide usage on the animals and plants, including crops, on a 300-hectare arable farm in eastern England. The aim was to investigate large-scale, long-term effects of pesticides under conditions as close as possible to farm conditions. It was a major project, lasting 7 years in its main phase, involving many scientific man–years. The account here is based on Greig-Smith *et al.* (1992b). Three regimes of pesticide usage were investigated. These were:

1. *full insurance*, in which relatively large amounts of pesticides were applied in advance of possible pest outbreaks, as 'insurance';
2. *supervised*, in which pesticides were applied only when needed, as assessed by monitoring pest, weed and disease levels;

3. *integrated*, which was similar to supervised, but incorporated some additional features of 'integrated pest management'.

A map of the farm is shown in figure 12.21. Because large-scale effects were to be investigated, the farm was divided into three large areas, applying one treatment to each area. Because long-term effects were of interest, there was no swapping of treatments between areas.

These features of the experimental design mean that the experiment is unreplicated. For this reason, little assessment can be made of whether the variation between areas is such as might have occurred by chance in the absence of pesticide treatments. The lack of replication precludes the application of the statistical tests normally used in agricultural trials. This could have been remedied by allocating pesticide regimes to fields on, for example, a random basis. Other designs are possible that allow for factors known to cause variation between fields, such as soil type, or the amount of non-crop habitat.

The lack of replication dissipates much of the power of the experimental approach to at-

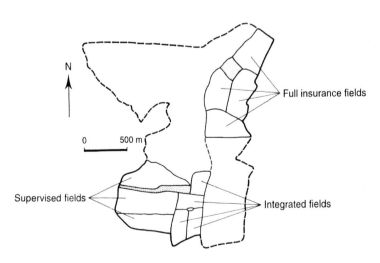

FIGURE 12.21 *Map of the farm at which the Boxworth project was conducted showing the location of the fields treated with each pesticide regime. Reproduced from Greig-Smith and Hardy (1992) with permission from HMSO.*

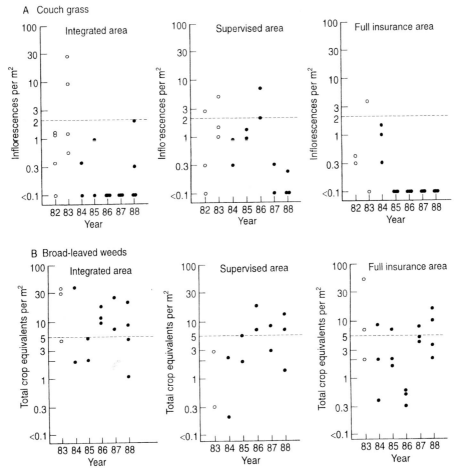

FIGURE 12.22 *Efficacy of the Boxworth pesticide regimes on the densities per m² of (A) the weed grass couch* (Elymus repens) *in July and (B) the broad-leaved weeds in spring; 1982 and 1983 were 'baseline' years before the pesticide regimes were applied. The dotted lines show 'spray decision thresholds' used in deciding whether to apply pesticides. Reproduced from Marshall (1992) with permission from HMSO.*

tribute pesticide cause to ecological effect. However, some idea of natural and baseline variation was obtained by monitoring the areas for 2 years before starting the trials. The pesticide regimes were applied for 5 years (1984–8), and some residual monitoring has continued since.

In the event, there was little difference between the supervised and integrated regimes (S + I henceforth). The analyses therefore generally consist in comparing the effects of full insurance with those of S + I.

Botanical monitoring showed that patterns

of grass weed densities reflected the efficiency of the weed control regimes. An example is shown in figure 12.22A. Full insurance was most effective, and held all grass weeds at low densities. By contrast, the densities of the broad-leaved weeds varied little between regimes (figure 12.22B). Further botanical comparisons were hampered by high variation between years, including the two baseline years.

Turning to the invertebrates, it appears that, overall, the density of herbivorous invertebrates was about 50% less under full insurance than under S + I. Worst affected were certain non-dispersing species. Carnivorous invertebrates (predators and parasites) showed a similar pattern, but detritivores were unaffected.

It proved particularly difficult to ascribe cause and effect in small mammals and birds, partly because of their mobility. The diet of common shrews (*Sorex araneus*) reflected the distribution of the invertebrates, described above, that form their diet. In particular, leatherjackets (crane-fly *Tipulidae* larvae) were less frequent in the stomachs of shrews caught in the full insurance area than in the supervised area. However, there was no evidence of long-term effects of the different pesticide regimes on any of the small mammal populations studied. Autumn application of slug pellets containing methiocarb had an immediate local impact on adult woodmice (*Apodemus sylvaticus*), but the population recovered quickly by immigration of juveniles from nearby woods, fields and hedges.

The only bird species affected by the full insurance regime was the starling (*Sturnus vulgaris*). Part of the observed reduction in numbers of starlings nesting in the full insurance area relative to the S + I area may have been due to the effects of the pesticides on leatherjackets.

The economics of the pesticide regimes were also evaluated. Yields of wheat in the full insurance fields were 0.92 ton ha^{-1} higher than in the supervised fields and 1.35 ton ha^{-1} higher than in the integrated area, although there was considerable variation from year to year. Grain quality also appeared better under full insurance. However, the extra costs inherent in full insurance meant that the supervised approach was as profitable as full insurance. The integrated regime used in the study was less profitable. However, truly integrated low-input systems might do better.

The Boxworth project shows clearly that the experimental study of pesticide effects in the field is a very expensive business. A replicated experimental design would have allowed stronger conclusions to be drawn, but as the farm would have been divided into smaller experimental plots mobility between areas would have increased and mobility was already a significant factor affecting the distributions of small mammals and birds. The lack of replication means that the Boxworth project was essentially a pilot study. It does nevertheless suggest which forms of wildlife were, and which were not, affected by high rates of application of pesticides.

12.7 *Summary*

The effect of a pollutant on a population is assessed by the effect it has on the population's growth rate. Population growth rate has usually been calculated from experiments in which the life history of the organism is recorded at a number of concentrations of the pollutant. Such experiments generally show that increasing the concentration of the pollutant has the effect of decreasing birth rate and/or increasing mortality rate. Decreasing birth rate and increasing mortality rate both decrease population growth rate. In consequence, an increase in the concentration of a pollutant generally results in a decrease in population growth rate.

Experiments of this type have usually been conducted at low population density. In the absence of stressors, populations at low population density undergo exponential population growth if per capita population growth rate is constant. Populations cannot expand indefinitely, however, because they eventually exhaust the available resources. As this starts to happen, population growth rate starts to decrease, until at very high population densities population growth rate is negative. In between, there is an equilibrium population density for which population growth rate is zero. This population density is called the carrying capacity of the environment in which the population lives. To date, there has been insufficient attention paid to the joint effects of pollutants and population density on population growth. Future work should among other things examine the effects of pollutants on carrying capacity.

Examination of three intensive case studies reveals the difficulties of establishing cause and effect in the field.

12.8 *Further reading*

BEGON, M. *et al*. (1996) *Population Ecology: A Unified Study of Animals and Plants*. Provides an introduction to population ecology, including the analysis of density dependence and interactions between species.

EVANS, P. R. (1990) Provides a useful discussion of the population effects of pesticides on birds and mammals.

FORBES, V. E. and CALOW, P. (1999) Reviews the literature and concludes that population growth rate is a better measure of responses to toxicants than are individual level effects.

LEVIN, L. *et al*. (1996) A classic recent example of an investigation of the effects of a pollutant on life history traits and population growth rate.

LINKE-GAMENICK, I. *et al*. (1999) Report of similar experiments which also examines the effects of density dependence.

SIBLY, R. M. (1999) and STARK, J. E. and BANKEN, J. A. O. (1999) Provide suggestions for the experimental design of population experiments.

13

Evolution of resistance to pollution

Evolutionary responses to pollution are referred to as resistance. It is implicit that resistance has a genetic basis, and what is known of its genetic inheritance is outlined in section 13.5. This chapter considers how genetic changes in resistance come about. Resistance represents an evolutionary response to environmental changes resulting from pollution, and the first sections of this chapter describe the general phenomenon of evolutionary response to environmental change. Pollution represents an environmental change for the worse, and resistance generally defends organisms against the deleterious consequences of pollution. Such defence may reduce an organism's mortality rate, but this is sometimes expensive, using energy and/or nutrients that could otherwise have been used for reproduction or somatic growth. Defence may, therefore, involve a trade-off between production and survival: increased survival may only be obtained at a cost of reduced growth or reproduction. As a result, resistance may have a fitness cost in an unpolluted environment. The possible physiological basis of this trade-off has been considered in Chap-

ter 8, the evolutionary implications are considered here in section 13.3. Four cases studies of evolutionary responses to pollution at the end of the chapter consider the evolution of pesticide resistance, of heavy metal tolerance in plants, of industrial melanism and of TBT resistance in dog whelks. These include some of the best documented studies of evolutionary responses to environmental changes. They show that the evolution of resistance generally consists in the selection of pre-existing genes rather than the appearance of new genes. The first sections of this chapter are quite theoretical, and an attractive approach for those encountering this material for the first time may be to begin by reading the case studies.

13.1 Chronic pollution is environmental change

Transient pollution by definition has only passing effects and so is unlikely to change gene frequencies. Chronic pollution, however, can have lasting effects because it changes the environment in which organisms live. In considering evolutionary responses to chronic pollution, we are, therefore, dealing with a particular case of the general phenomenon of evolutionary responses to environmental change. Before considering the effects of environmental change, however, we need to know what happens in unchanging environments.

13.2 *The evolutionary process in a constant environment*

Consider figure 13.1, which provides a simple graphical account of the main ideas. The account here is based on Sibly and Antonovics (1992). The evolutionary process is envisaged, put very simply, as consisting of the creation

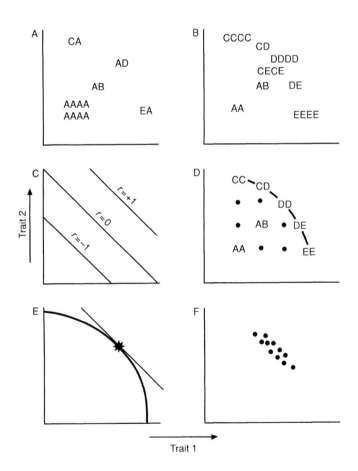

FIGURE 13.1 *Simple example of an evolutionary process. Axes represent two life history traits. Note that the alleles far from the origin (C–E) have increased in numbers between graph A and graph B, whereas those near the origin (A) have decreased. Graph C shows the 'per copy' rates of increase, i.e. fitnesses, here labelled* r. *Graph D shows the genetic options set with genotypes obtainable by recombination represented as dots. The boundary of the options set is the trade-off curve (thick line). Graph E shows the optimal strategy (evolutionary outcome), starred. Graph F shows the genetic options that may persist in the population at the end of the evolutionary process. See text for further details.*

(by mutation) of new alleles, which either displace or are displaced by their counterparts. In figure 13.1, alleles A–E affect two life history traits of carriers. These might, for example, be juvenile growth rate and juvenile survival. If the constant environment assumed here is a polluted environment, the alleles C–E might represent resistant alleles that confer improved survival (e.g. allele C) or growth (e.g. allele E). Alleles A and B would then represent non-resistant alleles, whose carriers on average have reduced survival or growth rate. Note that most individuals carry the A allele in figure 13.1A and so have small values of both traits, but a few individuals have larger values so that overall there is a small positive correlation between individuals. If there is not much effect of environmental variation, this reflects a positive genetic correlation (e.g. the two traits are correlated because they are determined by alleles that have an effect on the magnitude of both traits).

Figure 13.1B represents a hypothetical situation at some later time in the selection process. Now most of the small-trait A alleles have disappeared but the numbers of the large-trait alleles (C–E) have increased. Furthermore, the genetic correlation between individuals is now negative, whereas earlier it was positive.

Clearly, all depends on whether or not an allele spreads in the population, i.e. on the rates of increase of the alleles. The per capita (or more correctly per copy) rate of increase of an allele is here called its **fitness**. Because the rates of increase of alleles depend on their effects on life histories, the rates of increase (fitnesses) can be plotted out in the space of figure 13.1, as shown in figure 13.1C. In our example, the small-trait alleles have negative rates of increase because they are declining in the population. For these alleles, fitness is negative (e.g. $r = -1$ in figure 13.1C). On the other hand, the large-trait alleles have positive rates of increase because they are spreading. For them, fitness is positive

(e.g. $r = +1$). Evolutionary change can also occur, however, if both small- and large-trait alleles increase but at different rates.

In general, as selection proceeds, the cloud of points in figure 13.1 changes shape. In the absence of environmentally caused variation, the shape of the cloud is measured by genetic correlations and variances, and as the cloud changes shape the genetic correlations and variances change accordingly (cf. figure 13.5). In the absence of further mutation, where would this process end up?

In considering the eventual outcome of this selection process it must be remembered that we are here restricting attention to a constant environment. It is important to realize that the environment of an individual depends not only on physical features (e.g. temperature, rainfall, concentration of a pollutant) and biotic features determined by other species (e.g. food availability, predation) but also has characteristics determined by conspecifics, such as territory size, availability of mates, competition for food and so on.

In this environment, many alleles will affect life history components. Plotting out all these genetically codable options (including all possible recombinants) in a space such as that of figure 13.1A gives us a set of points that we shall call the **genetic options set** (figure 13.1D). Although the exposition here is in terms of two-dimensional examples, the concepts all have natural generalizations to three or more dimensions (Sibly and Antonovics, 1992).

Of particular interest, because it limits selection, is the boundary of the options set. We shall call this the **trade-off curve** (figure 13.1D). This trade-off curve represents the best that this organism can achieve genetically in the study environment.

Putting together the information about fitness (figure 13.1C) with the information on options sets (figure 13.1D), the **optimal strat-**

egy is readily identified (figure 13.1E) as that having the highest fitness in the study environment. This point represents the eventual outcome of selection in this environment.

In this section, a distinction has been made between the process and the outcome of selection. The process is modelled by **quantitative genetics**. This determines the short-term evolutionary trajectories within local populations from knowledge of the fitness surfaces together with the genetic options set, characterized by genetic correlations and variances. The outcome of selection can be identified by ecological optimality theory given knowledge of the shapes of the trade-off curve and the fitness contours. Note that the outcome of selection may be a number of alleles having similar fitness, i.e. a thin oval set of genetic options on or near the trade-off curve in the neighbourhood of the optimal strategy. In the example of figure 13.1F, these genetic options would be characterized by a negative genetic correlation. In this way, genetic correlations can provide evidence about the shapes of trade-off curves.

If there are no trade-offs, so that life history traits can be genetically altered independently of each other, then there is always selection to increase fecundity, decrease mortality rate and breed early. One way to achieve early breeding is through faster growth and development, so a corollary of selecting for early breeding is that there is always selection for faster growth and development – in the absence of trade-offs.

In this section, we have seen that the fitness of an allele can be measured by its rate of increase or, more specifically, by its 'per capita' rate of increase. In population genetics, it is usual to make the definition of fitness relative to the rate of increase of the most successful allele or genotype, but that approach is not followed here. The advantage of the present approach is that the fitness of an allele can be related directly to the life cycle of its carriers. Fitness is increased by reductions in mortality

rate, increases in fecundity or by breeding earlier. Thus, alleles are selected which reduce their carriers' mortality rate, increase their fecundity or make them breed earlier. It is important to note that the fitness of an allele depends on the environment in which its carriers live. If the environment changes, the fitness of the allele may change.

In these terms, an allele can be defined as **resistant** if it increases the fitness of its carriers in a polluted environment. Non-resistant alleles are also known as susceptible alleles. Thus, resistant alleles are favoured in polluted environments. What happens in unpolluted environments? If resistant alleles are then outperformed by susceptibles, the resistant alleles are said to have a **fitness cost**. Fitness costs are discussed further in section 13.4.

13.3 The evolution of resistance when there is a mortality–production trade-off

It follows from the above that if alleles exist that affect production rate but not mortality rate selection acts to maximize production rate. Conversely, if alleles affect mortality rate but not production rate, selection acts to minimize mortality rate. However, it may not be possible to alter one life history trait without affecting the other, if mortality and production are involved in a trade-off. A decrease in mortality rate can then only be achieved at the expense of a decrease in production rate. Possible physiological reasons for such a trade-off were described in Chapter 8 (section 8.5).

The action of selection on such a trade-off has been analysed by Sibly and Calow (1989). They focused on the juvenile phase, although the analysis can equally well be applied to adults. For the purposes of the analysis, they defined 'somatic growth rate' as the reciprocal of development period. This definition makes most biological sense if egg size and adult size are not subject to selection, at least for the period of time analysed.

For the reasons given in Chapter 8, it is likely that a decrease in mortality rate can only be achieved at the cost of a decrease in somatic growth rate. Protective mechanisms that decrease mortality rate are often thought of as defences. Resources allocated to these defences are not available for somatic growth, and it follows that increasing allocation to defence implies reduced somatic growth. Defensive structures and processes therefore have two key characteristics, from an evolutionary point of view. These characteristics are their effects reducing (i) mortality rate and (ii) growth rate. Plotting out the different genetic possibilities reveals the form of the trade-off curve (figure 13.2A, cf. figure 8.14).

It is important to remember that the form of the trade-off curve depends on the organism's environment. What is an effective defence in one environment may have no impact in another – defences against pollution are of no use in unpolluted environments. Thus, trade-offs are environment dependent.

The fitness contours for this trade-off are straight lines, and the most interesting contour, that on which fitness is zero (stable population), goes through the origin (figure 13.2B). Superimposing figures 13.2A and 13.2B reveals the position of the 'optimal strategy' (cf. figure 13.1E) representing the evolutionary outcome in the studied environment (figure 13.2C). This is the strategy towards which selection drives the population. It represents the optimal trade-off between mortality rate and growth. It indicates the optimal amount that the organism should spend on defence.

This analysis depends, as has been emphasized, on the constancy of the environment. What happens if the environment changes, as when it becomes polluted?

13.4 *Evolutionary responses to environmental change*

Pollutants may affect the shape and/or the position of the genetic options set. Changes in position will be discussed first. Pollution may damage organisms, either with lethal effect or with some detriment to production rate. Such

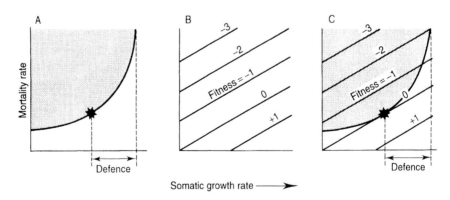

FIGURE 13.2 *(A) Genetic options set (shaded) and trade-off curve. (B) Fitness contours. Note that the zero fitness contour goes through the origin. (C) Superimposing (A) and (B) allows identification of the evolutionary outcome as the allele achieving highest fitness.*

processes increase mortality rate or reduce somatic growth rate, moving the options set either vertically upwards or horizontally to the left.

The outcome of the evolutionary response to such changes in the position but not the shape of the genetic options set is shown in figure 13.3. The analysis is straightforward. As in the last section, attention can be restricted to steady-state outcomes, in which the final value of fitness is zero. As before, the zero-fitness contours are straight lines going through the origin. Inspection of figure 13.3 shows that the evolutionary response is less defence if either mortality is increased, moving the options set vertically upwards, or if production is reduced, moving the options set horizontally to the left. In other words, either mortality stress or production stress elicits the same evolutionary response – less defence.

This prediction – less defence in polluted environments – appears paradoxical, and it should be emphasized that it applies only if the shape of the trade-off curve remains unchanged when its position is shifted. If more defence evolves in polluted environments then the inference must be that the shape of the trade-off curve has changed. A hypothetical example is

shown in figure 13.4. In this example, genes for optimal defence are selected in the polluted environment and genes for no defence are selected in the unpolluted environment, as shown by the stars in figure 13.4.

In general, pollution does change the shape of the options set. This may happen because pollution elicits the expression of genes that would not otherwise have been expressed. For instance, enzymes may be induced or rates of pumping or moulting may increase with consequent effects on life histories. Genetic variation that was of little consequence in the unpolluted environment may now distinguish survivors from non-survivors. Survival may depend on having alleles that increase relatively impermeable exterior membranes (e.g. Oppenoorth, 1985; Little *et al.*, 1989), more frequent moults (and consequent removal of toxicant in shed skin, e.g. Bengtsson *et al.*, 1985), a more comprehensive immune system or detoxication enzymes (Terriere, 1984; Oppenoorth, 1985), as described in Chapters 4, 5 and 7. Many examples are given in this book.

Figure 13.4 shows that if genes are expressed in polluted environments that before were silent or absent then evolutionary outcomes may

FIGURE 13.3 *The evolutionary outcomes of long-term mortality and production stresses. The straight lines are zero fitness contours.*

FIGURE 13.4 *If the trade-off curves in (A) polluted and (B) unpolluted environments have different shapes, then the evolutionary outcomes (starred) may involve more defence in polluted environments, as shown.*

differ between polluted and unpolluted environments and their populations may be genetically distinct. Such genetic differentiation can be documented and used experimentally in a **transplant experiment**. The method involves transplanting individuals from a number of environments to a common environment in which their life history components are then measured. Care must be taken that there are no maternal or other residual non-genetic effects carried over from the previous environment. For instance, mothers of measured individuals must be in equivalent condition because maternal quality can affect offspring performance ('maternal effect'). Since in a transplant experiment all individuals are assessed in a common environment, effects of source population reflect genetic differences. If the transplanted populations are genetically distinct, it is unlikely that the transplanted populations, carrying alleles that evolved in other environments, will be superior in the study environment to the population that evolved there. Hence, in each environment, the population that evolved there should outperform the others. This prediction has been confirmed by reciprocal transplant experiments (especially in plants) showing that resident populations outperform aliens (Sibly and Antovonics, 1992). Examples comparing resistant and susceptible strains in polluted and unpolluted environments are given in table 13.1. The three studies shown all indicate that, as

predicted, resistant strains are fitter in polluted environments and that susceptible strains are fitter in unpolluted environments. When this happens there is a **fitness cost of resistance**, meaning that resistant alleles, which are fitter in the polluted environment, are less fit than susceptibles in the unpolluted environment. Note, however, that whereas resistant strains are often much fitter than susceptibles in the polluted environments, the fitness of the resistants in the unpolluted environment is sometimes not much less than that of susceptibles (e.g. 0.23–0.62 in table 13.1). In general, the fitness costs of resistance depend on the resistance mechanism involved. Examples are discussed in the case studies at the end of this chapter. One experimental problem that arises in practice is that if one is interested in the fitness benefits and costs of a single allele then it is desirable to study the allele and its alternate on a common genetic background.

In general, then, chronic pollution results in a change in the shape of the genetic options set. Whereas before the environmental change the genetic options set may have been lined up on or alongside the trade-off curve and parallel to a fitness contour (figure 13.1F), after the environmental change, with genes expressed that before were silent, the genetic options set is quite likely to bulge out and to lie some way from the new trade-off curve. These predictions can be tested by measuring quantitative genetics parameters, which give some idea of the

TABLE 13.1 *Fitness advantage of resistant strains in environments with or without pesticides**†

Species	Pesticide	Fitness advantage with pesticide	Fitness advantage without pesticide
Mosquito (*Anopheles culifacies*)	DDT	0.34	−0.23
Mosquito (*Anopheles culifacies*)	Dieldrin	1.50	−0.49
Rat (*Rattus norvegicus*)	Warfarin	>>0	−0.62

*Fitness advantage = fitness$_{resistants}$ − fitness$_{susceptibles}$. Fitness cost is the negative of fitness advantage.
†Data from Bishop (1981) and Curtis *et al.* (1978).

shape of the genetic options set. The degree to which the points are spread out in figure 13.1F is measured by *additive genetic variance*. The prediction here is that additive genetic variance will increase if new genes are expressed in response to environmental pollution. The correlation between the points in figure 13.1F is measured by genetic correlation, and because in the unpolluted environment we argued that the genetic correlation would be tight (close to −1, figure 13.1F) any loosening of the correlation will move it away from ± 1.

Holloway *et al.* (1990) tested these predictions by transplanting a population of weevils from the environment in which it had evolved into a new toxin-rich environment. The new toxin-rich environment consisted of yellow split pea and the old environment of wheat, on which substrate the population had been maintained for 50 generations. Measurement of quantitative genetic parameters was achieved in a 'full-sib/half-sib' breeding experiment, using over 50 males and 250 females. This illustrates the gen-

eral point that quantitative genetic experiments are very demanding in terms of the numbers of animals needed. Because the population declined initially after transfer to the toxin-rich environment, genetic analysis could not be carried out until the fifth generation after transfer. The results are shown in figure 13.5. Additive genetic variance in development period and mortality rate increased as predicted after introduction to the new toxin-rich environment. Genetic variance in oviposition rate did not change. Changes in genetic correlation were quite large in two cases (shown in figure 13.5B). The genetic correlation between oviposition and juvenile mortality rate changed as predicted from a tight correlation close to +1 to a looser correlation of 0.4 in the toxin-rich environment (figure 13.5B). A correlation of +1, not −1 as in figure 13.1F, was predicted in the unpolluted environment because although the association between fitness and oviposition rate is positive that with mortality rate is negative. There was no change in the correlation of mortality and development period.

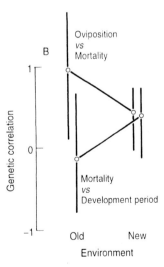

FIGURE 13.5 *Additive genetic variance in development period and mortality rate increased as predicted after introduction of rice weevils* Sitophilus oryzae *to a new toxin-rich environment, although the changes were not significant. Units are eggs day⁻¹ female⁻¹ (oviposition rate), 4 days (development period) or 4 × 10⁻⁵ (mortality rate). (B) Changes in genetic correlations. Vertical bars represent 95% confidence intervals. Reproduced from Holloway* et al. *(1990) with permission from Blackwell Science Ltd.*

This example shows how the genetic options set, measured by quantitative genetic parameters, may change when the environment becomes polluted. If the environment remains chronically polluted, selection will then favour alleles close to the trade-off curve in the neighbourhood of the new optimum. A new evolutionary process begins, but it is a process of the same type as that described in section 13.2.

13.5 *Resistance is often monogenic*

Alleles selected in polluted environments, with positive fitnesses, are said to be resistant, or, in some contexts, tolerant. Individuals that are homozygous resistant are referred to as resistant individuals. The relative resistance of heterozygotes measures the degree of dominance of the resistant allele. An example is shown in figure 13.6. Pesticide resistance often shows simple inheritance, with resistant genes being semidominant (i.e. heterozygotes half way between the homozygotes).

The degree of dominance affects the speed of spread of an allele, but not the final outcome of the evolutionary process (except when the allele is overdominant, i.e. the heterozygotes outperform both homozygotes). Advantageous dominant alleles spread faster initially than recessive alleles.

Complications arise if more than one locus is involved in resistance. To tell how many loci are involved, it is generally necessary to carry out a breeding experiment lasting several generations. For such experiments, homozygous strains are required. These are usually obtained by mass selection in the laboratory of field-collected strains. A breeding experiment starts by crossing a homozygous resistant strain with a homozygous susceptible strain. The offspring are necessarily heterozygous at all loci. These offspring are 'back-crossed' with the parental strains. Half the offspring of these back-crosses are expected to be heterozygous if only one locus is involved. On the other hand, if more than one locus is involved, less than half these offspring show resistance. The exact prediction depends on the degree(s) of dominance of the resistant allele(s). Statistical techniques are available for estimating the number of genes

FIGURE 13.6 *Dose–response curves for the mosquito* Culex quinquifasciatus *tested with permethrin (NRDC 167). Percentage mortality is plotted on a probit scale. SS shows the response of homozygous susceptible individuals, RS of heterozygotes and RR of homozygous resistant individuals. Reproduced from Taylor (1986) with permission from the Linnean Society of London.*

involved, but these make assumptions whose validity has to be checked. When discrimination between genotypes is difficult, as it can be in resistance studies, further experiments may be necessary. Repeated back-crossing can be useful. Genetic markers that map the positions of the resistant genes on the chromosomes have also proved particularly useful in difficult cases.

The general conclusion of studies of this type is that major genes (i.e. genes with large effects) are found in most cases of resistance, including resistance to insecticides, acaricides, fungicides, herbicides and heavy metals. Resistance is not always monogenic, but in most cases where resistance causes problems in pest control resistance appears to be largely controlled by one, or occasionally two, loci (Roush and Daly, 1990). Minor genes and modifier genes may, however, still have some small effects.

13.6 *Case studies*

Examples of evolutionary responses to pollution include some of the best known studies of the evolutionary process in the field. The reason for this is not hard to find. To study the evolutionary process it is usually necessary to find a population exposed to an environmental change, and in recent centuries most environmental changes have been effected by man. Many of these are cases of episodic or chronic pollution.

13.6.1 EVOLUTION OF PESTICIDE RESISTANCE

The evolutionary response to pesticides has been staggeringly varied and successful. The number of species in which resistance is known rose from 30 in the 1950s to over 450 in the 1980s. The account here is based on Taylor (1986), Mallet (1989) and especially Roush and Daly (1990).

Most insecticides now in use belong to one of four classes: (i) organochlorines, such as DDT, γ-HCH and dieldrin; (ii) organophosphorous compounds (OPs), such as malathion and dimethoate; (iii) carbamates, such as carbaryl; and (iv) synthetic pyrethroids, related to natural pyrethrins, and synthetics such as permethrin. Insects may be resistant to more than one insecticide, and often to insecticides in more than one class. When resistance to more than one insecticide is achieved by a single mechanism, this is true **cross resistance**, but when several resistance mechanisms are involved this is called **multiple resistance**. In some cases, the situation may be uncertain and these terms are then used more loosely.

All members of these four classes of insecticide act on the nervous system, although in different ways. The organochlorines and pyrethroids interfere with electrical conduction along the axons, whereas the OPs and carbamates act as inhibitors of acetylcholinesterase (Chapter 7).

The mechanisms by which insects have evolved resistance to these chemicals are summarized in table 13.2. Behavioural mechanisms include avoidance of an insecticide after low-level exposure to it. More important in practice are the various types of metabolic resistance that increase the rate at which the insecticide is broken down. A third type of resistance mechanism involves the animal becoming less sensitive to an insecticide. Often this is due to the site of action being insensitive to the chemical in the resistant strain (table 13.2). Last, the cuticle may be relatively impermeable to insecticides in resistant strains.

All of these resistance mechanisms could involve fitness costs in areas where there is no insecticide. Avoidance of certain microhabitats could result in reduced nutrient uptake, increased detoxication may be energetically expensive and reduced sensitivity or penetrance of insecticides may involve a costly reduction

TABLE 13.2 *Primary mechanisms of resistance**

Mechanisms	Insecticides affected
1. *Behavioural*	
Increased sensitivity to insecticide	DDT
Avoidance of treated microhabitats	Many
2. *Increased detoxication*	
DDT – dehydrochlorinase	DDT
Microsomal monooxygenase	Carbamates
	Pyrethroids
	Organophosphorous compounds
Glutathione transferase	Organophosphorous compounds
Hydrolases, esterases	Organophosphorous compounds
3. *Decreased sensitivity of target site*	
Decreased sensitivity of acetylcholinesterase	Organophosphorous compounds
	Carbamates
Decreased nerve sensitivity (K_{dr})	DDT
	Pyrethroids
Decreased nerve sensitivity (GABA receptor)	Dieldrin and other cyclodienes, γ-HCH
4. *Decreased cuticular penetration*	Most insecticides

*Modified from Taylor (1986).

in sensitivity or penetrance of other substances. Although the fitness costs of resistant alleles in untreated areas have rarely been studied, it is known that they can be large (table 13.1). Roush and Daly (1990) conclude that the most serious and consistent fitness costs are those associated with general esterases (e.g. carboxylesterases that hydrolyse naphthyl acetate). By contrast, in some studies little or no reproductive disadvantage has been found associated with malathion-specific carboxylesterases, increased oxidative detoxication, 'mutant' acetylcholinesterase and knock-down-resistance-like mechanisms.

Insecticide resistance is characterized by rapid evolution under strong selection. However, resistant alleles do not generally spread to fixation, completely displacing competitor alleles, in the field. This is probably because pesticide treatments are stopped when they become ineffective. However, inability to spread

to fixation would also result if resistant alleles have a fitness cost in untreated environments. Untreated populations mix with treated populations by dispersal. Immigration from untreated populations can considerably delay the increase of resistance, depending on the relative sizes of the treated and untreated populations and on the degree of migration.

One of the most spectacular examples of the evolution of insecticide resistance is that shown in figure 13.7. As part of a resistance management strategy, pyrethroid use was restricted to a short period ('stage II') each year in the middle of the cotton-growing season. Endosulfan use was permitted in stages I and II; other insecticides could be used at any time. Figure 13.7 shows that the proportion of individuals that were resistant rose each year during and immediately after pyrethroid application. The apparent rise immediately after pyrethroid application ceased is because of a time lag in measurement; affected individuals were not tested themselves, instead the test was carried out on their field-collected eggs. There was thus a delay while individuals matured and reproduced before the test was performed. After pyrethroid application the proportion of resistant individuals declined. This could be the result of a fitness cost of pyrethroid resistance, but it could also be because of immigration of susceptibles from adjoining untreated areas. Note that overall the frequency of pyrethroid resistance rose during the course of the 7-year programme.

13.6.2 EVOLUTION OF HEAVY METAL TOLERANCE IN PLANTS

The mining of heavy metals inevitably leads to pollution of the soils around mines, and spoil heaps are often rich in copper, zinc, lead or arsenic. The spoil heaps of Devon Great Consols, for example, which was the richest copper mine in Europe in the late nineteenth century, contain more than 1% copper and 5% arsenic in places. Such concentrations are highly toxic to most plants. Yet, plants can often be found growing on spoil heaps, as a result of the evo-

lution of metal tolerance (Macnair, 1987; Shaw, 1989; Schat and Bookum, 1992). These plants are genetically more tolerant than plants from neighbouring populations. Methods of quantifying metal tolerance in plants are described in section 6.3.4.

As an example, consider the wind-pollinated grass *Agrostis tenuis*, copper-tolerant genotypes of which grow on the spoil heaps of copper mines. Taking a transect through such a mine, as in figure 13.8, one finds copper tolerance on the copper-rich soils of the mine, but no tolerance outside it in the upwind direction. Some tolerance, is however, found immediately downwind as a result of pollen or seeds of tolerant parents being blown off the mine. Because tolerance is found on the mine but for the most part not off it, it is clear that tolerant alleles have increased in numbers on the mine. This is what is meant by saying that tolerant alleles are fitter in the polluted environment. Conversely, tolerant alleles seem to be outcompeted off the mine, indicating that there they are less fit In other words, there appears in this case to be a fitness cost of tolerance.

Figure 13.8 also contrasts the copper tolerance of individuals grown from seeds (white

FIGURE 13.8 *Copper tolerance in the grass* Agrostis tenuis *along a transect on the surface of a copper mine. The copper-impregnated part of the mine is shaded. Data of McNeilly (1968) redrawn by Macnair (1981), reproduced with permission from Academic Press.*

bars) with that of adults (black bars). Comparison of the two shows how selection works at different places on the transect. Thus, adults on the mine are more tolerant than seeds, indicating selection for tolerance on the mine. Conversely, downwind of the mine adults are less tolerant than seeds, indicating selection against tolerance off the mine.

It is generally believed that there is a fitness cost of tolerance, i.e. that tolerant genes are disadvantageous under normal conditions. If this were not the case, then tolerance genes would occur widely, at least in some species, given their competitive edge in polluted environments. However, there is disagreement as to the severity of the cost of tolerance. Some authors consider that they may be great (see Baker, 1987). Others suggest that they may be small on the basis of studies which carefully distinguish metal tolerance from other evolved attributes of mine populations (Macnair, 1987). For example, tolerant strains may be those which grow more successfully in the nutrient-poor conditions often found in mine waste (Ernst *et al.*, 1992).

Ernst (1976) suggested that the fitness costs of tolerance are a result of the energy needed by the mechanisms of tolerance. Energy spent on detoxication is not available for growth, and

Ernst suggested that this is why many tolerant plants grow more slowly and produce less biomass than non-tolerant conspecifics. This is the situation illustrated in figure 13.2A.

Why do some species evolve metal tolerance and others not? Presumably, similar selection pressures apply to all. Selection alone, however, is not enough, there must also be present some suitable 'tolerance genes' on which selection can act. Table 13.3 shows that normal populations of species able to evolve metal tolerance contain tolerance genes at low frequencies. Theory shows that whether or not tolerance genes exist in normal populations depends on mutation rate, i.e. the rate of creation of tolerance genes by mutation, on the fitness cost of tolerance and on the population size. Tolerance is more likely to evolve if the mutation rate is high, the fitness cost of tolerance is low and the species is common.

13.6.3 EVOLUTION OF INDUSTRIAL MELANISM

The blackening of industrialized parts of the countryside was a result of the Industrial Revolution which began in Britain in the eighteenth century. One result was that an area west of

TABLE 13.3 *The percentage of copper-tolerant individuals found in normal populations of various grass species that are commonly found near mines in Britain in relation to whether copper-tolerant populations of these species have been found on copper mines**

Species	Occurrence of tolerant individuals in normal populations (%)	Presence (+) or absence (–) of tolerant populations on copper mines
Holcus lanatus	0.16	+
Agrostis capillaris	0.13	+
Festuca ovina	0.07	–
Dactylis glomerata	0.05	+
Deschampsia flexuosa	0.03	+
Anthoxanthum odoratum	0.02	–
Festuca rubra	0.01	+
Lolium perenne	0.005	–
Poa pratensis	0.0	–
Poa trivialis	0.0	–
Phleum pratense	0.0	–
Cynosurus cristatus	0.0	–
Alopecurus pratensis	0.0	–
Bromus mollis	0.0	–
Arrhenatherum elatius	0.0	–

*Data of C. Ingram reported in Macnair (1987).

Birmingham became known as the Black Country. Blacker ('melanistic') forms of animals are better camouflaged in such environments and may thus be better protected against predators. This may give melanistic forms a survival advantage. Melanism is found in many species of arthropod, but most examples are in moths. The account here is based on Brakefield (1987) and Moriarty (1999).

The incidence of melanism has risen steadily since about 1850. Over 100 of the 780 species of larger moths in Britain now commonly include melanic forms. Similar changes have occurred in Europe and North America, usually in industrial areas. In many cases, the melanic forms have spread very rapidly and have become predominant.

The first and best studied example is the peppered moth, *Biston betularia* (Kettlewell, 1973). J. W. Tutt suggested in 1896 that the non-melanic form is well concealed ('cryptic') when at rest on the pale bark of trees in rural areas, where it resembles thalli of foliose lichens. By contrast, it was conspicuous on the completely black surfaces of trees and walls in industrial Britain. The blackening was mostly due to soot, but lichens were also killed, mainly by sulphur dioxide. Tutt suggested that in this environment the non-melanic form was conspicuous to avian predators, conversely melanic forms were cryptic.

The first melanic specimen was caught in Manchester in 1848, and by 1895 98% of the moths in that area were melanic. This corresponds to a 50% increase in the chances of survival of the melanic forms, according to calculations by J. B. S. Haldane (see Brakefield, 1987). Increases in melanic forms were recorded during this period in a number of industrialized zones. The phenomenon has been of major interest to evolutionary biologists since the 1920s.

Kettlewell organized surveys between 1952 and 1970 of the geographical distribution of melanism in *B. betularia* in Britain (figure 13.9). Three forms of the moth were recognized. The non-melanistic form, *typica*, shows a range of coloration from heavily speckled individuals to others that are almost white with fine, granular black markings. The common melanic form, *carbonaria*, produced by a dominant allele at a single locus, is usually all black, but sometimes has some light-coloured spots or patches. A third intermediate form, *insularia*, is also recognized. Figure 13.9 shows that in general the melanic forms occurred in the industrialized regions of Britain and in the areas downwind (the predominant winds in Britain are from the south-west). Kettlewell suggested that the downwind occurrence of the melanistic forms might be because soot had blackened the downwind areas, but it could also be the result of passive wind dispersal of larvae. When eggs hatch, the very small larvae suspend themselves on silk threads and are dispersed by air currents. It is possible that the high incidence of melanism in East Anglia is the result of long-distance dispersal of larvae from industrial zones in London and the English Midlands.

Interestingly, there has been a further environmental change in recent yeas consequent on

FIGURE 13.9 *The relative frequencies of the normal and two melanic forms of the peppered moth* Biston betularia *in Britain. The results are based on more than 30 000 records collected from 1952 to 1970 at 83 sites. Reproduced from Kettlewell (1973) with permission from Oxford University Press.*

clean air legislation and the creation of smoke-less zones in the late 1950s. This led to a rapid fall in emissions of smoke and sulphur dioxide. Tree surfaces became lighter and the relative frequency of melanics fell (figure 13.10). This phenomenon has been described as 'evolution in reverse'! The time lag between the reduction in pollution and the evolutionary response is probably a result of slow change in the composition of the lichens covering the tree branches on which the moths rest (Cook *et al.*, 1999).

Although it is generally accepted that melanism in *Biston betularia* is a result of atmospheric pollution, the detailed working of the selective mechanisms is still not fully understood. The fitness advantages of the melanic form in industrial areas and of the non-melanic forms in rural areas were demonstrated by Kettlewell in mark–release–recapture experiments. Marked individuals of melanic and normal forms were released in rural and industrial areas (table 13.4). More melanic individuals were recaptured in the industrial area, and more non-melanic in the rural areas.

To go further and to estimate the predation rate by birds it is necessary to have a detailed knowledge of the time budgets of adult moths of different ages and both sexes during the reproductive period. Peppered moths emerge shortly before dusk. After a dispersal flight, females rapidly attract mates and pair shortly after dusk. Thereafter, they do not fly and only walk short distances. They remain *in copula* for nearly 24 h. Many moths rest during the day underneath or on the side of small branches in the tree canopy. Studies of visual predation on the melanic and non-melanic forms have however generally been carried out on tree trunks. Although the findings are qualitatively in keeping with the camouflage protection hypothesis, the experiments need repeating on more realistic substrates.

13.6.4 EVOLUTIONARY RESPONSE OF DOG WHELKS, *NUCELLA LAPILLUS*, TO TBT CONTAMINATION

The use of dog whelks as biological monitors of TBT pollution is described in Chapter 15. Briefly, female dog whelks suffer imposex, a condition involving the growth of male organs which can prevent successful reproduction.

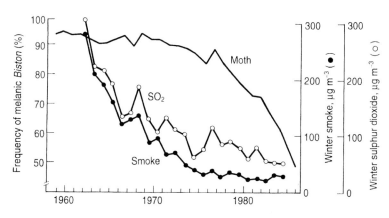

FIGURE 13.10 *The decline in the frequency of the melanic form of the peppered moth near Manchester after clean air legislation reduced emissions of smoke and sulphur dioxide. Reproduced from Brakefield (1987) with permission from Elsevier Trends Journals.*

TABLE 13.4 *The relative recoveries of marked individuals of two forms of the peppered moth,* Biston betularia *(typica and* carbonaria), *from two sites, one rural and one industrial**

Site	Form	No. released	No. recaptured	Recapture (%)
Rural Dorset	typica	496	62	12.5
	carbonaria	473	30	6.3
Industrial Birmingham	typica	201	32	15.9
	carbonaria	601	205	34.1

*From Moriarty (1999), presenting data from Kettlewell (1955, 1956).

Because migration is very limited in this species, population declines have been reported in many areas adjacent to harbours. One such area is the north coast of Kent, where dog whelks are known to have been abundant in pre-TBT times. Recently, however, Gibbs (1993) discovered that a population in this area at Dumpton Gap had evolved modified genitalia that allowed it to persist. The account here is based on Gibbs's paper.

The Dumpton Gap population is characterized by the absence of a penis or an undersized penis in about 10% of the males (absence of a penis in males is unknown elsewhere). The vas deferens and prostate are incompletely developed in affected males. Gibbs has labelled these abnormalities the 'Dumpton syndrome'. Laboratory-bred animals display the same characteristics, suggesting that the character is genetically determined. It cannot be due to some aberrant feature of the Dumpton environment such as predators or parasites as neither predators nor trematode parasites were present in the laboratory. About 75% of the Dumpton Gap females showed little or no imposex, whereas usually all females are expected to show imposex at the TBT levels experienced at Dumpton Gap. This phenomenon has also recently been reported in France (Huet et al., 1996).

The evidence therefore suggests that the 'Dumpton syndrome' is the result of a genetic mutation which reduces imposex in females,

allowing them to breed successfully even though it prevents breeding in some 10% of males. Overall, the gene must confer a marked fecundity advantage in environments contaminated with TBT. In TBT-free environments, however, the gene would be disadvantageous because a significant proportion of males are infertile (compare table 13.1). It seems, therefore, that the 'Dumpton syndrome' carries a fitness cost, and so would be selected against in TBT-free environments.

13.7 *Summary*

The case studies presented above are good examples of the evolutionary process because they report evolutionary responses to known environmental changes. In each case, the change is chronic man-made pollution which started at most a few centuries ago. For evolutionary responses to occur, genetic variation has to be present on which selection can act. Evolutionary responses occur when the change to the environment alters the relative fitnesses of different alleles. In each case, there was strong selection for resistant alleles in polluted environments, and evolutionary responses to pollution occurred within tens of generations. Where the genetic mechanisms have been investigated, they consist for the most part of one or occasionally two major genes. Although it is extraordinarily hard to study, it seems likely that resistance generally entails a fitness cost. In other words, resistant alleles are favoured at polluted sites, but selected against at unpolluted sites. Last, it should be emphasized that although our outline understanding seems secure, in each case there is still some uncertainty as to the detailed working of the evolutionary process.

13.8 *Further reading*

The account of the evolutionary process given here is based on Sibly and Antonovics (1992). This approach was selected because it is accessible and gives insight into how selection acts on life history characters such as juvenile growth rate, fecundity and survival. Those wanting a population genetic text should consult Hartl and Clark (1989); Bishop and Cook (1981) remains a useful review of the genetic consequences of man-made environmental changes.

FORBES, V. E. (ed.) (1999) *Genetics and Ecotoxicology*. Contains a number of case studies of adaptations to polluted environments.

MACNAIR, M. R. (1987) Gives a short but useful introduction to the evolution of heavy metal tolerance in plants.

MAJERUS, M. (1998) *Melanism: Evolution in Action*. Contains an extended discussion of the peppered moth story.

MALLET, J. (1989) Gives a brief but accessible introduction to the evolution of pesticide resistance.

ROUSH, R. T. and DALY, D. C. (1990) Provides a heavyweight authoritative view of resistance research, invaluable to anyone starting work on any aspect of resistance.

SHAW, A. J. (1999) Provides a recent review of this field.

CHAPTER 14

Changes in communities and ecosystems

14.1 Introduction

In the two preceding chapters, effects upon individual species were given particular attention. This is the approach usually followed when dealing with larger species. By contrast, microbiologists are particularly concerned with effects on communities and ecosystems, a subject area sometimes termed 'synecology'. The measurement of effects of pollutants upon ecosystems has certain strengths and limitations. It has the advantage of being a holistic approach, which can take into account the overall functional state of an ecosystem (Freedman, 1989).

There are two distinct approaches to studying changes in communities or ecosystems. One is structural, the other functional. Structural changes relate to changes in composition. In the extreme case, species may disappear altogether from communities and ecosystems in which they are usually found. Examples include the disappearance of the dog whelk from many coastal areas of southern England because of the effects of TBT (section 15.3) and the disap-

pearance of the sparrowhawk from large areas of eastern England because of the effects of organochlorine insecticides (see Chapter 12). Changes as severe as these indicate a reduction in biological diversity. More commonly, however, changes are only shifts in the balance of species in defined habitats. Changes on this scale may be evident from the use of biotic indices such as RIVPACS (section 11.2.2). Monitoring such changes provides an early warning system; it may give an early indication that pollution is causing changes that will eventually cause a serious reduction in biological diversity if remedial action is not taken.

In contrast to these indications of structural changes in community composition, a functional approach can provide a simple measure of the state of a community or ecosystem. It is relatively straightforward to measure the operation of the carbon cycle or the nitrogen cycle in soils and to determine whether it is adversely affected by pollutants. Although such a holistic approach gives no information on the status of individuals in a community or ecosystem, there are advantages in combining it with a structural approach. It makes sense to relate structure to function in communities as well as in individual organisms!

An ecosystem approach has been used with some success to study the effects of pollutants upon soils. The features of soil communities that make them amenable to scientific study are that (i) the boundaries of communities can be easily and clearly defined and (ii) similar communities exist in vast numbers in the field. These features make it possible to investigate the factors determining their field distribution and to carry out field experiments.

These useful attributes of soil communities are not shared with most larger-scale communities and ecosystems. Exceptions are the aquatic communities inhabiting lakes and rivers. Lakes exist in enormous numbers in Canada, Siberia and Scandinavia. The major types of pollution are acidification and heavy metal pollution, and their effects can be readily observed in terms of loss of species and diversity (see section 14.3). Field experiments are in principle a possibility, although because of their scale they are expensive.

Other types of terrestrial and aquatic communities are not so amenable to study. This may be because either their constituent species are highly mobile, so that their boundaries are hard to define, or because no two pollution events are the same. For instance, severe oil pollution rarely affects the same type of community twice, and the boundaries of the affected marine communities are ill-defined. Similar consideration applies to air pollution or to radioactive pollution as a result of nuclear warfare or accidents at nuclear power stations.

In the extreme case of an ecosystem analysis, the composition of the atmosphere can be taken as an index of the state of health of the entire planet. The recent increase in CO_2 levels in the atmosphere gives evidence of pollution on the global scale. This presents the threat of global warming because of the so-called 'glass-house' effect. Likewise, the disappearance of the ozone layer above Antarctica has been attributed to the appearance of CFC gases in the stratosphere. Reduction of the ozone layer brings an increase in ultraviolet radiation reaching the earth's surface and a consequent threat of cellular damage to living organisms (e.g. skin cancer in humans). In both of these examples there is a suggestion that ecological damage on the global scale may result from changes in natural processes caused by pollutants – namely the carbon cycle and the oxygen–ozone cycle. This view of pollution in terms of 'global' ecology is discussed by Lovelock in his elaboration of the 'Gaia' theory (see section 14.4).

14.2 *Soil processes: the functional approach*

Soil communities are complex associations between a variety of micro- and macroorganisms, minerals and non-living organic materials (refer to Chapter 5). The carbon cycle and the nitrogen cycle operate in soils and both provide indices of the function of soil communities. Stages in the cycles, e.g. CO_2 production (carbon cycle) and nitrification (nitrogen cycle), can be measured in the presence and absence of pollutants.

The basic carbon cycle, shown in figure 14.1, is an example of a nutrient cycle. Organisms which obtain their carbon from organic compounds are termed 'heterotrophs'. Animals and many microorganisms fall into this category. Organisms which obtain their carbon from carbon dioxide are termed 'autotrophs'. Autotrophs include green plants, green algae and certain bacteria and play a vital role in fixing atmospheric carbon dioxide into organic compounds. The heterotrophs then complete the cycle by converting organic compounds to carbon dioxide. Thus, the operation of the cy-

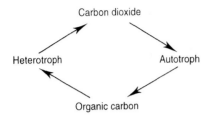

FIGURE 14.1 *The carbon cycle.*

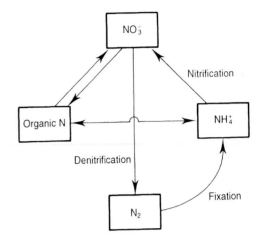

FIGURE 14.2 *The nitrogen cycle.*

cle may be affected by the action of pollutants upon autotrophs, heterotrophs or both.

A simplified version of the nitrogen cycle in soil is shown in figure 14.2. Atmospheric nitrogen can be 'fixed' by certain bacteria, including both freeliving species (e.g. *Azotobacter* spp.) and symbiotic bacteria (e.g. *Rhizobium* spp., which are found in the root nodules of leguminous plants). Fixation involves conversion of molecular nitrogen to ammonia, which then forms ammonium ions (NH_4^+) in soil water. NH_4^+ can be oxidized sequentially to nitrite ions (NO_2^-) and nitrate ions (NO_3^-) by soil bacteria (nitrification). Plants and bacteria can take up NH_4^+ and/or NO_3^- and use them for the biosynthesis of organic nitrogen compounds (e.g. amino acids and purines). Heterotrophs can then utilize organic nitrogen compounds as sources of nitrogen and can release NH_4^+ and NO_3^- from them. Finally, some microorganisms can convert NO_3^- to nitric oxide (NO), nitrous oxide (N_2O) and nitrogen, a process termed denitrification.

Microorganisms have an important role in the operation of the carbon and nitrogen cycles. Under normal environmental conditions, these cycles are subject to the influence of environmental factors such as temperature, pH and soil water content. The degree to which these environmental factors vary in time and space is dependent upon geographical location and the inherent properties of the soil (e.g. clay content, organic matter and $CaCO_3$ content). In evaluating the effects that pollutants may

have on soil processes, it is important to see them in relation to the fluctuations that occur in unpolluted sites. Effects of pollutants can be regarded as serious and significant if they go beyond the normal variations in operation of the cycles.

A widely used method of estimating hazards of chemicals to microorganisms involves determination of the *rate of carbon dioxide formation* in soil samples. Usually, a source of carbon (e.g. plant or horn meal) is added to soil to stimulate CO_2 formation, and a comparison is made between soils with and without the test chemical (figure 14.3).

The rate of production of nitrite and nitrate from soil ammonium (*nitrification*) can also be readily measured (table 14.1). This is regarded as a valuable method of testing the effects of pollutants because of the importance of nitrification in relation to soil fertility. A further example is measurement of the rate of nitrogen fixation by root nodule bacteria in a leguminous crop, which may be affected by soil pollutants.

Another factor to be considered is the *duration* of any effect produced by a chemical. Pollutants can change the composition of communities of soil microorganisms (Chapter

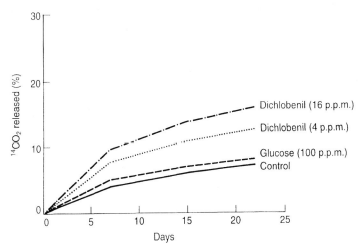

FIGURE 14.3 *Effect of a herbicide (dichlobenil) on the rate of CO_2 production in soil. {^{14}C}-Glucose was added to soil as a carbon source for microorganisms. The addition of the herbicide dichlobenil caused an increase in the rate of release of $^{14}CO_2$ (derived from {^{14}C}-glucose) over a 22-day period. The rate of $^{14}CO_2$ release from soil is expressed as the percentage added ^{14}C which appears in this form. From Somerville and Greaves (1987).*

5). When a chemical has an immediate effect on soil processes as a result of chemical toxicity, there may be a subsequent population growth in the species/strains of microorganisms which can metabolize it and use it as a nutrient source. Thus, the effects of an organic chemical will be relatively short lived because the increase in numbers of these microorganisms will lead to a more rapid breakdown of the compound in soil. It has been suggested that effects of chemicals lasting up to 30 days be regarded as 'normal' (1), up to 60 days as 'tolerable' (2) and beyond 60 days as critical (3). In reviewing effects of pesticide on soil microflora, some 90% of all cases fell into the first category.

A general problem when performing soil tests to evaluate the effects of chemicals is choosing the type of soil to use and the appropriate operating conditions. Attempts have been made to define standard soils – but the trouble here is that these are only representative of particular geographical and climatic areas. Also, a standard soil may not represent a 'worst case' scenario – it may not be the soil most likely to show the effect of a particular chemical.

Soil tests of the type described are of par-ticular interest with regard to the testing of pesticides. Herbicides, fungicides and insecticides are chemicals with high biological activity which may be expected to have effects on soil organisms and consequently upon soil fertility.

14.3 *Changes in communities in response to pollution*

Responses of communities to pollution in terrestrial and aquatic ecosystems have been covered briefly in Chapter 11 as part of the discussion on type 1 *in situ* biological monitoring (see section 11.2 for an introduction to some of the principles involved). In fresh waters, methods such as the river invertebrate prediction and classification scheme (RIVPACS) are well established. In marine systems, studies such as those by Clarke and Warwick (1998) and Warwick and Clarke (1998) have shown how a variety of indices (in their case a 'taxonomic distinctiveness index' utilizing nematodes) can be used to assess the 'health' of sediments af-

TABLE 14.1 *Effects of pollutants upon nitrification in soil**†

Treatment	mg NO$_3^+$ N day^{-1}	% of control
Control	0.53	100
5 × normal application rate of a pesticide	0.54	102
1 p.p.m. Nitrapyrin	0.14	26

*Horn meal was added to the soil as a nitrogen source. The rate of production of nitrate was unaffected by a high application of a pesticide. It was, however, strongly inhibited by nitrapyrin, a bactericide, used as a positive control.
†From Somerville and Greaves (1987).

fected by pollution. The marine planktonic community is more difficult to monitor; however, some success has been achieved by maintaining organisms within outdoor enclosures. Jak *et al.* (1998) were able to show that tributyl tin (TBT) reduced grazing activities of zooplankton (principally copepods) resulting in an algal 'bloom'.

'Before and after' studies are rare in ecotoxicology because it is often impossible to predict where a pollution event is likely to occur. Furthermore, it is difficult to get long-term funding for basic ecological monitoring. Environmental impact assessment of, for example, the site of a new industrial development may have to rely on a brief survey rather than a more extensive data set gathered over several seasons. Many of the most successful 'pollution and recovery' examples rely on data collected by natural historians which give a baseline with which the impact can be compared (e.g. the status of dog whelks before TBT, eggshells before DDT, lichens before air pollution).

Fortunately, enough of such data existed to enable accurate assessment of the environmental impact of the *Sea Empress*, a tanker which spilled 70 000 ton of crude oil into Milford Haven, Wales, in February 1996 (Crump *et al.*, 1999). Permanent quadrats had been established before the spill on rocky shores which became heavily smothered with oil. These were not sprayed with detergents and were allowed to clean up naturally. Normally, percentage cover of the rocks with seaweed is low because

of the grazing activities of **limpets** (mainly *Patella vulgata*). After the spill, limpet mortalities were very high and resulted in a dramatic green phase of the seaweeds *Enteromorpha* followed by a flush of *Porphyra* and ultimately a brown phase of *Fucus*. Limpets subsequently recolonized the rocks (probably recruited from small individuals which survived the pollution in deep crevices) and will eventually restore the natural ecological balance by renewed grazing.

Similar examples for soil ecosystems are much harder to come by. One of the reasons for this discrepancy is the lack of user-friendly identification keys for some of the most speciose groups (e.g. mites, enchytraeid worms). Nevertheless, progress is being made towards a soil prediction and classification scheme ('SOILPACS'), although it will be some years before this can be considered as routine as RIVPACS and the like. In the remainder of this section, examples will be given of a few of the methods which have been used to assess the impact of aerial deposition of metals on soil and leaf litter communities. It does not pretend to be comprehensive; further details can be found in the excellent reviews of Posthuma (1997), Van Straalen (1997, 1998) and Van Straalen and Kammenga (1998).

One of the characteristic features of terrestrial ecosystems contaminated by aerial deposition of metals is an accumulation of undecomposed **leaf litter**. This is attributed to a reduction in the populations of invertebrates which feed on dead plant material. Their faeces provide a more favourable substrate for microbial decay than intact leaves. Oligochaetes are particularly susceptible to the pollution as the moist body surface allows metals dissolved in the soil pore water to diffuse into their body tissues. In coniferous forests, enchytraeids are usually the dominant macroinvertebrates. However, in the vicinity of a copper–nickel smelter in Finland, numbers of **enchytraeids** were substantially lower than in more distant sites and there was a much thicker layer of undecomposed pine needles on the soil surface (Haimi and Siira-Pietikainen, 1996).

In temperate grassland and deciduous woodland, lumbricid **earthworms** dominate the invertebrate decomposer community in terms of biomass and consumption of leaf litter. At Avonmouth, UK, deposition of aerial emissions of cadmium, copper, lead and zinc from a smelting works have heavily contaminated soils in the vicinity of the factory (for further details, see Hopkin, 1989). About 600 m from the plant (sites 1 and 2 on figures 14.4 and 14.5; table 14.2), dead grass lying on the soil surface contains more than 2% **lead** and nearly 4% **zinc** on a dry weight basis. Detailed sampling by Spurgeon and Hopkin (1999) during the spring, summer, autumn and winter revealed a complete lack of earthworms from the two closest sites (< 0.6 km) and significantly reduced numbers compared with controls at a further five sites (< 3 km) (figure 14.4). Species richness was lowest at sites near to the factory (table 14.2). Worms such as *Aporrectodea caliginosa* and *Allolobophora chlorotica* that were dominant at relatively clean sites further from the smelter were absent from the most contaminated soils. **Multivariate cluster analysis** (figure 14.5) indicated that sites could be split into three groups based on relative species composition. Studies such as these provide strong circumstantial evidence that the accumulation of **organic material** (and the concomitant disruption of nutrient

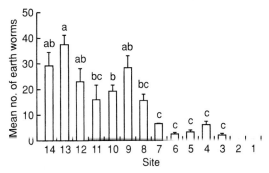

FIGURE 14.4 *Mean abundance of earthworms collected from six 25 × 25 cm quadrats taken at 13 sites along a gradient of contamination in the Avonmouth area and a control (site 14) 100 km from the smelting works (error bars indicate SE values) in April 1996. Worms are absent from the most heavily contaminated sites (1 and 2), and are present in reduced numbers at sites with a medium level of contamination (sites 3–7) in comparison with relatively uncontaminated localities (sites 8–14). Sites sharing the same letter indicate no significant differences at P > 0.05 as given by Tukey's test for the multiple comparison of means. Reproduced from Spurgeon and Hopkin (1999) with permission from the British Ecological Society.*

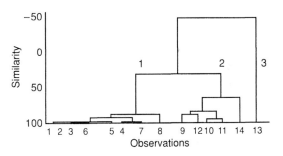

FIGURE 14.5 *Dendrograms of earthworm communities at Avonmouth sampled in April 1996 from the sites described in figure 14.4 ordered by cluster analysis using Euclidean distance and Ward's minimum variance method. Reproduced from Spurgeon and Hopkin (1999) with permission from the British Ecological Society.*

cycling) is related to the absence of key invertebrates which act as decomposition 'catalysts'.

According to Van Straalen (1997), a combination of ecophysiology (i.e. understanding the biology of the organisms involved) and

TABLE 14.2 *Shannon–Weiner diversity indices of earthworm communities at Avonmouth sampled during the period from spring 1996 to winter 1997 from the sites described in figure 14.4*

Site no.	Shannon–Weiner diversity			
	Spring	**Summer**	**Autumn**	**Winter**
14	1.83	1.55	1.37	1.56
13	1.1	1.46	1.22	1.76
12	1.65	0	1.38	1.69
11	1.46	1.3	1.56	1.68
10	1.45	0.9	1.13	1.56
9	1.66	1.39	1.41	1.53
8	1.95	1.36	1.44	1.49
7	0.31	0	0.67	0.93
6	1.03	1	0.83	0.24
5	0.80	0	0	0.87
4	0.88	0	0.68	0.85
3	0.64	0	0	0.64

Note that the diversity is invariably lower in sites closest to the smelting works (sites 3–7) in all seasons than in more distant sites (sites 8–14). Reproduced from Spurgeon and Hopkin (1999) with permission of the British Ecological Society.

multivariate statistics holds greatest promise for the future development of a 'SOILPACS' technique. At Avonmouth, this approach has already been adopted in a study of ground-running invertebrates (Read *et al.*, 1998). Principal components analysis and canonical correspondence analysis were used to show that the dominant factor accounting for differences in the species composition of woodlands was the degree of contamination with metals.

14.3.1 ACIDIFICATION OF LAKES AND RIVERS

An example of changes in a community caused by pollution is given by the consequences of the acidification of certain lakes as a result of acid rain. The pH of surface waters is determined by both abiotic and biotic factors. Such factors as the composition of the base rock and the precipitation of acids in rain or snow have a strong influence upon pH. Also important are the release of acids when organic residues are decomposed by microorganisms and the influence of neighbouring forest land. pH, in turn, influences the composition of aquatic communities. In particular, reduction of pH below 6 can be harmful to many species. Thus, the pH of surface waters is affected both directly *and indirectly* by pollutants (pollutants may affect pH of surface waters indirectly through action upon microorganisms or plants). In summary, the pH of surface waters is dependent upon the operation of natural processes, as well as pollution, and provides an indication of the diversity and 'health' of aquatic ecosystems (Barth, 1987).

The deposition of acid rain has resulted in the acidification of weakly buffered surface waters in many areas, including Scandinavia, eastern Canada and north-eastern USA. For example, the pH of 21 water bodies in central Norway decreased from an average of 7.5 to 5.4–6.3 between 1941 and the early 1970s; the pH of 14 surface waters in south-western Sweden decreased from 6.5–6.6 before 1950 to 5.4–

5.6 in 1971; the average pH of seven rivers in Nova Scotia decreased from 5.7 in 1954–5 to 4.9 in 1973. Although doubts have been expressed about the accuracy of older pH data, there is a broad scientific consensus that there has been recent acidification of many surface waters. The material in this section is based in part on Freedman (1995).

The effects of acidification on aquatic communities are sometimes dramatic, as in the loss of fish populations from a number of highly acidified waters. For instance, commercially valuable salmonids have been lost from many surface waters in Scandinavia (example in figure 14.6). A survey of more than 2000 lakes in southern Norway showed that one-third have lost their fish populations since the 1940s. The fishless lakes generally had a pH < 5.0. The local salmonids vary in their sensitivity to pH; requirements range from pH > 4.5 for non-

FIGURE 14.6 *The declining catch of Atlantic salmon in acidified rivers (– – –) compared with less polluted rivers (——). (A) In southernmost Norway: seven acidified rivers compared with the rest of the country (modified from Leivestad* et al. *(1976). (B) In Nova Scotia: 12 rivers with pH > 5.0 compared with 10 with pH ≤ 5.0 in 1980 (from Watt* et al. *1983). Reproduced with permission from the National Research Council of Canada.*

migratory brown trout (*Salmo trutta*) to 5.0–5.5 for Atlantic salmon (*Salmo salar*). Juveniles are generally more sensitive than adults, but adult deaths are sometimes recorded in the spring when water pH is lowest. Similar results involving more species have been recorded in many studies in Canada and north-eastern USA.

Field experiments are possible with lakes, just as they are with soil communities, but the costs are greater. The best example is the experimental acidification of lake 223, a 27-ha oligotrophic lake in the 'experimental lakes area' of north-western Ontario. The lake was extensively studied in 1974 and 1975 to provide baseline data, and then the lake was progressively acidified so that pH fell from 6.5–6.8 to 5.0–5.1 by 1981, after which pH was maintained in the range 5.0–5.1 until 1983. This was achieved by adding sulphuric acid, 27 400 l had been added by 1983. Experimental acidification has subsequently been carried out at two other lakes, one in the experimental lakes area, the other in northern Wisconsin. These lakes have different physical settings and originally had different assemblages of plants and animals, including fish.

Despite their initial differences, the responses of these three lakes to acidification were in many respects remarkably similar, particularly with regard to biogeochemical processes and effects on the lower trophic levels (Schindler *et al.*, 1991). Although acidification disrupted nitrogen cycling in all three lakes, each generated some buffering capacity internally.

The phytoplankton communities of all three lakes had originally been dominated by chrysophyceans and cryptophyceans. Acidification changed the dominant species and decreased diversity. Phytoplankton production and standing crop were somewhat increased, probably because light penetration increased. The littoral zones of the lakes, however, became dominated by a few species of filamentous green algae. These formed mats or clouds of algae which changed the entire character of the littoral zones when the pH fell below 5.6.

Acidification also changed the zooplankton communities, with cladocerans becoming increasingly dominant. As acidity increased, *Daphnia catawba* took over from other *Daphnia* species; the identity of the dominant rotifer species changed, and several sensitive zooplankton species disappeared.

Acidification also caused the loss of several species of large benthic crustaceans. The responses of fish species, however, were varied and appeared to depend on the sensitivity of key organisms lower down the food chains.

When acidification was reversed in lake 223 between 1984 and 1989, there was rapid partial recovery of the lacustrine communities, although the assemblages of phytoplankton and chironomids retained an acidophilic character.

Overall, acidification consistently reduced diversity, through loss of species and through the increased dominance of a small number of acidophilic taxa. These responses are similar to those found in atmospherically acidified lakes.

14.4 *Global processes*

The atmosphere enshrouds the whole planet and provides a link between oceans and continents. The air masses that constitute the lower atmosphere are in a constant state of circulation, individual gaseous components moving by diffusion processes through the air masses (Chapter 3). The composition of the atmosphere is influenced by living processes at the earth's surface. Levels of carbon dioxide, oxygen and nitrogen are dependent upon living processes. Thus, the effects of pollutants on ecosystems can, at least in theory, influence the operation of these processes globally and thus alter composition of the atmosphere. Some scientists conceive of the entire Earth as being a single living entity (as, for instance, in the Gaia theory of Lovelock).

Looking at things in this way, pollutant effects on living organisms may be direct or indirect. Inhibitors of photosynthesis are examples of pollutants which act directly on plants, slowing down the use of CO_2 and the release of O_2. The action of CFC gases is indirect. Because of a reduction of the ozone layer, more ultraviolet radiation reaches the Earth, causing damage to animals and plants.

Over the last 40+ years, the level of CO_2 in the atmosphere has increased by *c.* 10%, and there is continuing debate about the causes and the environmental effects. It is thought that most of this increase is due to the combustion of fossil fuels (coal, oil, etc.). It has also been argued that a substantial reduction in the area of forest land has also contributed to this – because trees are responsible for 'locking up' considerable quantities of the gas. The concern over the rise of atmospheric CO_2 is that it is having a 'glasshouse effect' – retaining more heat at the earth's surface and thereby causing global warming.

The issues raised here are long-term ones which are unlikely to go away. The effects of pollutants on the global scale can influence the state of ecosystems over the entire planet and are likely to be of increasing concern throughout this century.

14.5 *Summary*

Earlier chapters of this book have considered the biochemical, physiological and behavioural effects of pollutants on individual organisms. In the Introduction, we defined 'ecotoxicology' as 'the study of harmful effects of chemicals on *ecosystems*'. Thus, Chapter 14 examines the most important aspect of ecotoxicology, the manifestation of the damaging effects of pollution on communities of organisms, and wider eco-logical processes. Absence of sensitive species may not just result in lower overall biodiversity but also in disruption of essential ecological processes. For example, the reduction in populations of earthworms in soils contaminated with metals may lead to an accumulation of undecomposed leaf litter. This would normally be converted rapidly to faeces for more rapid microbial decay. Nutrients are locked up in the dead leaves which would otherwise be returned to the soil for new plant growth. In the aquatic environment, acidification of lakes results in lower overall diversity, although there are a few species which can take advantage of the reduced competition to reach higher population densities than in unacidified waters. One of the themes which runs throughout such studies is the need for long-term monitoring of natural uncontaminated ecosystems, and a good understanding of the biology of organisms, if we are to be able to recognize when ecosystems depart from their normal state.

14.6 *Further reading*

FREEDMAN, B. (1995) *Environmental Ecology*. Excellent coverage of the effects of pollution on ecosystems, and, in particular, useful for the effects of acidification on aquatic ecosystems.

LOVELOCK, J. (1989) *The Ages of Gaia*. An exposition of the Gaia hypothesis.

NEWMAN, M. C. (1998) *Fundamentals of Ecotoxicology*. Useful text on ecological effects of pollutants.

PANKHURST, C. E. *et al.* (eds) (1997) *Biological Indicators of Soil Health*, and VAN STRAALEN, N. and LØKKE, H. (eds) (1997) *Ecological Risk Assessment of Contaminants in Soil*. Include several useful chapters concerning effects of pollutants on soil communities.

SCHINDLER, D. W. *et al.* (1991) A review of the experimental acidification of lakes.

WOOD, M. (1995) *Environmental Soil Biology*, 2nd edn. Gives an account of the soil processes that may be affected by pollutants.

15

Biomarkers in population studies

It could be argued that the most crucial task for ecotoxicologists is to ensure that the structure and the function of ecosystems are preserved. It is also the most difficult. The linkages between biochemical, physiological, individual, population and community responses to pollutants are shown diagrammatically in the Introduction. The dilemma is that as the importance of a change increases so does the difficulty of measuring it and relating it to a specific cause. In this chapter, a number of specific examples will be considered to illustrate the approaches that have been used to tackle this important problem of ecotoxicology, involving the use of biomarkers. These examples are as follows:

1. DDE-induced eggshell thinning in raptorial and fish-eating birds;
2. reproductive failure of fish-eating birds on the Great Lakes of North America;
3. reproductive failure of molluscs caused by tributyl tin;
4. the forest spray programmes of eastern Canada.

The first three are investigations of unexpected adverse side-effects of the use of pesticides and/or industrial chemicals, and the fourth is an investigation of a large-scale operational use of pesticides.

15.1 DDE-induced eggshell thinning in raptorial and fish-eating birds

In the early 1950s, Derek Ratcliffe of the British Nature Conservancy found what he considered to be an abnormal number of broken eggs in the eyries of the peregrine in several regions of the British Isles. Later in that decade, representations were made to the British Home Office by racing pigeon enthusiasts concerning the losses caused by peregrines preying on their birds. These persons were lobbying for a change in the protected status of the peregrine because they claimed that the peregrine population had greatly increased. As a result of this pressure, the Nature Conservancy was asked to undertake a nationwide survey. This revealed that the peregrine population had declined greatly and that less than one-fifth of the birds successfully raised young in 1961 and 1962 (see figure 12.15).

This paper had an immediate impact in the USA. A team surveyed peregrines in eastern North America, travelling 22 500 km and visiting 133 known eyries in 1964 and found every single eyrie deserted. This finding was the spur for a meeting to examine the population changes of the peregrine and other birds of prey. The meeting was held at Madison, Wisconsin, in 1965 (Hickey, 1969) and was attended by persons from many parts of the world interested in the peregrine. The data presented at the meeting showed that the species was extirpated in eastern North America and had decreased markedly in many countries in Europe. In Scandinavia, the population was reduced to only 5% of its pre-war population by the early 1970s. In Germany, the number of breeding peregrines had decreased from 400 pairs to 40 pairs by 1975.

Based on the finding that egg breakage had become more common, it was likely that the thickness of eggshells had become thinner. Ratcliffe examined the temporal variation in peregrine and sparrowhawk eggshells collected in the UK since 1900. His findings that a pronounced decrease in eggshell thickness occurred in both species in the mid-1940s were published in *Nature* in 1967. A replotting of Ratcliffe's data for the peregrine over the critical period of change is shown in figure 12.15. The first sign of change was in 1946, although the mean was not statistically significant. In 1947, the change was clear and statistically significant, and in the decade that followed eggshells as thick as the pre-war norm were a rarity.

The most dramatic case of eggshell thinning was found in the brown pelican (*Pelecanus occidentalis*) off the coast of California in 1969.

The colony on Anacapa Island showed almost complete reproductive failure, with only four chicks raised from some 750 nests. Most nests were abandoned and remains of crushed eggshells were found throughout the colony (see figure 15.1A). On average, the thickness of these broken and crushed eggshells was only half of the normal value. The reproductive failure of the colony continued for the next few years. The productivity of double-crested cormorant (*Phalacrocorax auritus*) colonies was also close to zero and, again, the main cause was the breakage of the eggs. The most detailed studies on eggshell breakage were made by the Canadian Wildlife Service on colonies of cormorants on Lake Huron in 1972 and 1973 (Weseloh *et al.*, 1983). These workers visited all the known colonies of cormorants in Lake Huron. They found that 79% of the eggs were 'lost' within 8 days of laying and that by the end of the normal incubation period only 5% of the eggs remained in the nests. In about half the cases of 'lost' eggs, eggshell fragments were found in or around the nest. The eggshells averaged 24% thinner than pre-war values; although this is not as severe as that found in California, it was enough to cause almost complete reproductive failure. The subsequent recovery of this population is detailed in section 15.2.

The importance of the North American findings was that there clearly was a linkage between reproductive failure and eggshell thinning (Risebrough and Peakall, 1988). Analytical work revealed that DDT (and its metabolites) and PCBs were the only contaminants present in appreciable amounts. In this, the situation is quite different from that in the UK where dieldrin (see Chapter 12) was found to be a major factor in population declines of peregrine and sparrowhawk.

It was thought that a pesticide was responsible for eggshell thinning on the grounds that the effect occurred at the time that synthetic

A

B

FIGURE 15.1 *(A) Crushed eggs in the nest of a brown pelican, Anacapa Island, California, 1970. (B) Double-crested cormorant with deformed bill from a colony on Lake Michigan, USA.*

pesticides were introduced on a large scale and the declines in population were greatest in areas where pesticides were most heavily used. The analysis of residues of organochlorine pesticides was becoming more widespread by this time and by the beginning of the 1970s it was possible to show a correlation between the levels of DDE in the egg contents and the thickness of the shell (figure 15.2). This particular dose–response curve shows that, initially, a low concentration of DDE causes a large response and that the response tends to 'flatten out' with increasing dose.

These findings established a correlation between the concentration of DDE and the degree of eggshell thinning. Experimental studies with the American kestrel established a cause-and-effect relationship. Further, it was demonstrated that the relationship between the degree of eggshell thinning and DDE residue levels in the egg based on laboratory studies was the same as that found in the field. These studies also showed that PCBs did not cause eggshell thinning.

There was a major legal battle over DDT in 1971. It was the first pesticide over which en-

vironmental regulations were made and the producers feared, with some justification, that if they lost this battle they would lose others. The hearing that led to the ban of DDT in the USA was the most extended and bitterly fought ever conducted on an environmental contaminant. It ran from August 1971 to March 1972 and a total of 125 expert witnesses produced over 9000 pages of testimony. There were four batteries of lawyers. On one side, there were those representing the chemical industry (27 companies acting together as group petitioners) and the lawyers of the US Department of Agriculture; on the other side, those representing the US Environmental Protection Agency and the Environmental Defence Fund.

Observations of eggshell thinning linked to DDT were important evidence in this hearing. Despite the verdict going against DDT (and other similar ones in other countries), there was criticism of the finding. Because eggshell thinning had occurred so rapidly after the introduction of the insecticide, before it was widely used, it was argued that DDT could not have caused this effect. It was indeed true that all the work was carried out some 10–20 years after thinning had first been shown to occur; the first measurements were made on the levels of DDE in peregrine eggs in 1962, whereas eggshell thinning started in 1946. However, so stable is DDE and so sensitive are analytical techniques that it proved possible to extract enough DDE from the dried membranes of eggs collected over the critical period (1946–7) to demonstrate that enough DDE was present to have caused eggshell thinning.

It was soon found that the phenomenon of eggshell thinning in peregrines was global and that there was a close correlation between the degree of eggshell thinning and the health of the peregrine population. Studies from Australia, Europe and North America showed that when eggshell thinning exceeded 17–18% population declines followed. The relationship

FIGURE 15.2 *Relationship of eggshell thickness index to DDE residue levels in peregrine eggs collected from Alaska and northern Canada. From Peakall (1993), reprinted with permission from* Environmental Reviews.

between eggshell thinning and the status of populations is shown in figure 15.3. Four extirpated populations had eggshell thinning of 18–25%, declining populations had over 17%, whereas stable populations had less than 17% thinning.

As would be expected, there was considerable interest in the mechanism whereby DDE caused eggshell thinning. A wide variety of mechanisms have been proposed. However, the finding that eggshell thinning of up to 50% occurred in the brown pelican indicates that the site of impairment is in the shell gland, as reduction of 50% of calcium levels throughout the organism would cause profound physiologi-

cal changes. This hypothesis is supported by the finding that the circulating level of calcium in the blood is normal in birds laying eggs with thin shells. There is now general agreement that it is transport of calcium across the eggshell gland mucosa that is affected by DDE. Decreased activity of Ca-ATPase and effects on prostaglandins have been proposed as the key mechanisms. There is no reason to suppose that these mechanisms are mutually exclusive. However, no comprehensive answer to the mechanistic basis of species variation of DDE-induced eggshell thinning has been put forward.

The evidence that DDT and dieldrin was causing widespread decreases in several

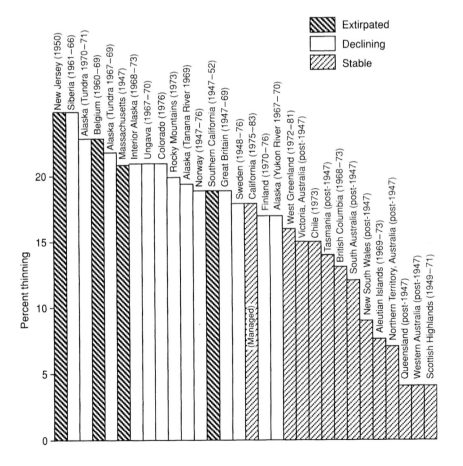

FIGURE 15.3 *Relationship between degree of eggshell thinning and the status of various populations of peregrines. From Peakall (1993), reprinted with permission from* Environmental Reviews.

populations of the peregrine and many other birds of prey was part of the evidence that led to bans and restrictions. The use of DDT and dieldrin was banned in the Scandinavian countries over the period 1969–72, and in Holland dieldrin was restricted in 1968 and DDT was banned in 1973. In Great Britain, there were voluntary restrictions on both DDT and dieldrin starting in 1962, although some small-scale uses were not banned until 1986. The approach in Great Britain was the opposite to that in the USA, where the ban on DDT imposed in 1972 followed a long court hearing.

In the eastern United States and southern Canada, the peregrine had disappeared completely by the early 1960s. In these areas, reintroduction was the only feasible means of restoring the population. Large-scale breeding programmes, with the intention of raising birds for reintroduction, were started by the Canadian Wildlife Service at Wainwright, Alberta, in 1972 and by the Peregrine Fund at Cornell University, New York, at about the same time. The first few captive-raised young were released in 1975 and the numbers released soon increased. The first breeding attempt of released birds in the wild was in 1979 and the progress has been rapid since then. We have now reached the point where releases are needed only in a few specific areas.

In other parts of the world, some breeding stock remained and reintroduction programmes were not essential. Recovery has occurred in Alaska, western Canada, the continental United States and Mexico, starting in the later 1970s. In the British Isles, where the most detailed data are available, recovery was described as 'virtually complete in overall numbers, though not in precise distribution'. In other parts of Europe, the populations have increased, but are still low compared with pre-war numbers, notably in the Czech Republic, Slovakia, East Germany and Poland, which formerly held large

populations. These, and other, recoveries of the peregrine are detailed in Cade *et al.* (1988).

An interesting aspect of the phenomenon of eggshell thinning is the wide difference in the sensitivity of different avian species. Most sensitive are the raptors and the fish-eating birds; this is unfortunate as it is these species that are exposed to the highest dose as a result of the bioaccumulation of DDE up the food chain. The raptors, in addition to the peregrine, that have shown marked effects include the osprey (*Pandion haliaetus*) and the bald eagle (*Haliaetus leucocephalus*) in North America and the sparrowhawk in Europe. In contrast, the species most commonly used in *in vivo* toxicity tests (quail, pheasant, chicken) are almost completely insensitive; and others (such as the duck) are only moderately sensitive. Thus, it is unlikely that even the studies conducted now before registration of a new pesticide would have detected this particular adverse effect of DDT. However, other negative aspects of DDT, such as its strong tendency to biomagnify up the food chain and its toxicity to fish, should prevent registration.

No story is quite as tidy as one would like. Although in many parts of the world it seems clear that DDT-induced eggshell thinning was the main cause of population declines of raptorial birds, in some areas other pesticides were certainly also involved. In the UK, direct mortality caused by dieldrin, another organochlorine that is magnified up the food chain, is considered to have been the most important factor (see Chapter 12), whereas in North America the levels of dieldrin recorded were well below the critical value.

The eggshell thinning story is one of the most comprehensive environmental investigations in the short history of environmental toxicology. An important feature of eggshell thinning and the subsequent collapse of the populations of the peregrine and other birds of prey lies in the fact that the effects occurred over a wide area, much of it remote from the place of application of the pesticide.

15.2 *Reproductive failure of fish-eating birds on the Great Lakes of North America*

The North America Great Lakes are, collectively, the largest body of fresh water in the world. The very vastness of the Great Lakes was strangely one of the causes of their pollution. They are so large that 'dilution seemed the solution to pollution'. The waterway of the lakes was a major factor in the opening up and development of this part of the New World. Hardly surprisingly, towns and industry sprang up on their shores. The harnessing of the Niagara Falls to generate electricity led to major industrial development along the Niagara River. Now some 36 million people live within the Great Lakes basin and over 13 000 manu-

facturing and industrial plants are located there. The result was that, by the 1960s, serious wildlife problems were occurring on these inland seas.

Investigations of the problems with wildlife of the Great Lakes had two separate and very distinct beginnings. One was proactive and the other reactive. The proactive approach was led by Joe Hickey, Professor of Wildlife Ecology at the University of Wisconsin. Concern over DDT was rising at this time and the study was started 'to determine what pesticide residues, if any, are present in different trophic levels in a Lake Michigan ecosystem' and 'to understand the biological significance, if any, of pesticide residues encountered in various layers of the Lake Michigan animal pyramid'. The bioaccumulation of DDT residues up the animal pyramid was clearly demonstrated and the findings are illustrated in figure 15.4. Although bioaccumulation of DDT was known, the importance of the Lake Michigan study lay in the fact that it demonstrated that these problems could occur in a large lake system.

The second starting point was more dra-

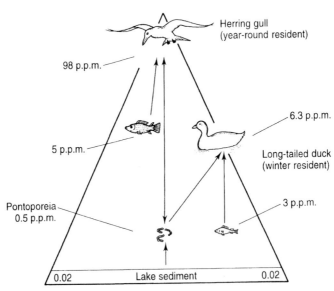

FIGURE 15.4 *Accumulation of DDT up the Lake Michigan food chain. Values are total DDT on wet weight basis. Reproduced after Hickey et al. (1966) with permission from the University of Wisconsin Press.*

matic. Michael Gilbertson, of the Canadian Wildlife Service, visited tern colonies on Lake Ontario in 1970. He described his visit as follows. 'As I walked about one of these islands, many birds whirled around my head and swooped down upon me again and again to prevent me from approaching their nests. Their shrill piercing cries rang in my ears. As I wandered about, I soon noticed that something was fundamentally wrong with the colony. While some young of varying age were found in the nests, the eggs in most had failed to hatch. On examining one of these eggs, I found that the young chick had died before it could completely crack open the shell. Several other eggs contained dead embryos. At the edge of a grass tussock, I also noticed an abnormal 2-week old chick, its upper and lower bill crossing over without meeting – a deformity which would result in certain starvation.'

The herring gull became the key indicator species in the studies of pollution on the North American Great Lakes. The reasons for this choice are given in box 15.1.

Although the first studies, which had shown high embryonic mortality, were carried out in the mid-1960s, it was not until the 1970s that systematic studies were set-up by the Canadian Wildlife Service. The serious nature of the problem was shown by the fact that the reproductive success of herring gull on Lake Ontario was very low, with only one young being produced by ten pairs of birds, whereas normal reproduction would be more than one young per pair. The hypothesis that these severe effects over a wide area were caused by pollutants, notably the organochlorines, was immediately put forward. A monitoring programme for residue levels in herring gulls was set up in 1974, and the birds were collected from 13 colonies throughout the Great Lakes. This programme has continued to the present day and some of the data from it are shown in figure 15.5.

One interesting study aimed to establish the relative importance of effects of pollutants in the adult birds which could cause behavioural changes (extrinsic effects) compared with effects of the pollutants in the egg of the developing embryo (intrinsic effects). To distinguish between these two causes of embryo mortality, an egg exchange experiment was devised. The basis of this experiment was to move eggs from a highly contaminated (dirty) colony and place them under adults in a relatively uncontaminated (clean) colony and vice versa. Additionally, it is possible to incubate eggs from both 'clean' and 'dirty' colonies artificially to examine the effects of residues on the embryo. The outline of this experiment is shown in table 15.1.

In theory, this is a simple enough experiment; in practice, it was not. Transportation did not

BOX 15.1 *Reasons for the herring gull being a key indicator species on the Great Lakes.*

1. The herring gull feeds at the top of the food chains of the Great Lakes.
2. The adult herring gull is a year-round resident of the Great Lakes, with comparatively little movement from lake to lake.
3. The herring gull nests colonially. Colonial birds are probably the only groups of organisms for which it is possible to count the entire breeding population. Collection of eggs and assessment of reproduction are also easier in a colonial species.
4. The herring gull breeds throughout much of Europe and North America. This allows comparison between contaminant levels, reproductive success, etc. on the Great Lakes and those in coastal and European populations.

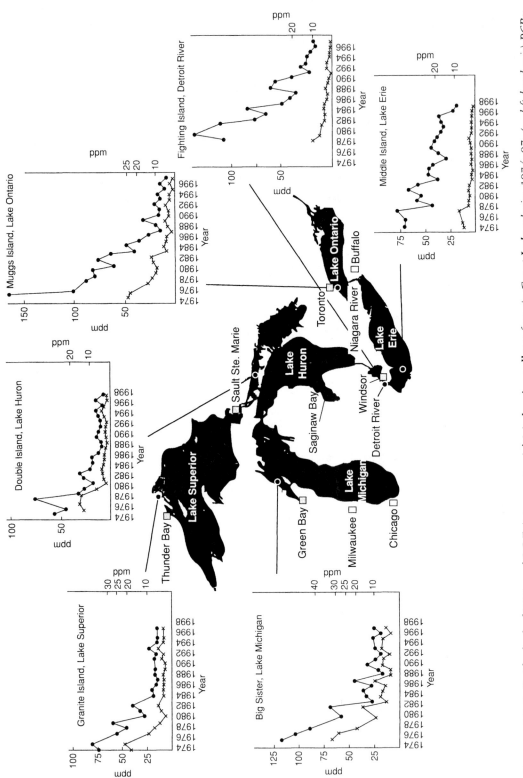

FIGURE 15.5 *Residue levels of **PCBs** and **DDE** (p.p.m., wet weight) in **herring gull** eggs from six **Great Lake** colonies, 1974–97. (•, left-hand axis) PCBs; (×, right-hand axis) DDE. Data supplied by the Canadian Wildlife Service.*

TABLE 15.1 *Outline of egg exchange experiments*

Adult	Egg	Information obtained
Clean	Clean	Normal reproduction in clean environment
Clean	Dirty	Effect of pollutants in egg on the embryo
Dirty	Clean	Effect of pollutants on reproduction via behavioural changes of adult
Dirty	Dirty	Impaired reproduction in dirty environment

affect the viability of the eggs, but to find colonies in different parts of the country with fresh eggs at the same time and move them rapidly from place to place was a logistical nightmare. Nevertheless, the results obtained from them in 1975 were quite clear: there was a major effect due to the pollutants in the egg and a significant effect of the pollutant on the behaviour of the adult. We were foolish enough to repeat the experiment in 1976, when the results were less clear-cut, and again in 1977, when no differences were found at all. In fact, what we were seeing was a sudden marked improvement in the reproductive success of the herring gull in Lake Ontario.

While the increased breeding success of birds on the Great Lakes was good news, it did make the scientific investigation of the impact of pollutants very difficult. The cause of the change was the decreased inputs of some of the organochlorines into the environment. In 1972, the use of DDT was banned in the USA and its use was curtailed in Canada. In the same year, Monsanto introduced its voluntary ban on the 'open circuit uses' of PCBs. Thus, the intensive phases of the studies on the herring gull were carried out against a background of decreasing inputs.

The effect of these restrictions on the residue levels of PCBs and DDE can be seen in figure 15.5. Although the patterns vary somewhat, they are characterized by a steep decline in the 1970s (especially in areas where the initial levels were highest), followed by slower declines in the 1980s and comparatively little

change in the 1990s. It is clear that we now have chronic low-level pollution.

The adverse effects on fish-eating birds were not confined to gulls. There was a severe decline of other species, such as the cormorant and the bald eagle. The best estimate of the population of cormorants on Lake Ontario in the 1940s and 1950s was 200 pairs; this had fallen to only three pairs by 1973. The main cause of failure was the breaking of eggshells. It was found at this time that 95% of the eggs broke before hatching. The recovery of the cormorant following the banning of DDT was rapid. By 1980, there were 375 pairs, which had risen to 6700 pairs by 1990, 13 100 by 1995 and to 18 800 pairs by 1998. Obviously, density-dependent factors such as food supply and nesting habitat (see Chapter 12) must come into play at some stage, but this has not been reached yet. An interesting question is why the population increased so much, way beyond the size of the initial population, once the pressure was released? The best explanation is a change in the fish stocks. Marked decreases in the population of the top predatory fish – lake trout and salmon – has allowed increases in the population of smaller fish such as alewife and smelt. The causes are complex, including both overfishing and the effects of pollutants on reproduction of the larger, long-lived fish. However, when the reproduction improved as DDT levels decreased, the species were able to exploit the increased food base of small fish.

Abnormal young have been found in several species throughout the Great Lakes. The best information is available on the cormorant. In this case, the aid of bird-watchers who ring these birds has been enlisted. They were asked to record the number of young with abnormalities, such as crossed beaks (figure 15.1B) during the procedure of banding the young. The highest incidence of abnormalities was found in certain areas of Lake Michigan, where one young in a hundred was defective. The rate of occurrence was considerably lower in other parts of the Great Lakes. Thus, although the occurrence of abnormalities was a potent reason for starting studies, the actual impact of these on populations of fish-eating birds has been slight.

On the Great Lakes, the scene had changed by the early 1980s. By this time, the reproductive problems of fish-eating birds were now confined to a few specific areas, rather than occurring on a lake-wide basis as they had a decade before. Other changes had occurred since the early 1970s. In hindsight, one problem that we had then was that neither the analytical chemistry nor the experimental toxicology was sophisticated enough. The confirmation that the chlorinated dioxins (PCDDs) and chlorinated dibenzofurans (PCDFs) were present in the Great Lakes was not made until 1980. The chemistry of these compounds is discussed in section 1.2.2.

Nowadays, analytical chemists routinely report on the levels of as many as 100 different organochlorines compared with a dozen a decade or so ago. Studies with specific congeners of PCBs, PCDFs and PCDDs have shown that the toxicity of specific compounds varies by several orders of magnitude. Toxicologists have demonstrated that 2,3,7,8-tetrachlorodibenzo-p-dioxin is one of the most toxic compounds known, but molecular biologists have found it useful for their work on the isolation and characterization of receptors. The identification of the Ah receptor in the mid-1970s by means of its strong specific binding to dioxin was a key finding in bringing molecular biology into the realm of toxicology. The Ah receptor is responsible for the control of the monooxygenase enzyme systems which are responsible for an important defence mechanism against foreign compounds (see Chapters 7 and 9).

It was found that the ability of individual compounds to induce the monooxygenase system is greatly influenced by the degree of chlorination and the chlorine substitution pattern. The most toxic PCBs are those that have no chlorine atoms next to the central bond, which allows the molecule to assume a coplanar configuration (see Chapter 1). Studies on the structure–activity relationship of a large number of organochlorines show that the toxic effects and strength of binding to the Ah receptor are related.

The application of this fundamental biochemistry of field investigations has involved expressing the complex mixtures of organochlorines as 'dioxin equivalents'. In the present case, this is based on the relationship found between the concentration required to induce a specific monooxygenase activity (aryl hydrocarbon hydroxylase) and the concentration required for toxic effects for a large number of PCBs, PCDFs and PCDDs. When calculating 'dioxin equivalents', the activity of the most powerful compound, 2,3,7,8-TCDD, is set at one and the potency of the other compounds are calculated from their ability to induce the monooxygenase relative to that of 2,3,7,8-TCDD (see section 9.2 for further explanation). These are called the **toxic equivalent factors** of the compounds. For each compound, this number can then be multiplied by the concentration and the toxic equivalence (TEQ) in terms of dioxin calculated. Although the potency of other compounds, such as individual PCBs, is much lower than dioxin, the concentrations are often much higher and therefore they frequently contribute more to the total 'dioxin equivalent' than dioxin itself. For example, recent studies in Lake Michigan suggest that 90% of the dioxin equivalents in the eggs of fish-eating birds is caused by two spe-

TABLE 15.2 *Toxic equivalency factor and dioxin equivalent of three compounds*

Compound	Concentration (pg g^{-1})	Toxic equivalency factor	Dioxin equivalent
2,3,7,8-TCDD	2	1	2
3,3',4,4',5-PCB	3000	2×10^{-2}	60
3,3',4,4',5,5'-PCB	25 000	5×10^{-4}	12.5

cific PCBs. An example of the calculation of dioxin equivalents is given in table 15.2. The approach is now used in reverse, the degree of induction of the enzyme is measured and then converted into dioxin equivalents. This bioassay approach is rapid and inexpensive compared with the conventional chemical analysis by gas chromatography – mass spectrometry.

The dioxin equivalents of egg samples from fish-eating birds collected from colonies of cormorants and terms in Michigan and Ontario have been determined. When the reproductive success of double-crested cormorant and Caspian tern (*Hydroprogne caspia*) colonies was plotted against dioxin equivalents of eggs from each colony, a high degree of correlation was found (figure 15.6).

FIGURE 15.6 *Relationship between dioxin equivalents and reproductive success in the Caspian tern on the North American Great Lakes. Reproduced from Peakall (1992) with permission from Chapman & Hall.*

The reproductive failure of fish-eating birds in the North American Great Lakes is an example in which two initially entirely different lines of research – molecular biology and investigations by field biologists – eventually blended to provide an answer to the problem.

15.3 Reproductive failure of molluscs caused by tributyl tin

The tributyl tin (TBT) compounds have the general formula $(\text{n-C}_4\text{H}_9)_3\text{-Sn-X}$, where X is an anion such as chloride or carbonate. TBTs have been used as molluscicides on boats, quays and other marine structures and as biocides for cooling systems, in pulp and paper mills, etc. Their use as active ingredients in marine antifouling paints began in the mid-1960s, and for the next decade their popularity increased as it became widely recognized as being extremely effective. The worldwide production of organotins was estimated in 1980 to be about 30 000 tons annually; about one-tenth of this was used in antifouling paints and a similar amount as wood preservatives.

Problems began to be recognized in the late 1960s with population declines of oysters and whelks in France and in southern England and in marine snails in Long Island Sound in the USA. Investigations into the decline of populations of a mollusc, the dog whelk (*Nucella lapillus*), in Plymouth Sound showed that these declining populations exhibited a high degree of imposex. Imposex (females developing male characteristics) is typified by the development in females of a small penis close to the right tentacle. A superficial vas deferens grows between the genital papilla and the penis. In the most extreme cases, this occludes the papilla, thus preventing egg liberation and reproduction.

Detailed laboratory and field studies, including transplant experiments, have been carried out by the Plymouth Marine Laboratory since the first finding of imposex in the dog whelk in 1969. These studies, reviewed in Matthiessen and Gibbs (1998), show abundant proof of the linkage of TBT with this irreversible sexual abnormality in female gastropods. A broader survey around the south-west peninsula of England revealed that imposex was widespread, with the most marked effects along the Channel coast.

One of the first findings was that there was a good relationship between the degree of imposex and the proximity of the affected population to harbours and marinas. This suggested a pollutant associated with the boating industry. Another suggestive finding was the marked increase of imposex after its discovery in 1969. These findings suggested that the causative agent was TBT. These correlations were confirmed by laboratory experiments which allowed a cause-and-effect linkage to be made. These studies determined that a few nanograms per litre (ng l^{-1}) were enough to sterilize young whelks (Bryan *et al.*, 1988).

Studies have shown that imposex is caused by elevated levels of testosterone that masculinize TBT-exposed females. The precise mechanism is not clear, but the weight of the evidence suggests that TBT acts as a competitive inhibitor of cytochrome P_{450}-mediated aromatase (Matthiessen and Gibbs, 1998).

Imposex has been widely reported in marine gastropods associated with marinas and harbours around the world. Beside the studies already referred to in the UK, France and eastern USA, records include western USA and Canada, Alaska, south-east Asia and New Zealand. Initially, the problems seemed to be restricted to areas close to marinas and harbours. However, in the North Sea, imposex was found in the common whelk (*Buccinum undatum*) and the frequency correlated well to shipping traffic intensities. It is now clear that imposex is a global phenomenon, with 72 species of 49 genera of gastropods shown to be affected.

Laboratory experiments and *in situ* transfer experiments indicate that imposex may be initiated in dog whelks with concentrations of TBT as low as 1 ng l^{-1}. The no observed effect concentration (NOEC) for development of imposex is given as less than 1.5 ng l^{-1} in the International Programme on Chemical Safety Environmental Health Criteria Document published by the World Health Organization in 1990. Sterilization of some females occurs at localities with 1–2 ng l^{-1} and is total in areas averaging 6–8 ng l^{-1}. Affected populations suffer reproductive failure and local extinction has occurred around marinas (Gibbs and Bryan, 1986).

In France, restrictions on the use of TBT were initially brought in 1982 and have subsequently been extended to a complete ban of the use of organotin compounds in antifouling paints. In the UK, the use of TBT-containing antifouling paints on small boats and aquaculture cages was banned in 1987 and the environmental quality standard set at 2 ng l^{-1} TBT. Vessels longer than 25 m are still able to use TBT paints, based on the premise that because they are in open waters much of the time the released TBT will be effectively diluted. Similar restrictions have been made in many other countries. In the USA, a number of states have imposed restrictions or bans, some of which are based on leaching rates. The leaching rate of TBT from paints has been considerably improved in recent years by incorporating the TBT in a co-polymer matrix.

Being moderately lipophilic, **TBT** would be expected to bioaccumulate. The ability of TBT to bioaccumulate in sediment and various organisms is shown in figure 15.7. Concentration factors are greatest for organisms which lack efficient degradation systems. TBT is rapidly degraded by fish, birds and mammals and thus is not considered a risk to top-level consumers.

In the decade after legislation to restrict TBT, imposex indices (Box 11.2) have improved in many surviving populations in the UK. However, populations which were close to collapse have shown little evidence of recovery and some have continued to extinction. The situation is made more difficult since *Nucella* does not have a planktonic larval stage to aid dispersal and thus recolonization depends on chance incursions of egg-laying adults.

15.4 *Forest spraying in eastern Canada to control spruce budworm*

The spraying programme in eastern Canada to control (or, as some of its critics would say, to attempt to control) spruce budworm

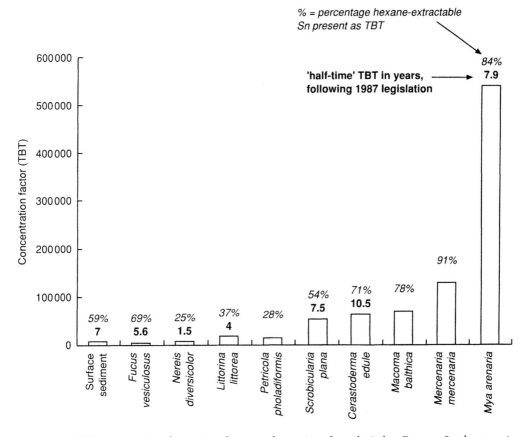

FIGURE 15.7 *TBT concentration factors in sediment and organisms from the Itchen Estuary, Southampton (water concentration 67 ng TBT l⁻¹). Figures above bars indicate per cent organotin as TBT and half-life of the latter following 1987 legislation; concentration factor = TBT concentration in tissue or sediment (dry weight) divided by TBT in water. Langston (1996).*

(*Choristoneura fumigerana*) has been the longest and largest programme of its kind in the world. In total, some 17 million kg of pesticides were sprayed on 67 million ha of forest. The total area sprayed is approximately half the area of England and Wales, but some areas would have been sprayed annually or even twice annually. Even so, at the peak of the spraying operation (1974–6), well over half of the province of New Brunswick was sprayed at least once annually (figure 15.8). The aerial spraying operation essentially stopped in the early 1990s, and by 1996 only trivial amounts of chemical pesticides were used; even the use of the microbial agent *Bacillus thuringiensis* was being phased out. The only chemical being sprayed on the forests of New Brunswick in any appreciable amount (112 000 kg) in 1996 was the herbicide glyphosphate. The total amounts of pesticides used in this programme are given in table 15.3.

What have been the ecological consequences of this spraying? Impact studies showed that

TABLE 15.3 *Total amounts of pesticides (kg) used in New Brunswick in forest spray operations, 1952–96*

DDT (1952–64)	5 745 000
Phosphamidon (1963–77)	771 000
Fenitrothion (1969–93)	9 696 906
Aminocarb (1972–92)	551 762
Trichlorofon (1977–8)	289 000

birds, especially canopy-living birds, were the most vulnerable members of the fauna. Mammals, living on the forest floor, appeared unaffected; amphibians and reptiles rather insensitive. Fish were severely affected by DDT, but not by organophosphorous or carbamate pesticides.

The impact of the organochlorine DDT was quite different from that of organophosphorous and carbamate pesticides, which act by inhibiting cholinesterase. The most important, from a commercial point of view, was the mortality of young salmon. It was discovered that entire year classes of salmon had been eliminated in some important salmon rivers of maritime Canada. Less important commercially, but just as dramatic, was the loss from the area of the fish-eating hawk the osprey. Also, the DDT used in forest spraying made a significant contribution to the DDT that caused the extinction of the peregrine falcon in eastern North America, south of the Arctic. This impact, mediated through reproductive failure caused by eggshell thinning, has already been discussed in some detail earlier in this chapter.

The first pesticide to replace DDT in the spray programmes in eastern Canada was phosphamidon. It turned out to be a most effective avicide; the effect was quite different from DDT. DDT, accumulating through the food chain, caused reproductive failure to those species that were sensitive to eggshell thinning, but did not cause outright mortality at the dosages used in forest spraying operations. The amounts of DDT that caused widespread mor-

FIGURE 15.8 *Map of New Brunswick, Canada, showing the extent of forest spraying (shaded) in 1976. Reproduced from Pearce et al. (1976) with permission from the Canadian Wildlife Services.*

tality of American robins after attempts to stop the spread of Dutch elm disease were much higher. In these programmes, some 250–550 g of DDT was sprayed on individual trees.

The main tool for the calculation of the impact of pesticides on forest birds is the line transect of singing males (further details in box 15.2). Another useful tool is the determination of cholinesterase levels (see also Chapter 7). This technique has the advantage that inhibition of cholinesterase is clearly related to exposure to organophosphorous or carbamate pesticides. Further experimental studies have determined that chronic inhibition of 50% or acute inhibition of 80% are associated with mortality. Thus, a bird found dead with 80% inhibition of its brain cholinesterase can be diagnosed with certainty as having been killed by a spray opera-

tion. Nevertheless, under field conditions, there are several limitations to the approach. These limitations are as follows.

1. The time-course of AChE depression varies with different pesticides and probably varies with different species. It is often difficult to collect enough specimens of a given species at any one time interval after the spray operation.
2. Sampling may be influenced by birds coming into the area after spraying has occurred, so that their only exposure is secondary.
3. Birds with marked AChE inhibition are inactive, unwilling to fly or to be flushed and are thus less likely to be captured or collected.

BOX 15.2 *Line transect to measure impact of pesticides.*

Repeated counts of singing male birds were made along roads and trails through the forest before and after spray treatment. The routes were 5 km long, took about 3 h and were carried out in the early morning. This method was considered to be more effective than the intensive study of small plots as in the latter method the numbers of birds recorded were often too small for statistical analysis to be valid. Another problem with the small-plot approach is that in operational spraying the coverage is far from uniform, so that a small plot may be under- or oversprayed or even missed completely. The line transect, especially if undertaken at right-angles to the line of spray, does not have this weakness. Nevertheless, there are limitations to the approach, the most important of which are as follows.

1. The counts may underestimate any decrease in vocal output as songs become more easily heard as their total number decreases.
2. The mortality of birds caused by the spray may be masked by immigration from unsprayed areas. The huge blocks of forest sprayed in eastern Canada at the height of the spray programme make this less of a problem, although long-distance migration may occur.
3. The method does not prove that the cause of the change was the pesticide, although a marked decline immediately after the spray does strongly suggest that the pesticide was the cause. The most likely confounding factor is mortality caused by bad weather.
4. Even though a considerable number of transits are run and the results averaged, the degree of extrapolation involved is large.

Using the time transect technique, it was estimated that nearly 3 million birds were killed in New Brunswick in 1975, largely because of phosphamidon. The impact of fenitrothion was considered to be much less. These calculations, despite being open to criticism, do suggest that very considerable mortality occurred and were a major reason for the phasing out of phosphamidon in forest-spraying programmes.

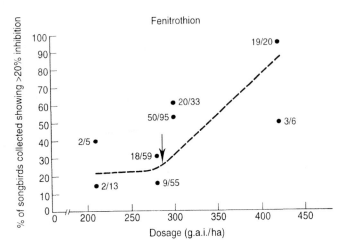

FIGURE 15.9 *Relationship between inhibition of cholinesterase in songbirds and the dosage of fenitrothion applied to the forest. The arrow marks the dosage above which effects are seen on songbirds as judged by transect analysis.*g.a.i./ha, grams active ingredient per hectare.*

All these factors bias the data in favour of underestimating the degree of AChE inhibition. If the application rate is plotted against the degree of AChE inhibition (figure 15.9), then one finds that the application rate known (from transient surveys or other population studies) to cause effects occurs at 35–40% inhibition (Mineau and Peakall, 1987).

It is clear that pesticides used in forest spraying can cause mortality, sometimes considerable mortality, at operational dosages. The dosage of fenitrothion commonly used in Canadian spray programmes (210 g ha⁻¹) does not have any appreciable safety margin. A considerable number of studies have been carried out, some showing effects, others not. The safety margins for malathion and aminocarb are greater and several studies with these pesticides have not revealed adverse effects. Fenitrothion was in use in New Brunswick for 25 years (1969–93). The question may fairly be asked 'what are the effects of the persistent use of a non-persistent pesticide?'. Regrettably, the answer to that interesting question is that we do not know. An assessment on the environmental effects of fenitrothion used in forestry (Ernst *et al.*, 1989) failed to reach any definite conclu-

sion. It is, indeed, a difficult question to answer. First of all, the accurate measurement of long-term trends of avian (or indeed any wildlife) populations are difficult and expensive to determine. Even if this effort were made, it is difficult, if not impossible, to assign a cause to the changes that have been established. For example, in the case of North American songbirds, declines have most frequently been attributed to the destruction of wintering habitat in Latin America.

To collect enough information on populations, it is necessary to have a large number of persons to carry out a large number of surveys for a long period of time. Virtually the only way to do this is to enlist amateurs (the word 'amateur' is used in its old sense of a person who does the work for the love of it, rather than any suggestion of the work being second-rate), as has been done by the British Trust for Ornithology (BTO). Some of the surveys currently being run by the BTO have been given in tables 12.3 and 12.4. In addition, there are many more specialized surveys, including the heronies survey, seabird colonies register, national wildfowl counts and the estuaries enquiry.

15.5 *Summary*

Specific examples of linkages between biochemical, physiological, individual, population and community responses are presented. The examples given are DDE-induced eggshell thinning in predatory birds, effects of contaminants on fish-eating birds in the Great Lakes of North America, reproductive failure of molluscs caused by tributyl tin and the forest-spraying programmes in eastern Canada. In the first three examples, population changes were well documented; in the last, although considerable mortality of songbirds was found, population effects were not demonstrated.

15.6 *Further reading*

BRYAN, G. W. *et al.* (1986) The first major paper documenting the decline of the dog whelk and linkage of this decline to tributyl tin.

CADE, T. J. *et al.* (eds) (1988) *Peregrine Falcon Populations: Their Management and Recovery*. A detailed account of the recovery of the peregrine and several other birds of prey throughout the world.

GILBERTSON, M. *et al.* (1998) A series of papers on trends in levels and effects of toxic substances in the Great Lakes.

HOFFMAN, D. J. *et al.* (eds) (1995) *Handbook of Ecotoxicology*. Gives a number of other case studies.

IPCS (INTERNATIONAL PROGRAMME ON CHEMICAL SAFETY) (1990) *Tributyl Tin Compounds*. A formal document that reviews the analytical methods, environmental levels and effects on organisms (ranging from microorganisms to man).

MATTHIESSEN, P. and GIBBS, P. E. (1998) A critical review of the effects of TBT on molluscs.

MINEAU, P. and PEAKALL, D. B. (1987) An evaluation of the methods – transects counts, number of singing male birds and cholinesterase levels – available to assess the impact of OP and carbamate pesticides used in forest spraying.

PEAKALL, D. B. (1993) A personal account of the discovery of eggshell thinning in the peregrine and its relationship to population changes.

PEAKALL, D. B. and BART, J. R. (1983) A detailed review of the amounts of pesticides used, the areas sprayed and the effects seen in the spraying programmes in eastern Canada and the north-eastern USA.

Appendix: Introduction to population projection matrices

A simple example will show how population projection matrices (also known as Leslie matrices) are used. Suppose the study organism lives for 4 years, and that:

the proportion of 1-year-olds that live to be 2 is 0.1;
the proportion of 2-year-olds that live to be 3 is 0.8;
the proportion of 3-year-olds that live to be 4 is 0.6;
the proportion of 4-year-olds that live to be 5 is 0.

Suppose that the organism breeds at ages 2, 3 and 4 years, producing, respectively, five, 10 and three offspring that live to be 1 year old. These numbers are tabulated in a matrix, which we designate M1, as follows:

$$M1 = \begin{pmatrix} 0 & 5 & 10 & 3 \\ 0.1 & 0 & 0 & 0 \\ 0 & 0.8 & 0 & 0 \\ 0 & 0 & 0.6 & 0 \end{pmatrix}$$

This is an example of a 'population projection matrix'. Note that the age-specific fecundities are entered in order in the columns of the top row, and that the survivorships appear, in order, one per column, in the lower rows of the matrix. Thus, survivorship of the 1-year-olds appears in the second row in the first column, of 2-year-olds in the third row in the second column, and so on.

The reason for this peculiar method of tabulation is that it allows one to use matrix algebra in the following way. Suppose that initially there are in the population 1000 1-year-olds, 50 2-year-olds, 50 3-year-olds and 10 4-year-olds. This is the initial **age distribution** of the population and is tabulated as a vector (column) that we shall refer to as C1, as follows:

$$C1 = \begin{pmatrix} 1000 \\ 50 \\ 50 \\ 10 \end{pmatrix}$$

The age distribution 1 year later, which we shall call C2, can now be obtained by matrix algebra:

$$C2 = M1 \times C1$$

or, in full,

$$C2 = \begin{pmatrix} 780 \\ 100 \\ 40 \\ 30 \end{pmatrix} = \begin{pmatrix} 0 & 5 & 10 & 3 \\ 0.1 & 0 & 0 & 0 \\ 0 & 0.8 & 0 & 0 \\ 0 & 0 & 0.6 & 0 \end{pmatrix} \times \begin{pmatrix} 1000 \\ 50 \\ 50 \\ 10 \end{pmatrix}$$

Readers unfamiliar with matrix algebra might like to check that this gives the correct answer, i.e. that it calculates C2, the age distribution in year 2, correctly. Computers can be used to carry out the matrix algebra. Any computer software that performs matrix algebra may be used, e.g. MINITAB. Furthermore, the computer can be used to calculate the age distributions in later years, i.e. C3 = M1 × C2, C4 = M1 × C3, etc.

It will be found that after several years the population settles down and each age class then grows at the same rate each year. Specifically, each age class increases by a factor of 1.14 each year. Thus, the net reproductive rate λ = 1.14. When all age classes increase by the same factor each year, a population is said to be in **stable age distribution**.

The population growth rate $r = \log_e \lambda = 0.13$. When the population has reached its sta-

ble age distribution, each age class grows exponentially, with equation

$$a_x(t) = a_x(0)e^{rt}$$

where $a_x(t)$ represents the number of organisms in age class x, i.e. the number of age x at time t.

The value of λ eventually achieved does *not* depend on the initial age distribution. It satisfies the Euler–Lotka equation (equation 12.3 on p. 197) and can be found by the iterative approach described there. An alternative method uses the fact that λ is the 'dominant eigenvalue' of the matrix M1. Dominant eigenvalues of matrices are easily found by appropriate computer software.

Note: The notation here has been revised at reprint to achieve consistency with equation 12.3.

Glossary

AChE: Acetylcholinesterase.

acid rain: Rain made more acidic by the action of oxides of nitrogen and sulphur (pH below 5.5).

adducts: In the context of the present text, products of the linkage between a xenobiotic and an endogenous molecule, e.g. adducts formed between benzo(a)pyrene metabolites and DNA.

Ah receptor: aryl hydrocarbon receptor. A receptor located on a cytoplasmic protein to which bind coplanar molecules such as PAHs, coplanar PCBs and dioxins. Binding initiates induction of cytochrome P_{450} 1A.

ALAD: Aminolaevulinic acid dehydrase.

allele: One of a pair of genes that occupy the same relative position on homologous chromosomes and separate during meiosis.

anion: A negatively charged atom or radical.

antagonism: Where the toxicity of a mixture is less than the sum of the toxicities of its components.

anthropogenic: Generated by the activities of man.

anthropogenic organic enrichment factor: Measurement of the contribution of human activity to the global cycle of a pollutant.

ATPases: Adenosine triphosphatases.

autotroph: An organism which obtains its carbon from carbon dioxide.

bioaccumulation factor (BAF):

$$\frac{\text{concentration of a chemical in an animal}}{\text{concentration of the same chemical in its food}}$$

bioconcentration factor (BCF):

$$\frac{\text{concentration of a chemical in an organism}}{\text{concentration of the same chemical in the ambient medium}}$$

biomarker: Any biological response to a chemical at the individual level or below demonstrating a departure from normal status. Usually restricted to responses at the level of organization of the whole organism or below.

biotransformation: Conversion of a chemical into one or more products by a biological mechanism (nearly always by enzyme action).

birth rate: The number of offspring born to each reproductive female per year.

BOD (biochemical oxygen demand): The amount of dissolved oxygen used by microorganisms to oxidize 1 l of a sewage sample.

carboxylesterases: Esterases that hydrolyse organic compounds with carboxyester bonds. In the classification of enzymes by the International Union of Biochemistry, EC3.1.1.1 refers to carboxylesterases which are inhibited by OP compounds.

carrying capacity (of the environment in which a population lives): The population size to which the population is driven by density-dependent processes (see p. 196).

cation: A positively charged atom or radical.

cetaceans: Whales and dolphins.

CFC: Chlorofluorocarbon.

ChE (cholinesterase): A general term for esterases which hydrolyse cholinesters.

CMPP: 2-methyl-4-chloro-phenoxy propionic acid.

COD (chemical oxygen demand): The amount of oxygen required to achieve a complete chemical oxidation of 1 l of a sewage sample.

congener: A member of a group of structurally related compounds.

conjugate: In biochemical toxicology, a molecule formed by the combination of a xenobiotic (usually a phase I metabolite) with an endogenous molecule (e.g. glucuronic acid, glutathione or sulphate).

contaminant: See Introduction.

crankcase blowby: Leakage around pistons into the crankcase of an internal-combustion engine.

curie: A unit of measurement of radioactivity; 1 curie represents 3.7×10^{10} disintegrations per second.

cyclodiene insecticides: A group of organochlorine insecticides including aldrin, dieldrin, endrin and heptachlor.

cytochrome P_{450}: An iron-containing protein that catalyses many biological oxidations.

2,4-D: 2,4-Dichlorophenoxyacetic acid.

2,4-DB: 2,4-Dichlorophenoxybutyric acid

p,p'-**DDT:** *p,p'*-Dichlorodiphenyltrichloroethane.

density dependence: The phenomenon whereby factors vary in their effects with population density (see p. 200).

DNA adduct: See adduct.

DNOC: Dinitroorthocresol.

e: Base of natural logarithms (2.718).

EBI: Ergosterol biosynthesis inhibitor (fungicide).

EC(D)$_{50}$: Concentration (dose) that affects designated criterion (e.g. behavioural trait) of 50% of a population. Also known as median effect concentration (dose).

endogenous: Originating within an organism.

endoplasmic reticulum: Membranous network of cells which contains many enzymes that metabolize xenobiotics. Microsomes consist mainly of vesicles derived from endoplasmic reticulum.

epoxide hydrolase: Enzyme which converts epoxides to diols by the addition of water.

ester: An organic salt which will yield an acid and a base when hydrolysed.

eukaryotes: Organisms that contain their DNA within nuclei.

eutrophication: The stimulation of algal growth in surface waters caused by high levels of nitrates and phosphates. Such pollution may be caused by sewage or run-off from agricultural land treated with fertilizers from industrial waste.

eyrie: The nest of certain birds of prey, e.g. peregrine and golden eagle.

fitness: Used here to mean the population growth rate of an allele (note that this is different from the way fitness is defined in population genetics) (see p. 198).

fitness cost: The reduction in the fitness of resistant alleles relative to that of susceptible alleles in unpolluted environments (see p. 226).

free radicals: Chemical species (atoms, molecules or ions) that contain unpaired electrons. They are usually highly reactive.

fugacity: A measurement of the 'escaping tendency' of a molecule (see Chapter 3).

GABA: Gamma-aminobutyric acid.

genotoxic: Toxic to the genetic material of the organism.

glucuronyl transferases: A group of enzymes that catalyse the formation of conjugates between glucuronic acid and a foreign compound (usually a phase I metabolite).

gray (Gy): The SI unit of absorbed radiation dose corresponding to an energy absorption of 1 J kg^{-1} of matter (cf. sievert).

haemprotein: A protein containing haem as a prosthetic group, e.g. cytochrome P$_{450}$.

hazard: Potential of a chemical to cause harm.

hazard assessment: Comparison of ability to cause harm (see hazard) with expected environmental concentration.

HCB: Hexachlorobenzene.

hermaphrodite: An organism having both male and female characteristics.

HMO: Hepatic microsomal monooxygenase (see Chapter 5).

hydrophobic: 'Water hating'. Oils and non-polar solvents immiscible with water are examples of hydrophobic liquids.

immiscible: Refers to pairs of liquids which are unable to mix with one another, e.g. water and oil.

imposex: The imposition of male characters on females in prosobranch molluscs, principally the dog whelk, *Nucella lapillus*. A measure of imposex is the size of the female penis relative to that of the male. Within a population, imposex can be calculated with the following formula:

$$\text{level of imposex (\%)} = \frac{\left(\text{meanlength of female penis mm}\right)^3}{\left(\text{mean length of male penis mm}\right)^3} \times 100$$

induction: With reference to enzymes: an increase in activity due to an increase in cellular concentration. This may be a response to a xenobiotic and usually involves an increased rate of synthesis of the enzyme.

k-value: A measure of mortality (see p. 202).
K: Used generally as a biological constant; used in population ecology to mean carrying capacity.
K_{ow}: Octanol–water partition coefficient.

λ: Net reproductive rate – a measure of the rate of increase of the population, namely the factor by which population size is multiplied each year. $\lambda = e^r$ (see p. 197).

LC(D)$_{50}$: Concentration (dose) that kills 50% of the population observed. Also known as median lethal concentration (dose).
ligand: A substance which binds specifically.
lipophilic: Literally 'fat loving'. Lipophilic molecules have a high affinity for lipids and tend to move from water into membranes and fat depots.
lipoprotein: An association between lipids and proteins.
logistic equation: A simple equation showing how population growth rate may depend on population density (see p. 201).
logistic regression: A type of statistical regression that estimates the probability of an event occurring using one or more predictor variables.

macromolecule: A large molecule, e.g. proteins, DNA and polysaccharides.
MCPA: 2-methyl 4-chloro-phenoxyacetic acid.
melanism: Possessing the dark pigment melanin.
metallothionein: A metal-binding protein.
microcosm, mesocosm, macrocosm: 'Small', 'medium' or 'large' multispecies system in which physical and biological parameters can be altered and subsequent effects monitored. They may be field or laboratory based and are thought to mimic responses of organisms in the field more realistically than single-species test systems.
microsomes/microsomal: When tissue homogenates are subjected to differential centrifugation, the microsomal fraction is separated between approximately $10\,000\,g$ and $105\,000\,g$. It contains mainly vesicles derived from the endoplasmic reticulum in the case of most vertebrate tissues.

mitochondrion: A subcellular organelle in which oxidative phosphorylation occurs, leading to the generation of ATP.

MO: Monooxygenase. An enzyme system found in the endoplasmic reticulum of many cells. It contains cytochrome P_{450} as its active centre and catalyses the oxidation of many lipophilic organic compounds.

mutualism: A relationship between two species in which each benefits from the presence of the other.

NOEC(D): No observed effect concentration (dose).

NOEL: No observed effect level. See NOEC(D).

OC: Organochlorine compound.

OP: Organophosphorous compound.

oxyradical: An unstable form of oxygen containing an unpaired electron.

ozonosphere: Layer of the atmosphere in which ozone is concentrated.

PAH: Polycyclic aromatic hydrocarbon.

parthenogenetic: Relating to 'virgin births'.

partition coefficient: See Chapter 3. K_{ow} is an example of a partition coefficient.

PCB: Polychlorinated biphenyl.

PCDD: Polychlorinated dibenzodioxin.

PCDF: Polychlorinated dibenzofuran.

PEC: Predicted environmental concentration.

photochemical smog: A complex pollution arising as a consequence of photochemical reactions occurring in certain areas where there is strong solar radiation and high levels of aerial pollutants (e.g. from car exhaust fumes).

phytotoxic: Toxic to plants.

pinnipeds: Seals, sea lions and walruses.

PNEC: Predicted no effect concentration (see Chapter 6).

poikilotherms: Organisms that are unable to regulate their body temperatures.

polar: With reference to chemicals: molecules which have charge.

pollutant: See Introduction.

population growth rate, r: Per capita rate of increase of the population (see p. 196).

porphyrins: Chemical structures which constitute prosthetic groups of certain proteins, e.g. cytochrome P_{450}, haemoglobin.

potentiation: With reference to toxicity: the situation where the toxicity of a combination of compounds is greater than the summation of the toxicities of its individual components.

probit analysis: A statistical procedure used to analyse data arising from toxicity tests. The response measured in the test is transformed into 'probit' values.

pyrethroids: Synthetic insecticides having a structural resemblance to the naturally occurring pyrethrins.

QSAR: Quantitative structure–activity relationships; relationships between parameters describing the structure of molecules and toxicity.

r, **population growth rate:** See p. 196.

rain-out: Removal of pollutants from air by incorporation into developing rain droplets of rain clouds.

reductase: An enzyme that performs reductions.

resistance: Reduced susceptibility to the toxicity of a chemical which is genetically determined (a characteristic of a resistant strain of animal or plant) (see Chapter 13).

resistant allele: An allele which increases the fitness of its carriers in polluted environments (see p. 223).

risk assessment: Probability of hazard being realized. Best expressed as fraction of a population (community) likely to be affected. In practice, however, hazard assessment rather than risk assessment is often carried out.

selective toxicity: A difference in the toxicity of a chemical between different species, sexes, strains or age groups. Expressed as a selectivity ratio, e.g.:

$$\frac{LD_{50}\text{to species A}}{LD_{50}\text{to species B}}$$

selectivity: See selective toxicity.

SFG: Scope for growth

SH: Sulphydryl group.

sievert (Sv): This SI unit is needed because the damage caused by radiation depends on the rate at which it is absorbed. Thus, a dose of relatively massive alpha particles of 20 Sv is typically equal to 1 Gy, whereas for the less damaging beta particles and gamma rays typically 20 Sv = 20 Gy.

sister chromatid exchange (SCE): Reciprocal exchange of DNA at loci between chromatids during replication of DNA.

somatic growth rate: The rate of increase of an individual's body mass.

SR: Synergistic ratio (see Chapter 10).

standard deviation: A measure of the variation in a sample.

standard error: A measure of the precision of an estimate.

stereochemistry: A branch of chemistry which is concerned with the three-dimensional structure of chemicals.

sublimation: volatilization of a solid.

sulphotransferases: Enzymes which catalyse the formation of conjugates between xenobiotics and sulphate.

survivorship: The proportion of animals surviving between two specified ages.

survivorship curve: Graph showing how survivorship from birth varies with age (see p. 199).

susceptible allele: Non-resistant allele.

synergism: Similar to potentiation (q.v.), but some authors use the term in a more restricted way, e.g. where one component of a mixture (synergist) would cause no toxicity if applied alone at the stated dose.

2,4,5-T: 2,4,5-Trichlorophenoxyacetic acid.
TBT: Tributyl tin.
TCDD: Tetrachlorodibenzodioxin.
tolerance: The ability of an organism to withstand the adverse effects of pollution.
toxic: Harmful to living organisms.
toxic equivalent: A value that expresses the toxicity of a mixture of chemicals relative to that of a toxic compound used as a standard (see p. 155).
toxicodynamics: Relating to the harmful effects of chemicals upon living organisms.
toxicokinetics: The uptake, metabolism, distribution and excretion of chemicals that express toxicity, i.e. concerned with the fate of chemicals in living organisms.
trade-off: Exchange of one advantageous character for another (see p. 222).

vitellogenin: A protein that forms part of the yolk of egg-laying vertebrates.

wash-out: Removal of air pollutants by falling rain or snow.

xenobiotic: 'Foreign compound'. A compound that is not part of the normal biochemistry of a defined organism. A foreign compound to one species may be a normal endogenous compound to another.

Bibliography

ADDISON, R. F. (1996) The use of biological effects monitoring in studies of marine pollution. *Environmental Reviews* 4, 225–37.

AGUILAR, A and BORRELL, A. (1994) Assessment of organochlorine pollutants in Cetaceans by means of skin and hypodermic biopsies. In FOSSI, M.C. and LEONZIO, C. *Non-destructive Biomarkers in Vertebrates.* Boca Raton, FL: Lewis

AHLBORG, U. G., BECKING, G. C., BIRNBAUM, L. S., BROUWER, A., *et al.* (1994) Toxic equivalency factors for dioxin – like PCBs. *Chemosphere* 28, 1049–67.

ÅHMAN, B. and ÅHMAN, G. (1994) Radiocesium in Swedish reindeer after the Chernobyl fallout: seasonal variations and long-term decline. *Health Physics* 66, 503–12.

ALDRIDGE, W. N. (1953) Serum esterases I and II. *Biochemical Journal* 53, 110–17, 117–24.

ALLEN, Y., SCOTT, A. P., MATTHIESSEN, P., HAWORTH, S., *et al.* (1999) Survey of estrogenic activity in United Kingdom estuarine and coastal waters and its effect on gonadal development of the flounder *Platicthys flesus. Environmental Toxicology and. Chemistry* 18, 1791–1800.

ALLOWAY, B. J. and JACKSON, A. P. (1991) The behaviour of heavy metals in sewage sludge-amended soils. *Science of the Total Environment* 100, 151–76.

ANDREWARTHA, H. G. and BIRCH, L. C. (1954) *The Distribution and Abundance of Animals.* Chicago: University of Chicago Press.

APPLEQUIST, H., ASBIRK, S. and DRABAEK, I. (1995) Variation in mercury content of guillemot feathers over 150 years. *Marine Pollution Bulletin* 16, 244–48.

ARCAND-HOY, L. D. and BENSON, W. H. (1998) Fish reproduction: an ecologically relevant indicator of endocrine disruption. *Environmental Toxicology and Chemistry* 17, 49–57.

ATCHISON, G. J., SANDHEINRICH, M. B. and BRYAN, M. D. (1996) Effects of environmental stressors on interspecific interactions of aquatic animals. In NEWMAN, M. C. and JAGOE, C. H. (eds) *Quantitative Ecotoxicology: A Hierarchical Approach.* Chelsea, MI: Lewis.

ATIENZAR, F. A., CONRADI, M., EVENDEN, A. J., JHA, A. N. and DEPLEDGE, M. H. (1999) Qualitative assessment of genotoxicity using random amplified polymorphic DNA: comparison of genomic template stability with key fitness parameters in *Daphnia magna* exposed to benzo(a)pyrene. *Environmental Toxicology and Chemistry* 18, 2283–8.

AXELSEN, J., HOLST, N., HAMERS, T. and KROGH, P. H. (1997) Simulations of the predator–prey interactions in a two species ecotoxicological test system. *Ecological Modelling* 101, 15–25.

BACCI, E. (1993) *Ecotoxicology of Organic Contaminants*. Boca Raton, FL: Lewis.

BAIRD, D. J., BARBER, I. and CALOW, P. (1990) Clonal variation in general responses of *Daphnia magna* Straus to toxic stress. I. Chronic life-history effects. *Functional Ecology* 4, 399–407.

BAKER, A. J. M. (1987) Metal tolerance. *New Phytologist* 106 (supplement), 93–111.

BAKER, A. J. M. and PROCTOR, J. (1990) The influence of cadmium, copper, lead and zinc on the distribution and evolution of metallophytes in the British Isles. *Plant Systematics and Evolution* 173, 91–108.

BAKER, A. J. M. and WALKER, P. L. (1989) Physiological responses of plants to heavy metals and the quantification of tolerance and toxicity. *Chemical Speciation and Bioavailability* 1, 7–12.

BALK, F. and KOEMAN, J. H. (1984) Future hazards from pesticide use. Commission on Ecology Papers no. 6. International Union for the Conservation of Nature and Natural Resources.

BALLANTYNE, B. and MARRS, T. C. (1992) *Clinical Experimental Toxicology of Organophosphates and Carbamates*. Oxford: Butterworth/Heinemann.

BALLS, M, BRIDGES, J. M. and SOUTHEE, J. (eds) (1991) *Animals and Alternatives in Toxicology: Present Status and Future Prospects*. Basingstoke: Macmillan.

BARTH, H. (ed.) (1987) *Reversibility of Acidification*. London: Elsevier.

BAUMANN, P. C. and HARSHBARGER, J. (1998) Long term trends in liver neoplasm epizootics of brown bullhead in the Black River, Ohio. *Environmental Monitoring and Assessment* 53, 213–23.

BEEBY, A. (1991) Toxic metal uptake and essential metal regulation in terrestrial invertebrates: a review. In NEWMAN, M. C. and MCINTOSH (eds) *Metal Ecotoxicology: Concepts and Applications*, pp. 65–89. Boca Raton, FL: Lewis.

BEGON, M., MORTIMER, M and THOMPSON, D. J. (1996) *Population Ecology: a Unified Study of Animals and Plants*, 3rd edn. Oxford: Blackwell Scientific Publications.

BEITINGER, T. L. (1990) Behavioral reactions for the assessment of stress in fishes. *Journal of Great Lakes Research* 16, 495–528.

BENGTSSON, G., GUNNARSSON, T. and RUNDGREN, S. (1983) Growth changes caused by metal uptake in a population of *Onychiurus armatus* (Collembola) feeding on metal polluted fungi. *Oikos* 40, 216–25.

BENGTSSON, G., GUNNARSSON, T. and RUNDGREN, S. (1985) Influence of metals on reproduction, mortality and population growth in *Onychiurus armatus* (Collembola). *Journal of Applied Ecology* 22, 967–78.

BENN, F. R. and MCAULIFFE, C. A. (eds) (1975) *Chemistry and Pollution*. London: Macmillan.

BERGGREN, D., BERGVIST, B., FALKENGREN-GRERUP, U., FOLKESON, L. and TYLER, G. (1990) Metal solubility and pathways in acidified forest ecosystems of South Sweden. *Science of the total Environment* 96, 103–14.

BERGMAN, Å., OLSSON, M. and REILAND, S. (1989) High frequency of skeletal deformities in skulls of the Baltic grey seal. In YABLOKOV, A. V. and OLSSON, M. (eds) *Influence of Human Activities on the Baltic Ecosystem*, pp. 87–95. <author to insert place of publication and name of publisher>

BERROW, S. D., LONG, S. C., MCGARRY, A. T., POLLARD, D., *et al.* (1998) Radionuclides (^{137}Cs and ^{40}K) in harbour porpoises *Phocoena phocoena* from British and Irish coastal waters. *Marine Pollution Bulletin* 36, 569–76.

BESSELINK, H. T., VAN SANTEN, E., VORSTMAN, W., VETHAAK, A. D., *et al.* (1997) High induction of cytochrome P4501A activity without changes in retinoid and thyroid hormone levels in flounder (*Platichthys flesus*) exposed to 2,3,7,8-tetrachlorodibenzo-p-dioxin (TCDD). *Environmental Toxicology and Chemistry* 16, 816–23.

BICKHAM, J. W., MAZET, J. A., BLAKE, J. SMOLEN, M. J., *et al.* (1998) Flow cytometric determination of genotoxic effects of exposure to petroleum in mink and sea otters. *Ecotoxicology* 7, 191–99.

BISHOP, J. A. (1981) A neoDarwinian approach to resistance: examples from mammals. In BISHOP, J. A. and COOK, L. M. (eds) *Genetic Consequences of Man-made Change*, pp. 37–51. London: Academic Press.

BISHOP, J. A. and COOK, L. M. (1981) *The Genetic Consequences of Man-made Change*. London: Academic Press.

BLOXHAM, M. J., WORSFOLD, P. J. and DEPLEDGE, M. H. (1999) Integrated biological and chemical monitoring: *In situ* physiological responses of freshwater crayfish to fluctuations in environmental ammonia concentrations. *Ecotoxicology* 8, 225–37.

BOON, J. P., EIJGENRAAM, F., EVERAARTS, J. M. and DUINKER, J. C. (1989) A structure–activity relationship (SAR) approach towards metabolism of PCBs in marine animals from different trophic levels. *Marine Environmental Research* 27, 159–76.

BORG, H., ANDERSSON, P. and JOHANSSON, K. (1989) Influence of acidification on metal fluxes in Swedish forest lakes. *Science of the Total Environment* 87/88, 241–53.

BORIO, R., CHIOCCHINI, S., CICIONI, R., ESPOSTI, P. D., *et al.* (1991) Uptake of radiocesium by mushrooms. *Science of the Total Environment* 106, 183–90.

BOSVELD, A.T. C. and VAN DEN BERG, M. (1994) Effects of polychlorinated biphenyls, dibenzo-*p*-dioxins and dibenzofurans on fish-eating birds. *Environmental Reviews* 2, 147–66.

BOWER, J. S., BROUGHTON, G. F. J., STEDMAN, J. R. and WILLIAMS, M. L. (1994) A winter NO$_2$ smog episode in the UK. *Atmospheric Environment* 28, 461–75.

BRAKEFIELD, P. M. (1987) Industrial melanism: do we have the answers? *Trends in Ecology and Evolution* 2, 117–22.

BRITISH ECOLOGICAL SOCIETY (1990) River water quality. Ecological issues no. 1, Field Studies Council, Preston Montford.

BROOKS, G. T. (1974) *Chlorinated Insecticides* (2 vols). Cleveland, OH: CRC Press.

BROUWER, A. and VAN DEN BERG, K. J. (1986) Binding of a Metabolite of 3,4,3′,4′-tetrachlorobiphenyl to transthyretin reduces serum vitamin A transport by inhibiting the formation of the protein complex carrying both retinol and thyroxin. *Toxicology and Applied Pharmacology* 85, 301–12.

BROUWER, A., REIJINDERS, P. J. H. and KOEMAN, J. H. (1989) Polychlorinated biphenyl (PCB)-contaminated fish induces vitamin A and thyroid hormone deficiency in the common seal (*Phoca vitulina*). *Aquatic Toxicology* 15, 99–105.

BROUWER, A., MURK, A. J. and KOEMAN, J. H. (1990) Biochemical and physiological approaches in ecotoxicology. *Functional Ecology* 4, 275–81.

BRYAN, G. W. and LANGSTON, W. J. (1992) Bioavailability, accumulation and effects of heavy metals in sediments with special reference to United Kingdom estuaries: a review. *Environmental Pollution* 76, 89–131.

BRYAN, G. W., GIBBS, P. E., HUMMERSTONE, L. G. and BURT, G. R. (1986) The decline of the gastropod *Nucella lapillus* around South-West England: evidence for the effect of tributyltin from antifouling paints. *Journal of the Marine Biological Association of the United Kingdom* 66, 611–40.

BRYAN, G. W., GIBBS, P. E., and BURT, G. R. (1988) A comparison of the effectiveness of tri-n-butyltin chloride and five other organotin compounds in promoting the development of imposex in the dog whelk *Nucella lapillus*. *Journal of the Marine Biological Association of the United Kingdom* 68, 733–44.

BUNCE, N. (1991) *Environmental Chemistry*. Winnipeg: Wuerz.

BUSBY, D. G., WHITE, L. M. and PEARCE, P. A. (1990) Effects of aerial spraying of fenitrothion on breeding white-throated sparrows. *Journal of Applied Ecology* 27, 743–55.

BUTLER, J. D. (1979) *Air Pollution Chemistry*. London: Academic Press.

CADE, T. J., ENDERSON, J. H., THELANDER, C. G. and WHITE, C. M. (eds) (1988) *Peregrine Populations: Their Management and Recovery*. Boise, ID: The Peregrine Fund Inc.

CAFFREY, P. B. and KEATING, K. I. (1997) Results of zinc deprivation in Daphnid culture. *Environmental Toxicology and Chemistry* 16, 572–5.

CALABRESE, E. J. (1999). Evidence that hormesis represents an 'overcompensation' response to a disruption in homeostasis. *Ecotoxicology and Environmental Safety* 42, 135–7.

CALAMARI, D. and VIGHI, M. F. (1992) Role of evaluative models to assess exposure to pesticides. In TARDIFF, R. G. (ed.) *Methods to Assess Adverse Organisms (SCOPE)*, pp. 119–32. Chichester: John Wiley.

CALLAHAN, C. A., MENZIE, C. A., BURMASTER, D. E., WILBORN, D. C. and ERNST, T. (1991) On-site methods for assessing chemical impact on the soil environment using earthworms: a case study at the Baird and McGuire superfund site. *Environmental Toxicology and Chemistry* 10, 817–26.

CALOW, P. (1989) Ecotoxicology? *Journal of Zoology* 218, 701–4.

CALOW, P. (ed.) (1993) *Handbook of Ecotoxicology*, vol. 1. Oxford: Blackwell Science.

CALOW, P. (ed.) (1994) *Handbook of Ecotoxicology*, vol. 2. Oxford: Blackwell Science.

CALOW, P., SIBLY, R. M. and FORBES, V. (1997) Risk assessment on the basis of simplified life-history scenarios. *Environmental Toxicology and Chemistry* 16, 1983–9.

CAMPBELL, P. M. and HUTCHINSON, T. H. (1998) Wildlife and endocrine disruptors: requirements for hazard identification. *Environmental Toxicology and Chemistry* 17, 127–35.

CAMPLANI, A., SAINO, N. and MØLLER, A. P. (1999) Carotenids, sexual signals and immune function in barn swallows from Chernobyl. *Proceedings of the Royal Society of London* **266B**, 1111–16.

CARTER, L. J. (1976) Michigan's PBB incident: chemical mix-up leads to disaster. *Science* **192**, 240–3.

CASTELLINI, S., SAVVA, D., RENZONI, A. and MATTEI, E. N. (1996) Biomarkers of genotoxicity in two species of bivalves molluscs from coastal ecosystems of north Adriatic Sea. *Proceedings of the Italian Society of Ecology* **17**, 399–402.

CASWELL, H. (1989) *Matrix Population Models.* Sunderland, MA: Sinauer Associates Inc.

CATTANI, O., FABBRI, D., SALVATI, M., TROMBINI, C. and VASSURA, I. (1999) Biomonitoring of mercury pollution in a wetland near Ravenna, Italy, by translocated bivalves (*Mytilus galloprovincialis*). *Environmental Toxicology and Chemistry* **18**, 1801–5.

CHAMP, M. A. (1999) The need for the formation of an Independent, International Marine Coatings Board. *Marine Pollution Bulletin* **38**, 239–46.

CLARK, R. B. (1992) *Marine Pollution*, 3rd edn. Oxford: Clarendon Press.

CLARKE, B. (1975) The contribution of ecological genetics to evolutionary theory: detecting the direct effects of natural selection on particular polymorphic loci. *Genetics* **79**, 101–8.

CLARKE, K. R. and Warwick, R. M. (1998) A taxonomic distinctness index and its statistical properties. *Journal of Applied Ecology* **35**, 523–31.

COALE, K. H., JOHNSON, K. S., FITZWATER, S. E., GORDON, R. M. *et al.* (1996) A massive phytoplankton bloom induced by an ecosystem-scale iron fertilization experiment in the equatorial Pacific Ocean. *Nature* **383**, 495–501. (Twenty authors in total!)

COLEMAN, J. E. (1967) Metal ion dependent binding of sulphonamide to carbonic anhydrase. *Nature* **214**, 193–4.

COOK, L. M., DENNIS, R. L. H. and MANI, G. S. (1999) Melanic morph frequency in the peppered moth in the Manchester area. *Proceedings of the Royal Society of London* **266B**, 293–7.

CRICK, H. Q. P., BAILLIE, S. R., BALMER, D. E., BASHFORD, R. I., *et al.* (1998) Breeding birds in the wider countryside; their conservation status. Research report 198. British Trust for Ornithology. Thetford: BTO.

CROMMENTUIJN, T., BRILS, J. and VAN STRAALEN, N. M. (1993) Influence of cadmium on life-history characteristics of *Folsomia candida* (Willem) in an artificial soil substrate. *Ecotoxicology and Environmental Safety* **26**, 216–27.

CROSBY, D. G. (1998) *Environmental Toxicology and Chemistry*. Oxford: Oxford University Press.

CROSSLAND, N. O. (1988) A method for evaluating effects of toxic chemicals on fish growth rates. In ADAMS, J. A., CHAPMAN, G. A. and LANDIS, W. G. (eds) *Aquatic Toxicology and Hazard Assessment*, 10th vol. Philadelphia: American Society for Testing and Materials.

CROSSLAND, N. O. (1994) Extrapolating from mesocosms to the real world. *Toxicology and Ecotoxicology News* **1**, 15–22.

CROUAU, Y., CHENON, P. and GISLARD, C. (1999) The use of *Folsomia candida* (Collembola: Isotomidae) for the bioassay of xenobiotic substances and soil pollutants. *Applied Soil Ecology* 12, 103–11.

CROUT, N. M. J., BERESFORD, H. A. and HOWARD, B. J. (1991) The radiological consequences for lowland pastures used to fatten upland sheep contaminated with radiocaesium. *Science of the Total Environment* 103, 73–88.

CRUMP, R. G., MORLEY, H. S. and WILLIAMS, A. D. (1999) West Angle Bay, a case study. Littoral monitoring of permanent quadrats before and after the *Sea Empress* oil spill. *Field Studies* 9, 497–511.

CULBARD, E. B., THORNTON, I., WATT, J., WHEATLEY, M., *et al.* (1988) Metal contamination in British urban dusts and soils. *Journal of Environmental Quality* 17, 226–34.

CURTIS, C. F., COOK, L. M. and WOOD, R. J. (1978) Selection for and against insecticide resistance and possible methods of inhibiting the evolution of resistance in mosquitoes. *Ecological Entomology* 3, 273–87.

DALLINGER, R. (1993) Strategies of metal detoxification in terrestrial invertebrates. In DALLINGER, R. and RAINBOW, P. S. (eds) *Ecotoxicology of Metals in Invertebrates*, pp. 245–89. Chelsea, MI: Lewis.

DALLINGER, R. and RAINBOW, P. S. (eds) (1993) *Ecotoxicology of Metals in Invertebrates*. Chelsea, MI: Lewis.

DALLINGER, R., BERGER, B., HUNZIKER, P. and KÄGI, J. H. R. (1997) Metallothionein in snail Cd and Cu metabolism. *Nature*, 388, 237.

DANIELS, R. E. and ALLAN, S. D. (1981) Life table evaluation of chronic exposure to pesticide. *Canadian Journal of Fisheries and Aquatic Sciences* 38, 485–94.

DEININGER, R. A. (1987) Survival of Father Rhine. *Journal of American Water Works Association* 79, 78–93.

DELL'OMO, G. (ed.) (2000) *Behaviour in Ecotoxicology*. Chichester: Wiley.

DELL'OMO, G. and SHORE, R. F. (1996a) Behavioral and physiological effects of acute sublethal exposure to dimethoate on wood mice, *Apodemus sylvaticus*. I. Laboratory studies. *Archives of Environmental Contamination and Toxicology* 31, 91–7.

DELL'OMO, G. and SHORE, R. F. (1996b) Behavioral and physiological effects of acute sublethal exposure to dimethoate on wood mice, *Apodemus sylvaticus*. II. Field studies on radio-tagged mice in a cereal ecosystem. *Archives of Environmental Contamination and Toxicology* 31, 538–42.

DEMPSTER, J. P. (1975) Effects of organochlorine insecticides on animal populations. In MORIARTY, F. (ed.) *Organochlorine Insecticides: Persistent Organic Pollutants*, pp. 231–48. London: Academic Press.

DEPLEDGE, M. H. (1994) The rational base for the use of biomarkers as ecotoxicological tools. In FOSSI, C. M. and LEONZIO, C. (eds) *Nondestructive Biomarkers in Vertebrates*, pp. 271–95. Boca Raton, FL: Lewis.

DEPLEDGE, M. H. and ANDERSEN, B.B. (1990) A computer-aided physiological monitoring system for the continuous, long-term recording of cardiac activity in selected invertebrates. *Comparative Biochemistry and Physiology* **96A**, 473–7.

DEPLEDGE, M. H., AMARAL-MENDEL, J. J., DANIEL, B., HALBROOK, R. S., *et al.* (1993) The conceptual basis of the biomarker approach. In PEAKALL, D. B. and SHUGART, R. L. (eds) *Biomarkers. Research and Application in the Assessment of Environmental Health*, pp. 15–29. Berlin: Springer-Verlag.

DEVONSHIRE, A. L. and SAWICKI, R. M., (1979) Insecticide-resistant *Myzus persicae* as an example of evolution by gene duplication. *Nature* **280**, 140–1.

DIX, H. M. (1981) *Environmental Pollution: Atmosphere, Land, Water and Noise.* Chichester: John Wiley.

DONKER, M. H., EIJSACKERS, H. and HEIMBACH, F. (eds) (1994) *Ecotoxicology of Soil Organisms.* Boca Raton, FL: Lewis.

DROBNE, D. (1997). Terrestrial isopods – a good choice for toxicity testing of pollutants in the terrestrial environment. *Environmental Toxicology and Chemistry* **16**, 1159–64.

DROBNE, D. and HOPKIN, S. P. (1994) Ecotoxicological laboratory test for assessing the effects of chemicals on terrestrial isopods. *Bulletin of Environmental Contamination and Toxicology* **53**, 390–7.

EDWARDS, C. A. (1976) *Persistent Pesticides in the Environment*, 2nd edn. Boca Raton, FL: CRC Press.

EDWARDS, T. (1994) Chornobyl [sic!]. *National Geographic* **186**, 100–15.

EEVA, T., LEHIKOINEN, E. and RÖNKA, M. (1998). Air pollution fades the plumage of the Great Tit. *Functional Ecology* **12**, 607–12.

ELLIOTT, J. M. (1993) A 25-year study of production of juvenile Sea-trout, *Salmo trutta*, in an English Lake District stream. *Canadian Special Publication of Fisheries and Aquatic Sciences* **118**, 109–22.

ELDEFRAWI, M. E. and ELDEFRAWI, A. T. (1990) Nervous system-based insecticides. In HODGSON, E. and KUHR, R. J. (eds) (1990) *Safer Insecticides – Development and Use.* New York: Marcel Dekker.

ENDERSON, J. H., TEMPLE, S. A. and SWARTZ, L. G. (1972) Time-lapse photographic records of nesting Peregrine Falcons. *Living Bird* **11**, 113–28.

ENVIRONMENTAL HEALTH CRITERIA NO. 63 (1986) *Organophosphorus Insecticides – a General Introduction.* Geneva: WHO.

ENVIRONMENTAL HEALTH CRITERIA NO. 64 (1986) *Carbamate Pesticides – a General Introduction.* Geneva: WHO.

ENVIRONMENTAL HEALTH CRITERIA NO. 86 (1989) *Mercury – Environmental Aspects.* Geneva: WHO.

ENVIRONMENTAL HEALTH CRITERIA NO. 101 (1990) *Methylmercury.* Geneva: WHO.

ENVIRONMENTAL HEALTH CRITERIA NO. 116 (1990) *Tributyl Tin Compounds.* Geneva: WHO.

ERNST, W. (1976) Physiological and biochemical aspects of metal tolerance. In MANSFIELD, T. A. (ed.) *Effects of Air Pollutants on Plants.* Cambridge: Cambridge University Press.

ERNST, W. H. O. and PETERSON, P. J. (1994) The role of biomarkers in environmental assessment. (4) Terrestrial plants. *Ecotoxicology* **3**, 180–92.

ERNST, W. H. O., VERKLEIJ, J. A. C. and SCHAT, H. (1992) Metal tolerance in plants. *Acta Botanica Neerlandica* **41**, 229–48.

ERNST, W. R., PEARCE, P. A. and POLLOCK, T. L. (1989) Environmental effects of fenitrothion use in forestry. Environment Canada, Atlantic Region Report.

ETO, M. (1974) *Organophosphorus Insecticides: Organic and Biological Chemistry.* Cleveland, OH: CRC Press.

EVANS, P. R. (1990) Population dynamics in relation to pesticide use, with particular reference to birds and mammals. In SOMERVILLE, L. and WALKER, C. H. (eds) *Pesticide Effects on Terrestrial Wildlife*, pp. 307–17. London: Taylor and Francis.

EVERAARTS, J. M., SHUGART, L. R., GUSTIN, M. K., HAWKINS, W. E. and WALKER, W. W. (1993) Biological markers in fish: DNA integrity, hematological parameters and liver somatic index. *Marine Environmental Research* 35, 101–7.

EVERAARTS, J. M., DEN BESTEN, P. J., HILLEBRAND, M. Th. J., HALBROOK, R. S. and SHUGART, L. R. (1998) DNA strand breaks, cytochrome P-450-dependent monooxygenase system activity and level of chlorinated biphenyl congeners in the pyloric caeca of the seastar (*Asterias rubens*) from the North Sea. *Ecotoxicology* 7, 69–80.

FAIRBROTHER, A. (1994) Clinical biochemistry. In FOSSI, M. C. and LEONZIO, C. (eds) *Nondestructive Biomarkers in Vertebrates*, pp. 63–89. Boca Raton, FL: Lewis.

FAIRBROTHER, A., MARDEN, B. T., BENNETT, J. K. and HOOPER, M. J. (1991) Methods used in determination of cholinesterase activity. In MINEAU, P. (ed.) *Cholinesterase-inhibiting Insecticides – their Impact on Wildlife and the Environment*, pp. 35–71. Amsterdam: Elsevier.

FEDER, M. E., BENNETT, A. F., BURGGREN, W. W. and HUEY, R. B. (1987) *New Directions in Ecological Physiology.* Cambridge: Cambridge University Press.

FEST, C. and SCHMIDT, K.-J. (1982) *Chemistry of Organophosphorus Pesticides*, 2nd edn. Berlin: Springer-Verlag.

FISCHER, H. (1989) Cadmium in seawater recorded by mussels: regional decline established. *Marine Ecology Progress Series* 55, 159–69.

FORBES, V. E. (ed.) (1999) *Genetics and Ecotoxicology.* Philadelphia: Taylor and Francis.

FORBES, V. E. and CALOW, P. (1999) Is the per capita rate of increase a good measure of population-level effects on ecotoxicology? *Environmental Toxicology and Chemistry* 17, 1544–56.

FORRESTER, N. W., CAHILL, M., BIRD, L. J. and LAYLAND, J. K. (1993) Management of pyrethroid and endosulfan resistance in *Helicoverpa armigera* (Lepidoptera: Noctuidae) in Australia. *Bulletin of Entomological Research* (supplement 1), 1–132.

FOSSI, M. C. and LEONZIO, C. (ed.) (1994) *Nondestructive Biomarkers in Vertebrates.* Boca Raton, FL: Lewis.

FOX, G. A., KENNEDY, S. W., NORSTROM, R. J. and WINGFIELD, D. C. (1988) Porphyria in herring gulls: a biochemical response to chemical contamination of Great Lake food chains. *Environmental Toxicology and Chemistry* 7, 831–9.

FRANK, P. W., BOLL, C. D. and KELLY, R. W. (1957) Vital statistics of laboratory cultures of *Daphnia pulex* DeGeer as related to density. *Physiological Zoology* 30, 287–305.

FREEDMAN, B. (1989) *Environmental Ecology: the Impacts of Pollution and Other Stresses on Ecosystem Structure and Function.* San Diego: Academic Press.

FREEDMAN, B. (1995) *Environmental Ecology: the Environmental Effects of Pollution Disturbance and Other Stresses*, 2nd edn. San Diego: Academic Press.

FRYDAY, S. L., HART, A. D. M. and DENNIS, N. J. (1994) Effects of exposure to an organophosphate on the seed-handling efficiency of the House sparrow. *Bulletin of Environmental Contamination and Toxicology* 53, 869–76.

FUTUYMA, D. J. (1986) *Evolutionary Biology*, 2nd edn. Sunderland, MA: Sinauer Associates.

GIBBS, P. E. (1993) A male genital defect in the dog-whelk, *Nucella lapillus* (Neogastropoda), favouring survival in a TBT-polluted area. *Journal of the Marine Biological Association of the United Kingdom* 73, 667–78.

GIBBS, P. E. and BRYAN, G. W. (1986) Reproductive failure in populations of the dog whelk *Nucella lapillus*, caused by imposex induced by tributyltin from antifouling paints. *Journal of the Marine Biological Association of the United Kingdom* 66, 767–77.

GIBBS, P. E., PASCOE, P. L. and BURT, G. R. (1988) Sex change in the female dog whelk, *Nucella lapillus*, induced by tributyl tin from antifouling paints. *Journal of the Marine Biological Association of the United Kingdom* 68, 715–31.

GIBSON, G. and SKETT, P. (1986) *Introduction to Drug Metabolism.* London: Chapman & Hall.

GILBERTSON, M., FOX, G. A. and BOWERMAN, W. W. (1998) *Trends in Levels and Effects of Persistent Toxic Substances in the Great Lakes.* Dordrecht: Kluwer Academic Press.

GOUGH, J. J., McINDOE, E. C. and LEWIS, G. B. (1994) The use of dimethoate as a reference compound in laboratory acute toxicity tests on honey bees (*Apis mellifera* L.) 1981–1992. *Journal of Apicultural Research* 33, 119–25.

GRANT, A. and MIDDLETON, R. (1990) An assessment of metal contamination in the Humber Estuary, U.K. *Estuarine, Coastal and Shelf Science* 31, 71–85.

GRAY, J. S. (1981) *The Ecology of Marine Sediments.* Cambridge: Cambridge University Press.

GREIG-SMITH, P. W. and HARDY, A. R. (1992) Design and management of the Boxworth project. In GREIG-SMITH, P., FRAMPTON, G. and HARDY, T. (eds) *Pesticides, Cereal Farming and the Environment. The Boxworth Project.* London: HMSO.

GREIG-SMITH, P. W., BECKER, H., EDWARDS, P. J. and HEIMBACH, F. (eds) (1992a) *Ecotoxicology of Earthworms.* Andover: Intercept.

GREIG-SMITH, P. W., FRAMPTON, G. and HARDY, T. (1992b) *Pesticides, Cereal Farming and the Environment. The Boxworth Project.* London: HMSO.

GRUE, C. E., FLEMING, W. J., BUSBY, D. G. and HILL, E. F. (1983) Assessing hazards of organophosphate pesticides to wildlife. *Transactions North American Wildlife Conference* 48, 200–20.

GRUE, C. E., HOFFMAN, D. J., BEYER, W. N. and FRANSON, L. P. (1986) Lead concentrations and reproductive success in European starlings *Sturnus vulgaris* nesting within highway roadside verges. *Environmental Pollution* **42A**, 157–82.

GRUE, C. E., HART, A. D. M. and MINEAU, P. (1991) Biological consequences of depressed brain cholinesterase activity in wildlife. In MINEAU, P. (ed.) *Cholinesterase-inhibiting Insecticides – their Impact on Wildlife and the Environment*, pp. 151–210. Amsterdam: Elsevier.

GUILHERMINO, L., BARROS, P. SILVA, M. C. and SOARES, A. M. V. M. (1998) Should the use of inhibition of cholinesterases as a specific biomarker for organophosphate and carbamate pesticides be questioned? *Biomarkers* **3**, 157–63.

GUNTHER, A. J., DAVIS, J. A., HARDIN, D. D., GOLD, J. *et al.* (1999) Long-term bioaccumulation monitoring with transplanted bivalves in the San Francisco Estuary. *Marine Pollution Bulletin* **38**, 170–81.

GUPTA, S. K. and SUNDARARAMAN, V. (1991) Correlation between burrowing capability and AChE activity in the earthworm, *Pheretima posthuma*, on exposure to carbaryl. *Bulletin of Environmental Contamination and Toxicology* **46**, 859–65.

GUTHRIE, F. E. and PERRY, J. J. (eds) (1980) *Introduction to Environmental Toxicology*. New York: Elsevier.

HÅGVAR, S. and ABRAHAMSEN, G. (1990) Microarthropoda and Enchytraeidae (Oligochaeta) in naturally lead-contaminated soil: a gradient study. *Environmental Entomology* **19**, 1263–77.

HAIMI, J. and SIIRA-PIETIKAINEN, A. (1996) Decomposer animal communities in forest soil along heavy metal pollution gradient. *Fresenius Journal of Analytical Chemistry* **354**, 672–5.

HAMER, D. M. (1986) Metallothionein. *Annual Review of Biochemistry* **55**, 913–51.

HAMERS, T. and KROGH, P. H. (1997). Predator–prey relationships in a two species toxicity test system. *Ecotoxicology and Environmental Safety* **37**, 203–12.

HANDY, R. D. and DEPLEDGE, M. H. (2000) Physiological responses; their measurement and use as environmental biomarkers in ecotoxicology. *Ecotoxicology*

HARRIES, J. E., SHEAHAN, D. A., JOBLING, S., MATTHIESSEN, P., *et al.* (1997) Estrogenic activity in five United Kingdom rivers detected by measurement of vitellogenesis in caged male trout. *Environmental Toxicology and Chemistry* **16**, 534–42.

HARRIES, J. E., JANBAKHSH, A., JOBLING, S., MATTHIESSEN, P., *et al.* (1999) Estrogenic potency of effluent from two sewage treatment works in the United Kingdom. *Environmental Toxicology and Chemistry* **18**, 932–7.

HART, A. D. M. (1993) Relationships between behaviour and the inhibition of acetylcholinesterase in birds exposed to organophosphorus pesticides. *Environmental Toxicology and Chemistry* **12**, 321–36.

HART, L. G., SHULTICE, R. W. and FOUTS, J. R. (1963) Stimulatory effects of chlordane on hepatic microsomal drug metabolism in the rat. *Toxicology and Applied Pharmacology* **5**, 371–86.

HARTL, D. L. and CLARK, A. G. (1989) *Principles of Population Genetics*. Sunderland, MA: Sinauer Associates.

HARTWELL, S. I., CHERRY, D. S. and CAIRNS, JR, J. (1987) Field validation of avoidance of elevated metals by fathead minnows (*Pimephales promelas*) following *in situ* acclimation. *Environmental Toxicology and Chemistry* 6, 189–200.

HASSALL, K. A. (1990) *The Biochemistry and Uses of Pesticides*, 2nd edn. London: Macmillan.

HEDGECOTT, S. (1994) Prioritization and standards for hazardous chemicals. In CALOW, P. (ed.) *Handbook of Ecotoxicology*, pp 368–93, Oxford: Blackwell.

HEGDAL, P. L. and BLASKIEWICZ, R. W. (1984) Evaluation of the potential hazard to Barn Owls of talon (brodifacoum bait) used to control rats and house mice. *Environmental Toxicology and Chemistry* 3, 167–79.

HEGGESTAD, H. E. (1991) Origin of Bel-W3, Bel-C and Bel-B tobacco varieties and their use as indicators of ozone. *Environmental Pollution* 74, 264–91.

HELIÖVAARA, K., VÄISÄNEN, R., BRAUNSCHWEILER, H. and LODENIUS, M. (1987) Heavy metal levels in two biennial pine insects with sap-sucking and gall-forming life styles. *Environmental Pollution* 48, 13–23.

HENSBERGEN, P. J., DONKER, M. H., VAN VELZEN, M. J. M., ROELOFS, D., *et al.* (1999) Primary structure of a cadmium-induced metallothionein from the insect *Orchesella cincta* (Collembola). *European Journal of Biochemistry* 259, 197–203.

HICKEY, J. J. (1969) *The Peregrine Falcon Populations: Their Biology and Decline*. Madison, WI: University of Wisconsin Press.

HICKEY, J. J., KEITH, J. A. and COON, F. B. (1966) An exploration of pesticides in a Lake Michigan ecosystem. *Journal of Applied Ecology* 3 (supplement), 141–54.

HILL, E. F. and FLEMING, W. J. (1982) Anticholinesterase poisoning of birds: field monitoring and diagnosis of acute poisoning. *Environmental Toxicology and Chemistry* 1, 27–38.

HODGSON, E. and LEVI, P. (1994) *Introduction to Biochemical Toxicology*, 2nd edn. Norwalk, CT: Appleton and Lange.

HODGSON, E. and KUHR, R. J. (eds) (1990) *Safer Insecticides – Development and Use*. New York: M Dekker.

HOFFMAN, A. A. and PARSONS, P. A. (1991) *Evolutionary Genetics and Environmental Stress*. Oxford: University Press.

HOFFMAN, D. J., RATTNER, B. A., BURTON, JR, G. A. and CAIRNS, JR, J. (eds) (1995) *Handbook of Ecotoxicology*. Lewis Publishers, Boca Raton.

HOLLOWAY, G. J., SIBLY, R. M. and POVEY, S. R. (1990) Evolution in toxin-stressed environments. *Functional Ecology* 4, 289–94.

HOLMES, S. B. and BOAG, P. T. (1990) Inhibition of brain and plasma cholinesterase activity in zebra finches orally dosed with fenitrothion. *Environmental Toxicology and Chemistry* 9, 323–34.

HOPKIN, S. P. (1989) *Ecophysiology of Metals in Terrestrial Invertebrates*. Barking (UK): Elsevier Applied Science.

HOPKIN, S. P. (1990) Critical concentrations, pathways of detoxification and cellular ecotoxicology of metals in terrestrial arthropods. *Functional Ecology* **4**, 321–7.

HOPKIN, S. P. (1993a) *In situ* biological monitoring of pollution in terrestrial and aquatic ecosystems. In CALOW, P. (ed.) *Handbook of Ecotoxicology*, vol. 1, pp. 397–427. Oxford: Blackwell.

HOPKIN, S. P. (1993b) Deficiency and excess of copper in terrestrial isopods. In DALLINGER, R. and RAINBOW, P. S. (eds) *Ecotoxicology of Metals in Invertebrates*, pp. 359–82. Chelsea, MI: Lewis.

HOPKIN, S. P. (1993c) Ecological implications of '95% protection levels' for metals in soils. *Oikos* **66**, 137–41.

HOPKIN, S. P. (1995) Deficiency and excess of essential and non-essential metals in terrestrial insects. *Symposia of the Royal Entomological Society of London* **17**, 251–70.

HOPKIN, S. P. (1997) *Biology of the Springtails (Insecta: Collembola)*. Oxford: Oxford University Press.

HOPKIN, S. P. and MARTIN, M. H. (1984) Heavy metals in woodlice. *Symposia of the Zoological Society of London* **53**, 143–66.

HOPKIN, S. P. and SPURGEON, D. J. (2000) Forecasting the environmental effects of zinc, the metal of benign neglect in soil ecotoxicology. In RAINBOW, P. S., HOPKIN, S. P. and CRANE, M. (eds) *Forecasting the Fate and Effects of Toxic Chemicals*. John Wiley, Chichester.

HOPKIN, S. P., HARDISTY, G. N. and MARTIN, M. H. (1986) The woodlouse *Porcellio scaber* as a 'biological indicator' of zinc, cadmium lead and copper pollution. *Environmental Pollution* (series B) **11**, 271–90.

HOPKIN, S. P., HAMES, C. A. C. and DRAY, A. (1989) X-ray microanalytical mapping of the intracellular distribution of pollutant metals. *Microscopy and Analysis* **14**, 23–7.

HOVE, K., PEDERSON, O. GARMO, T. H., HANSEN, H. S. and STAALAND, H. (1990) Fungi; a major source of radiocesium contamination of grazing ruminants in Norway. *Health Physics* **59**, 189–92.

HUCKLE, K. R., WARBURTON, P. A., FORBES, S. and LOGAN, C. J. (1989) Studies on the fate of flucoumafen in the Japanese Quail (*Coturnix coturnix japonica*). *Xenobiotica* **18**, 51–62.

HUDSON, R. H., TUCKER, R. K. and HAEGELE, M. A. (1984) *Handbook of Toxicity of Pesticides to Wildlife*. US Fish and Wildlife Service Resource Publishers, 153.

HUET, M., PAULET, Y. M. and LE PENNEC, M. (1996). Survival of *Nucella lapillus* in a tributyltin-polluted area in West Brittany: A further example of a male genital defect (Dumpton syndrome) favouring survival. *Marine Biology* **125**, 543–9.

HUGGETT, R. J., KIMERLE, R. H., MEHRLE, JR, P. M. and BERGMAN, H. L. (eds) (1992) *Biomarkers. Biochemical, Physiological, and Histological Markers of Anthropogenic Stress*. Boca Raton, FL: Lewis.

HUTTON, M. (1980) Metal contamination of feral pigeons *Columba livia* from the London area. Part 2. Biological effects of lead exposure. *Environmental Pollution* **22A**, 281–93.

HYNES, H. B. N. (1960) *The Biology of Polluted Waters*. Liverpool: Liverpool University Press.

INOUE, K., OSAKABE, M., ASHIHARD, W. and HAMAMURA, T. (1986) Factors affecting abundance of the Kanzawa spider mite *Tetranychus kanzawai* on grapevine in a glasshouse; influence of pesticidal application on occurrence of the Kanzawa spider mite and its predators. *Bulletin of the Fruit Tree Research Station*, Series E, 103–16.

JAGOE, C. H., DALLAS, C. E., CHIESSER, R. K., SMITH, M. H. *et al.* (1998). Contamination near Chernobyl: radiocaesium, lead and mercury in fish and sediment radiocaesium from waters within the 10 km zone. *Ecotoxicology*, 7, 201–9.

JAK, R. G., CEULEMANS, M., SCHOLTEN, M. C. T. and VAN STRAALEN, N. M. (1998) Effects of tributyltin on a coastal North Sea plankton community in enclosures. *Environmental Toxicology and Chemistry* 17, 1840–7.

JANSSEN, P. A. H., FABER, J. H. and BOSVELD, A. T. C. (1998) *(Fe)male?* IBN Science Contribution 13.

JANSSEN, R. P. T., POSTHUMA, L., BAERSELMAN, R., DEN HOLLANDER, H. A., *et al.* (1997) Equilibrium partitioning of heavy metals in Dutch field soils. II. Prediction of metal accumulation in earthworms. *Environmental Toxicology and Chemistry* 16, 2479–88.

JENSEN, C. S., GARSDAL, L. and BAATRUP, E. (1997) Acetylcholinesterase inhibition and altered locomotor behavior in the carabid beetle *Pterostichus cupreus*. A linkage between biomarkers at two levels of biological complexity. *Environmental Toxicology and Chemistry* 16, 1727–32.

JOHNSTON, G. O. (1995) The study of interactive effects of pesticides in birds – a biomarker approach. *Aspects of Applied Biology* 41, 25–31.

JONES, K. C., SYMON, C., TAYLOR, P. J. L., WALSH, J. and JOHNSTON, A. E. (1991) Evidence for a decline in rural herbage lead levels in the U.K. *Atmospheric Environment* 25A, 361–9.

JORGENSEN, S. E. (ed.) (1991) *Modelling in Ecotoxicology*. Amsterdam: Elsevier.

JUKES, T. (1985) Selenium not for dumping. *Nature* 349, 438–40.

KAZAKOV, V. S., DEMIDCHIK, E. P. and ASTAKHOVA, L. N. (1992) Thyroid cancer after Chernobyl. *Nature* 359, 21.

KENNEDY, S. W. and JAMES, C. A.(1993) Improved method to extract and concentrate porphyrins from liver tissue for analysis by high-performance liquid chromatography. *Journal of Chromatography* 619, 127–32.

KETTLEWELL, B. (1973) *The Evolution of Melanism*. Oxford: Clarendon Press.

KINTER, W. B., MERKINS, L. S., JANICKI, R. H. and GUARINO, A. M. (1972) Studies on the mechanism of toxicity of DDT and polychlorinated biphenyls (PCBs): disruption of osmoregulation in marine fish. *Environmental Health Perspectives* 1, 169–73.

KOEMAN, J. H. and VAN GENDEREN, H. (1972) Tissue levels in animals and effects caused by chlorinated hydrocarbons, chlorinated biphenyls and mercury in the marine environment. In *Marine Pollution and Sea Life*, pp. 1–8. West Byfleet: Fishing New Books.

Koss, G., Schuler, E., Arndt, B., Siedel, J., *et al.* (1986) A comparative toxicological study on pike (*Esox lucius* L.) from the River Rhine and River Lahn. *Aquatic Toxicology* **8**, 1–9.

Kruckeberg, A. L. and Wu, L. (1992) Copper tolerance and copper accumulation of herbaceous plants colonizing inactive Californian copper mines. *Ecotoxicology and Environmental Safety* **23**, 307–19.

Kudo, A., Miyahara, S. and Miller, D. R. (1980) Movement of mercury from Minimata Bay into Yatsushiro Sea. *Progress in Water Technology* **12**, 509–24.

Kuhr, R. and Dorough, W. (1977) *Carbamate Insecticides*. Cleveland, OH: CRC Press.

Lacerda, L. D., Pfeiffer, W. C., Ott, A. T. and Silveira, E. G. (1989) Mercury contamination in the Madeira River, Amazon – Hg inputs to the environment. *Biotropica* **21**, 91–3.

Langston, W. J. (1996) Recent developments in TBT ecotoxicology. *Toxicology and Ecotoxicology News* **3**, 179–87.

Langston, W. J. and Bebianno, M. J. (eds) (1998) *Metal Metabolism in Aquatic Environments*. Chapman & Hall, London.

Leahey, J. P. (1985) *The Pyrethroid Insecticides*. London: Taylor and Francis.

Leivestad, H., Hendry, G., Muniz, I. P. and Snekvik, E. (1976) Effects of acid precipitation on freshwater organisms. In Brakke, F. H. (ed.) *Impact of Acid Precipitation on Forest and Freshwater Ecosystems in Norway*, pp. 87–111. Research Report FR 6/76. Oslo: SNSF Project.

Lepp, N. W. and Dickinson, N. M. (1994) Fungicide-derived copper in tropical plantation crops. In Ross, S. M. (ed.) *Toxic Metals in Soil-Plant Systems*, pp. 367–93. Chichester: John Wiley.

Levin, L., Caswell, H., Bridges, T., Dibacco, C., *et al.* (1996) Demographic responses of estuarine polychaetes to sewage, algal and hydrocarbon additions: life table response experiments. *Ecological Application* **6**, 1295–1313.

Lewis, D. F. V (1996) *Cytochromes P450: Structure, Function, and Mechanism*. Taylor and Francis: London.

Lewis, M. A. (1990) Are laboratory-derived toxicity data for freshwater algae worth the effort? *Environmental Toxicology and Chemistry* **9**, 1279–84.

Lima, S. L. and Dill, L. M. (1990) Behavioural decisions made under the risk of predation: a review and prospectus. *Canadian Journal of Zoology* **68**, 619–40.

Linke-Gamenick, I., Forbes, V.E. and Sibly, R. M. (1999) Density-dependant effects of a toxicant on life history traits and population dynamics of a capitellid polychaete. *Marine Ecology Progress Series* **184**, 139–148.

Little, E. J., McCaffery, A. R., Walker, C. H. and Parker, T. (1989) Evidence for an enhanced metabolism of cypermethrin by a monooxygenase in a pyrethroid-resistant strain of the tobacco budworm (*Heliothis virescens* F.). *Pesticide Biochemistry and Physiology* **34**, 58–68.

Livingstone, D. R. and Stegeman, J. J. (eds) (1998) *Forms and functions of cytochrome P450. Comparative Biochemistry and Physiology* **121C** (supplement).

Loganathan, B. G., Tanabe, S., Tanaka, H., Watanabe, S., *et al.* (1990) Comparison of organochlorine residue levels in the striped dolphin from

Western Northern Pacific, 1978–1979 and 1986. *Marine Pollution Bulletin* 21, 435–9.

LØKKE, H. and VAN GESTEL, C. A. M. (eds) (1998) *Handbook of Soil Invertebrate Toxicity Tests*. Chichester: John Wiley.

LOSEY, J. E., RAYOR L. S. and CARTER, M. E. (1999) Transgenic pollen harms monarch larvae. *Nature* 399, 214.

LOVELOCK, J. (1989) *The Ages of Gaia: a Biography of our Living Planet*. Oxford: Oxford University Press.

LOWE, V. P. W. (1991) Radionuclides and the birds at Ravenglass. *Environmental Pollution* 70, 1–26.

LUKASHEV, V. K. (1993) Some geochemical and environmental aspects of the Chernobyl nuclear accident. *Applied Geochemistry* 8, 419–36.

LUTGENS, F. K. and TARBUCK, E. J. (1992) *The Atmosphere: an Introduction to Meteorology*. Hemel Hempstead: Prentice Hall.

MACKAY, D. (1991) *Multimedia Environmental Models: the Fugacity Approach*. Chelsea, MI: Lewis.

MACKENZIE, A. B. and SCOTT, R. D. (1993) Sellafield waste radionuclides in Irish Sea intertidal and salt marsh sediments. *Environmental Geochemistry and Health* 15, 173–84.

MACKENZIE, A. B., SCOTT, R. D., ALLAN, R. L., BEN SHABAN, Y. A., *et al.* (1994) Sediment radionuclide profiles: implications for mechanisms of Sellafield waste dispersal in the Irish Sea. *Journal of Environmental Radioactivity* 23, 36–69

MACNAIR, M. R. (1981) Tolerance of higher plants to toxic materials, In BISHOP, J. A. and COOK, L. S. (eds) *Genetic Consequences of Man-made Change*. London: Academic Press.

MACNAIR, M. R. (1987) Heavy metal tolerance in plants: a model evolutionary system. *Trends in Evolution and Ecology* 2, 354–9.

MAJERUS, M. (1998) *Melanism: Evolution in Action*. Oxford: Oxford University Press.

MALINS, D. C. and COLLIER, T. K. (1981) Xenobiotic interactions in aquatic organisms: effects on biological systems. *Aquatic Toxicology* 1, 257–68.

MALLET, J. (1989) The evolution of insecticide resistance: have the insects won? *Trends in Evolution and Ecology* 4, 336–40.

MALTBY, L. and NAYLOR, C. (1990) Preliminary observations on the ecological relevance of the *Gammarus* 'scope for growth' assay: effect of zinc on reproduction. *Functional Ecology* 4, 393–7.

MALTBY, L., NAYLOR, C. and CALOW, P. (1990a) Effect of stress on a freshwater benthic detrivore: scope for growth in *Gammarus pulex. Ecotoxicology and Environmental Safety* 19, 285–91.

MALTBY, L., NAYLOR, C. and CALOW, P. (1990b) Field deployment of a scope for growth assay *Gammarus pulex* a freshwater benthic detrivore. *Ecotoxicology and Environmental Safety* 19, 292–300.

MANAHAN, E. E. (1994) *Environmental Chemistry*, 6th edn. Boca Raton, FL: Lewis.

MARKERT, B. and WECKERT, V. (1994) Higher lead concentrations in the environment of former West Germany after the fall of the 'Berlin Wall'. *Science of the Total Environment* 158, 93–6.

MARSHALL, J. (1992) Weeds. In GRIEG-SMITH, P., FRAMPTON, G. and HARDY, T. (eds) *Pesticides, Cereal Farming and the Environment. The Boxworth Project.* London: HMSO.

MARTIKAINEN, E. A. T. and KROGH, P. H. (1999) Effects of soil organic matter content and temperature on toxicity of dimethoate to *Folsomia fimetaria* (Collembola: Isotomidae). *Environmental Toxicology and Chemistry* 18, 865–72.

MARTIN, M. H. and BULLOCK, R. J. (1994) The impact and fate of heavy metals in an oak woodland ecosystem. In ROSS, S. M. (ed.) *Toxic Metals in Soil–Plant Systems*, pp. 327–65. Chichester: John Wiley.

MARTINEAU, D., LAGACÉ, A., BÉLAND, P., HIGGINS, C. R., *et al.* (1988) Pathology of stranded beluga whales (*Delphinapterus leucas*) from the St Lawrence estuary (Quebec, Canada). *Journal of Comparative Pathology* 38, 287–308.

MATTHIESSEN, P and GIBBS, P. E. (1998) Critical appraisal of the evidence for tributyl-mediated endocrine disruption in molluscs. *Environmental Toxicology and Chemistry* 17, 37–43.

McCARTHY, J. F. and SHUGART, L. R. (1990) Biological markers of environmental contamination. In McCARTHY, J. F. and SHUGART, L. R. (eds) *Biomarkers of Environmental Contamination*, pp. 3–14. Boca Raton, FL: Lewis.

McNEILLY, T. (1968) Evolution in closely adjacent plant populations. III. *Agrostis tenvis* on a small copper mine. *Heredity* 23, 99–108.

MEHLHORN, H., O'SHEA, J. M. and WELLBURN, A. R. (1991) Atmospheric ozone interacts with stress ethylene formation by plants to cause visible plant injury. *Journal of Experimental Botany* 42, 17–24.

MELLANBY, K. (1967) *Pesticides and Pollution.* London: Collins.

MERIAN, E. (ed.) (1991) *Metals and their Compounds in the Environment.* Weinheim: VCH.

MILLER, K. L., FERNANDES, T. F. and READ, P. A. (1999) The recovery of populations of dogwhelks suffering from imposex in the Firth of Forth 1987–1997/98. *Environmental Pollution* 106, 183–92.

MINEAU, P. (ed.) (1991) *Cholinesterase-inhibiting Insecticides: their Impact on Wildlife and the Environment.* Amsterdam: Elsevier.

MINEAU, P. and PEAKALL, D. B. (1987) An evaluation of avian impact assessment techniques following broad-scale forest insecticide spraying. *Environmental Toxicology and Chemistry* 6, 781–91.

MOORE, N. W. and WALKER, C. H. (1964) Organic chlorine insecticide residues in wild birds. *Nature* 201, 1072–3.

MORAN, P. J. and GRANT, T. R. (1991) Transference of marine fouling communities between polluted and unpolluted sites; impact on structure. *Environmental Pollution* 72, 89–102.

MORIARTY, F. (1999) *Ecotoxicology*, 3rd edn. London: Academic Press.

MORIARTY, F. (ed.) (1975) *Organochlorine Insecticides: Persistent Organic Pollutants.* London: Academic Press.

MORIARTY, F. and WALKER, C. H. (1987) Bioaccumulation in food chains – a rational approach. *Ecotoxicology and Environmental Safety* 13, 208–15.

MOSE, D. G., MUSHRUSH, G. W. and CHROSNIAK, C. E. (1992) A two-year study of seasonal indoor radon variations in southern Maryland. *Environmental Pollution* 76, 195–9.

MOUW, O., KALITIS, K., ANVER, M., SCHWARTZ, J., *et al.* (1975) Lead. Possible toxicity in urban vs rural rats. *Archives Environmental Health* 30, 276–80.

MURK, A. J., LEONARDS, P. E. G., BULDER, A. S., JONAS, A., *et al.* (1997) The CALUX (chemical-activated luciferase expression) assay, a sensitive in vitro bioassay for measuring TCDD-equivalents in blood plasma. *Environmental Toxicology and Chemistry* 16, 1583–9.

NAYAK, B. N. and PETRAS, M. L. (1985) Environmental monitoring for genotoxicity: *in vivo* sister chromatid exchange in the house mouse (*Mus musculus*). *Canadian Journal of Genetics and Cytology* 27, 351–6.

NAYLOR, C., MALTBY, L. and CALOW, P. (1989) Scope for growth in *Gammarus pulex*, a freshwater benthic detrivore. *Hydrobiologia*, 188/189, 517–23.

NEBERT, D. W. and GONZALEZ, F. J. (1987) P450 Genes: structure, evolution and regulation. *Annual Review of Biochemistry* 56, 945–93.

NELSON, D. R. (1998) Metazoan P450 evolution. In LIVINGSTONE, D. R. and STEGEMAN, J. J. (eds) *Forms and Functions of Cytochrome P450. Comparative Biochemistry and Physiology* 121C (supplement), 15–22.

NELSON, R. W. (1976) Behavioural aspects of egg breakage in peregrine falcons. *Canadian Field Naturalist* 90, 320–9.

NEWMAN, M. C. (1998) *Fundamentals of Ecotoxicology*. Chelsea, MI: Ann Arbor Press.

NEWTON, I. (1986) *The Sparrowhawk*, Calton: Poyser.

NEWTON, I. (1988) Determination of critical pollutant levels in wild populations, with examples from organochlorine insecticides in birds of prey. *Environmental Pollution* 55, 29–40.

NEWTON, I. and HAAS, M. B. (1984) The return of the sparrowhawk. *British Birds* 77, 47–70.

NEWTON, I. and WYLLIE, I. (1992) Recovery of a sparrowhawk population in relation to declining pesticide contamination. *Journal of Applied Ecology* 29, 476–84.

NIEBOER, E. and RICHARDSON, D. H. S. (1980) The replacement of the nondescript term 'heavy metals' by a biologically and chemically significant classification of metal ions. *Environmental Pollution* 1B, 3–26.

NRC (NATIONAL RESEARCH COUNCIL) (1985) *Oil in the Sea: Inputs Fates and Effects*. Washington, DC: National Academy Press.

OECD (1991) *The State of the Environment*. Paris: OECD.

OLLSON, M., ANDERSON, Ö., BERGMAN, Å., BLOMKIST, G., *et al.* (1992) Contaminants and diseases in seals from Swedish waters. *Ambio* 21, 561–2.

OPPENOORTH, F. J. (1985) Biochemistry and genetics of insecticide resistance. In KERKUT, G. A. and GILBERT, L. I. (eds) *Comprehensive Insect Physiology Biochemistry and Pharmacology*, 12. Oxford: Pergamon Press.

PAASIVIRTA, J. (1991) *Chemical Ecotoxicology*. Chelsea, MI: Lewis.

PAIN, D and PIENKOWSKI, M. (eds) (1997) *Farming and Birds in Europe*. New York: Academic Press.

PANKHURST, C. E., DOUBE, B. M. and GUPTA, V. V. S. R. (eds) (1997) *Biological Indicators of Soil Health*. Wallingford: CAB International.

PAYNE, J. F., FANCEY, L. L., RAHIMTULA, A. D. and PORTER, E. L. (1987) Review and perspective on the use of mixed-function oxygenase enzymes in biological monitoring. *Comparative Pharmacology and Physiology* 86C, 233–45.

PAYNE, J. F., MATHIEU, A., MELVIN, W. and FANCEY, L. L. (1996) Acetylcholinesterase, an old biomarker with a new future? Field trials in association with two urban rivers and a paper mill in Newfoundland. *Marine Pollution Bulletin* 32, 225–31.

PEAKALL, D. B. (1985) Behavioural responses of birds to pesticides and other contaminants. *Residue Reviews* 96, 45–77.

PEAKALL, D. B. (1992) *Animal Biomarkers as Pollution Indicators*. London: Chapman & Hall.

PEAKALL, D. B. (1993) DDE-induced eggshell thinning: an environmental detective story. *Environmental Reviews* 1, 13–20.

PEAKALL, D. B. and BART, J. R. (1983) Impacts of aerial application of insecticides on forest birds. *Critical Reviews in Environmental Control* 13, 117–65.

PEAKALL, D. B. and CARTER, N. (1997) Decreases in farmland birds and agricultural practices: a huge ecotoxicological experiment. *Toxicology and Ecotoxicology News* 4, 162–3.

PEAKALL, D. B. and SHUGART, L. R. (eds) (1993) *Biomarkers. Research and Application in the Assessment of Environmental Health*. Berlin: Springer-Verlag.

PEAKALL, D. B., WALKER, C. H. and MIGULA, P. (1999) *Biomarkers: A Pragmatic Basis for Remediation of Severe Pollution in Eastern Europe*. NATO Science Series, Vol. 54. NATO. Dordrecht: Kluwer Academic Publishers.

PEARCE, P. A., PEAKALL, D. B. and ERSKINE, A. J. (1976) Impact on forest birds of the 1975 spruce budworm spray operations in New Brunswick. Canadian Wildlife Series Progress Notes No. 62.

PERKINS, D. F. and MILLAR, R. O. (1987) Effects of airborne fluoride emissions near an aluminium works in Wales. Part 1. Corticolous lichens growing on broad-leaved trees. *Environmental Pollution* 47, 63–78.

PFEIFFER, W. C., DE LACERDA, L. D., MALM, O., SOUZA, C. M. M., *et al.* (1989) Mercury concentrations in inland waters of gold-mining areas in Rondonia, Brazil. *Science of the Total Environment* 87/88, 233–40.

PFEIFFER, W. C., DE LARCERDA, L. D., SALOMONS, W. and MALM, O. (1993) Environmental fate of mercury from gold mining in the Brazilian Amazon. *Environmental Research* 1, 26–37.

PICKERING, Q. C., HENDERSON, C. and LEMKE, A. E. (1962) The toxicity of organic phosphorus insecticides to different species of warmwater fishes. *Transactions of the American Fisheries Society* 91, 175–84.

PILLING, E. D., BROMLEY-CHALLENOR, K. A. C., WALKER, C. H. and JEPSON, P. C. (1995) Mechanisms of synergism between the pyrethroid insecticide *t*-cyhalothrin and the imidazole fungicide prochloraz in the honeybee (*Apis mellifera*). *Pesticide Biochemistry and Physiology* 51, 1–11.

PLENDERLEITH, R. W. and BELL, L. C. (1990) Tolerance of the twelve tropical grasses to high soil concentrations of copper. *Tropical Grasslands* 24, 103–10.

PORTEOUS, A. (1992) *Dictionary of Environmental Science and Technology*. Chichester: John Wiley.

POSTEL, S. (1984) Air pollution, acid rain and the future of forests. Worldwatch Paper 58.

POSTHUMA, L. (1997) Effects of toxicants on population and community parameters in field conditions, and their potential use in the validation of risk assessment models. In VAN STRAALEN, N. M. and LØKKE, H. (eds) *Ecological Risk Assessment of Contaminants in Soil*, pp. 85–123. London: Chapman & Hall.

POSTHUMA, L. and VAN STRAALEN, N. M. (1993) Heavy-metal adaptation in terrestrial invertebrates; a review of occurrence, genetics, physiology and ecological consequences. *Comparative Biochemistry and Physiology* 106C, 11–38.

POTTS, G. R. (1986) *The Partridge*. London: Collins

POTTS, G. R. (2000) The grey partridge. In PAIN, D. and DIXON, J. (eds) *Bird Conservation and Farming Policy in the European Union*. London: Academic Press.

PREIN, A. E., THIE, G. M., KOEMAN, J. H. and POELS, C. L. M. (1978) Cytogenic changes in fish exposed to water of the River Rhine. *Science of the Total Environment* 9, 287–91.

PURDOM, C. E., HARDIMAN, P. A., BYE, V. J., ENO, N. C., *et al.* (1994) Estrogenic effects of effluents from sewage treatment works. *Chemistry Ecology* 8, 275–85.

PYZA, E., MAK, P., KRAMARZ, P. and LASKOWSKI, R. (1997) Heat shock proteins (HSP70) as biomarkers in ecotoxicological studies. *Ecotoxicology and Environmental Safety* 38, 244–51.

RAMADE, F. (1992) *Précis d 'Ecotoxicologie*. Paris: Masson.

RATCLIFFE, D. (1993) *The Peregrine Falcon*, 2nd edn. Calton: T and D Poyser.

RATCLIFFE, D. A. (1967) Decrease in eggshell weight in certain birds of prey. *Nature* 215, 208–10.

READ, H. J., MARTIN, M. H. and RAYNER, J. M. V. (1998) Invertebrates in woodlands polluted by heavy metals – an evaluation using canonical correspondence analysis. *Water, Air and Soil Pollution* 106, 17–42.

RICHTER, C. A., DRAKE, J. B., GIESY, J. P. and HARRISON, R. O. (1994) Immunoassay monitoring of polychlorinated biphenyls (PCBs) in the Great Lakes. *Environmental Science and Pollution Research* 1, 69–74.

RISEBROUGH, R. W. and PEAKALL, D. B. (1988) The relative importance of the several organochlorines in the decline of Peregrine Falcon populations. In CADE, T. J., ENDERSON, J. H., THELANDER, C. G. and WHITE, C. M. (eds) *Peregrine Falcon Populations: their Management and Recovery*, pp. 449–68. Boise ID: The Peregrine Fund, Inc.

ROBINSON, J., RICHARDSON, A., CRABTREE, A. N., COULSON, J. C. and POTTS, G. R. (1967) Organochlorine residues in marine organisms. *Nature* 214, 1307–11.

RONDAY, R. and HOUX, N. W. H. (1996) Suitability of seven species of soil-inhabiting invertebrates for testing toxicity of pesticides in soil pore water. *Pedobiologia*, 40, 106–12.

RONIS, M. J. J. and WALKER, C. H. (1989) The microsomal monooxygenases of birds. *Reviews in Biochemical Toxicology* 10, 310–84.

ROSMAN, K. J. R., CHISHOLM, W., BOUTRON, C. F., CANDLESTONE, J. P. and GÖRLACH, U. (1993) Isotopic evidence for the source of lead in Greenland snows since the late 1960s. *Nature* 362, 333–5.

ROSS, S. M. (ed.) (1994) *Toxic Metals in Soil–Plant Systems*. Chichester: John Wiley.

ROUSH, R. T. and DALY, D. C. (1990) The role of population genetics in resistance research and management. In ROUSH, R. T. and TABASHNIK, B. E. (eds) *Pesticide Resistance in Arthropods*. New York: Chapman & Hall.

SAFE, S. (1990) Polychlorinated biphenyls (PCBs), dibenzo-p-dioxins (PCDDs), dibenzofurans (PCDFs), and related compounds: Environmental and mechanistic considerations which support the development of toxic equivalency factors (TEFs). *Critical Reviews in Toxicology* 21, 51–88.

SALMINEN, J. and HAIMI, J. (1997) Effects of pentachlorophenol on soil organisms and decomposition in forest soil. *Journal of Applied Ecology*, 34, 101–10.

SALOMONS, W., BAYNE, B. L., DUURSMA, E. K. and FÜRSTNER, V. (eds) (1988) *Pollution of the North Sea: an Assessment*. Berlin: Springer-Verlag.

SANDERS, B. M. (1993) The cellular stress response. *Environs* 16, 3–6.

SAVVA, D. (1998) Use of DNA fingerprinting to detect genotoxic effects. *Ecotoxicology and Environmental Safety* 41, 103–6.

SAWICKI, R. M. and RICE, A. D. (1978) Response of susceptible and resistant peach-potato aphid *Myzus persicae* (Sulz.) to insecticides in leaf-dip assays. *Pesticide Science* 9, 513–6.

SCHAT, H. and BOOKUM, W. M. T. (1992) Genetic control of copper tolerance in *Silene vulgaris*, *Heredity* 68, 219–29.

SCHEUHAMMER, A. M. (1989) Monitoring wild bird populations for lead exposure. *Journal of Wildlife Management* 53, 759–65.

SCHINDLER, D. W. (1996) Ecosystems and ecotoxicology: a personal perspective. In NEWMAN, M. and JAGOE, C. (eds) *Quantitative Ecotoxicology: a Hierarchical Approach*. Chelsea, MI: Lewis.

SCHINDLER, D. W., FROST, T. M., MILLS, K. H., CHANG, P. S. S., *et al.* (1991) Comparisons between experimentally and atmospherically-acidified lakes during stress and recovery. *Proceedings of the Royal Society of Edinburgh* 97B, 193–226.

SCHÜÜRMANN, G. and MARKERT, B. (eds) (1998) *Ecotoxicology*. Chichester: John Wiley, and Heidelberg: Spektrum.

SCHMITT, C. J. and BRUMBAUGH, W. G. (1990) National contaminant biomonitoring program: concentrations of arsenic, cadmium copper, lead, mercury, selenium and zinc in US freshwater fish, 1976–1984. *Archives of Environmental Contamination and Toxicology* 19, 731–47.

SCHWARZENBACH, R. P., GSCHWEND, P. M. and IMBODEN, D. M (1993) *Environmental Organic Chemistry*. New York: John Wiley.

SEAGER, J. and MALTBY, L. (1989) Assessing the impact of episodic pollution. *Hydrobiologia* 188/189, 633–40.

SHAW, A. J. (ed.) (1989) *Heavy Metal Tolerance in Plants: Evolutionary Aspects.* Boca Raton, FL: CRC Press.

SHAW, A. J. (1999) The evolution of heavy metal tolerance in plants: adaptations, limits, and costs. In FORBES, V. E. (ed.) *Genetics and Ecotoxicology.* Philadelphia: Taylor and Francis.

SHEPPARD, S., BEMBRIDGE, J., HOLMSTRUP, M. and POSTHUMA, L. (eds) (1998) *Advances in Earthworm Ecotoxicology.* Pensacola, FL: SETAC Publications.

SHUGART, L. (1994) Genotoxic responses in blood. In FOSSI, M. C. and LEONZIO, C. (eds) *Nondestructive Biomarkers in Vertebrates*, pp. 131–45. Boca Raton, FL: Lewis.

SIBLY, R. M. (1996) Effects of pollutants on individual life histories and population growth rates. In NEWMAN, M. C. and JAGGE, C. (eds) *Quantitative Ecotoxicology: a Hierarchical Approach.* Chelsea, MI: Lewis.

SIBLY, R. M. (1999) Efficient experimental designs for studying stress and population density in animal populations. *Ecological Applications* 9, 496–503.

SIBLY, R. M. and ANTONOVICS, J. (1992) Life-history evolution. In BERRY, R. J., CRAWFORD, T. J. and HEWITT, G. M. (eds) *Genes in Ecology.* Oxford: Blackwell Scientific Publications.

SIBLY, R. M. and CALOW, P. (1989) A life cycle theory of responses to stress. *Biological Journal of the Linnean Society* 37, 110–16.

SIBLY, R. M., NEWTON, I. and WALKER, C. H. (2000) Effects of dieldrin on population growth rates of sparrowhawks 1963–1986 *Journal of Applied Ecology* 37, 540–6.

SIBLY, R. M., WILLIAMS, T. D. and JONES, M. B. (2000) How environmental stress affects density dependence and carrying capacity in a marine copepod. *Journal of Applied Ecology* 37, 388–97.

SIBUET, M., CALMET, D. and AUFFRET, G. (1985) Reconnaissance photographique de conteneurs en place dans la zone d 'immersion des dechets faiblement radioactifs de l'Atlantique Nord-Est. *Compte Rendu Hebdomadaire des Seances de l'Academie des Sciences* Paris, Serie 111, 301, 497–502.

SIMKISS, K. (1993) Radiocaesium in natural systems – a UK coordinated study. *Journal of Environmental Radioactivity* 18, 133–49.

SINCLAIR, A. R. E. (1989) Population regulation in animals. In CHERRETT, J. M. (ed.) *Ecological Concepts.* Oxford: Blackwell Scientific Publications.

SMIT, C. E. and VAN GESTEL, C. A. M. (1998) Effects of soil type, prepercolation, and ageing on bioaccumulation and toxicity of zinc for the springtail *Folsomia candida. Environmental Toxicology and Chemistry* 17, 1132–41.

SMITH, P. J. (1996) Selective decline in imposex levels in the dogwhelk *Lepsiella scobina* following a ban on the use of TBT anti-foulants in New Zealand. *Marine Pollution Bulletin*, 32, 362–5.

SÖDERGREN, A. (1991) Environmental fate and effects of bleached pulp mill effluents. Swedish Environment Protection Agency Report 4031.

SOLOMONS, W. G. (1986) *Organic Chemistry*, 2nd edn. New York: John Wiley.

SOLONEN, T. and LODENIUS, M. (1990) Feathers of birds of prey as indicators of mercury contamination in Southern Finland. *Holarctic Ecology* 13, 229–37.

SOMERVILLE, L. and GREAVES, M. P. (1987) *Pesticide Effects on Soil Microflora.* London: Taylor and Francis.

SOMERVILLE, L. and WALKER, C. H. (eds) (1990) *Pesticide Effects on Terrestrial Wildlife.* London: Taylor and Francis.

SØRENSEN, F. F., WEEKS, J. M. and BAATRUP, E. (1997). Altered locomotory behaviour in woodlice (*Oniscus asellus* (L.)) collected at a polluted site. *Environmental Toxicology and Chemistry*, 16, 685–90.

SPEAR, P. A., MOON, T. W. and PEAKALL, D. B. (1986) Liver retinoid concentrations in natural populations of herring gulls *(Larus argentatus)* contaminated by 2,3,7,8-tetrachlorodibenzo-p-dioxin and in ring doves (*Streptopelia risoria*) injected with a dioxin analogue. *Canadian Journal of Zoology* 64, 204–8.

SPURGEON, D. J. and HOPKIN, S. P. (1996) Effects of variations of the organic matter content and pH of soils on the availability and toxicity to zinc of the earthworm *Eisenia fetida*. *Pedobiologia*, 40, 80–96.

SPURGEON, D. J. and HOPKIN, S. P. (1999) Seasonal variation in the abundance, biomass and biodiversity of earthworms in soils contaminated with metal emissions from a primary smelting works. *Journal of Applied Ecology*, 36, 173–83.

SPURGEON, D. J. and HOPKIN, S. P. (2000) The development of genetically inherited resistance to zinc in laboratory selected generations of the earthworm *Eisenia fetida*. *Environmental Pollution* 109, 193–201.

SPURGEON, D. J., HOPKIN, S. P. and JONES, D. T. (1994) Effects of cadmium, copper, lead and zinc on growth, reproduction and survival of the earthworm *Eisenia fetida* (Savigny): assessing the environmental impact of point-source metal contamination in terrestrial ecosystems. *Environmental Pollution* 84, 123–30.

STANLEY, Jr, T. R., SMITH, G. J., HOFFMAN, D. D. J., HEINZ, G. H. and ROSSCOE, R. (1996) Effects of boron and selenium on mallard reproduction and duckling growth and survival. *Environmental Toxicology and Chemistry* 15, 1124–32.

STARK, J. E. and BANKEN, J. A. O. (1999) Importance of population structure at the time of toxicant exposure. *Ecotoxicology and Environmental Safety* 42, 282–7.

STEIN, J. E., COLLIER, T. K., REICHERT, W. L., CASILLAS, E., HOM, T. and VARANASI, U. (1992) Bioindicators of contaminant exposure and sublethal effects: studies with benthic fish in Puget Sound, Washington. *Environmental Toxicology and Chemistry* 11, 701–14.

SUTER, G. W. (ed.) (1993) *Ecological Risk Assessment.* Boca Raton, FL: Lewis.

SVENDSEN, C., MEHARG, A. A., FREESTONE, P. and WEEKS, J. (1996) Use of an earthworm lysosomal biomarker for the ecological assessment of pollution from an industrial plastics fire. *Applied Soil Ecology* 3, 99–107.

TANFORD, C. (1980) *The Hydrophobic Effect: Formation of Micelles and Biological Membranes*, 2nd edn. New York: J. Wiley.

TARDIFF, R. (ed.) (1992) *Methods to Assess Adverse Effects of Pesticides on Non-target Organisms. Scope 49*. Chichester: John Wiley.

TAYLOR, C. E. (1986) Genetics and evolution of resistance to insecticides. *Biological Journal of the Linnean Society* 27, 103–12.

TERRIERE, L. C. (1984) Induction of detoxification enzymes in insects. *Annual Review of Entomology* 29, 71–88.

THEODORAKIS, C. W., BLAYLOCK, B. G. and SHUGART, L. R. (1997) Genetic ecotoxicology. I. DNA integrity and reproduction in mosquitofish exposed *in situ* to radionuclides. *Ecotoxicology* 6, 205–18.

THOMPSON, H. M. and HUNT, L. V. (1999) Extrapolating from honeybees to bumblebees in pesticide risk assessment. *Ecotoxicology*, 8, 147–66.

TIMBRELL, J. A. (1995) *Introduction to Toxicology*, 2nd edn. London: Taylor and Francis.

TIMBRELL, J. A. (1999) *Principles of Biochemical Toxicology*, 3rd edn. London: Taylor and Francis.

TOLBA, M. K. (1992) *Saving our Planet*. London: Chapman & Hall.

TUREKIAN, K. K. (1976) *Oceans*, 2nd edn. Hemel Hempstead: Prentice-Hall.

TURLE, R., NORSTROM, R. J. and COLLINS, B. (1991) Comparison of PCB quantitation methods: re-analysis of archived specimens of herring gull eggs from the Great Lakes. *Chemosphere* 22, 201–13.

TURTLE, E. E., TAYLOR, A., WRIGHT, E. N., THEARLE, R. J. P., *et al.* (1963) The effects on birds of certain chlorinated insecticides used as seed dressings. *Journal of Science in Food and Agriculture* 8, 567–76.

UNEP (1993) *Environmental Data Report 1993–94*. Oxford: Blackwell.

VAN DER GAAG, M. A., VAN DER KERKHOF, J. F. J., VAN DER KLIFT, H. W. and POELS, C. L. M. (1983) Toxicological assessment of river quality in bio-assays with fish. *Environmental Monitoring and Assessment* 3, 247–55.

VAN GESTEL, C. A. M. and HENSBERGEN, P. J. (1997) Interaction of Cd and Zn toxicity for *Folsomia candida* Willem (Collembola: Isotomidae) in relation to bioavailability in soil. *Environmental Toxicology and Chemistry* 16, 1177–86.

VAN STRAALEN, N. M. (1993) Soil and sediment quality criteria derived from invertebrate toxicity data. In DALLINGER, R. and RAINBOW, P. S. (eds) *Ecotoxicology of Metals in Invertebrates*, pp. 427–41. Chelsea, MI: Lewis.

VAN STRAALEN, N. M. (1996) Critical body concentrations: their use in bioindication. In VAN STRAALEN, N. M. and KRIVOLUTSKY, D. A. (eds) *Bioindicator Systems for Soil Pollution*, pp. 5–16. Dordrecht: Kluwer Academic.

VAN STRAALEN, N. M. (1997) Community structure of soil arthropods as a bioindicator of soil health. In PANKHURST, C. E., DOUBE, B. M. and GUPTA, V. V. S. R. (eds) *Biological Indicators of Soil Health*, pp. 235–64. Wallingford, CAB International.

VAN STRAALEN, N. M. (1998) Evaluation of bioindicator systems derived from soil arthropod communities. *Applied Soil Ecology* 9, 429–37.

VAN STRAALEN, N. M. and KAMMENGA, J. E. (1998) Assessment of ecotoxicity at the population level using demographic parameters. In SCHÜÜRMANN, G. and MARKERT, B. (eds) *Ecotoxicology*, pp. 621–44. Chichester: John Wiley.

VAN STRAALEN, N. M. and LØKKE, H. (eds) (1997) *Ecological Risk Assessment of Contaminants in Soil.* London: Chapman & Hall.

VAN STRAALEN, N. M. and VERHOEF, H. A. (1997) The development of a bioindicator system for soil acidity based on arthropod pH preferences. *Journal of Applied Ecology* 34, 217–32.

VAN ZUTPHEN, L. F. M. and BALLS, M. (eds) (1997) *Animal Alternatives, Welfare and Ethics.* Amsterdam: Elsevier.

VARANASI, U., REICHART, W. L. and STEIN, J. E. (1989) 32P-Postlabelling analysis of DNA adducts in liver of wild English sole (*Parophys vetulus*) and winter flounder (*Pseudopleuronectes americanus*). *Cancer Research* 49 1171–7.

VASQUEZ, M. D., POSCHENRIEDER, C., BERCELO, J., BAKER, A. J. M., *et al.* (1994) Compartmentation of zinc in roots and leaves of the zinc hyperaccumulator *Thlaspi caerulescens* J and C Presl. *Botanica Acta* 107, 243–50.

WAID, J. S. (ed.) (1985–7) *PCBs in the Environment*, vols I-III. Cleveland, OH: CRC Press.

WALKER, C. H. (1975) *Environmental Pollution by Chemicals*, 2nd edn. London: Hutchinson.

WALKER, C. H. (1980) Species differences in some hepatic microsomal enzymes that metabolize xenobiotics. *Progress in Drug Metabolism* 5, 118–64.

WALKER, C. H. (1983) Pesticides and birds: mechanisms of selective toxicity. *Agricultural Ecosystems and Environ* 9, 211–26.

WALKER, C. H. (1987) Kinetic models for predicting bioaccumulation of pollutants in ecosystems. *Environmental Pollution* 44, 227–40.

WALKER, C. H. (1998a) Avian forms of cytochrome P_{450}. In LIVINGSTONE, D. R. and STEGEMAN, J. J. (eds) *Forms and Functions of Cytochrome P450. Comparative Biochemistry and Physiology* 121C (supplement).

WALKER, C. H. (1998b) Alternative approaches and tests in ecotoxicology: a review of the present position and prospects for change taking into account ECVAMs duties, topic selection, and test criteria. *ATLA (Alternatives to Laboratory Animals)* 26, 649–77.

WALKER, C. H. (1990a) Kinetic models to predict bioaccumulation of pollutants. *Functional Ecology* 4, 295–301.

WALKER, C. H. (1990b) Persistent pollutants in fish-eating sea birds: bioaccumulation metabolism and effects. *Aquatic Toxicology* 17, 293–324.

WALKER, C. H. (1992) Biochemical responses as indicators of toxic effects of chemicals in ecosystems. *Toxicology Letters* 64/65, 527–32.

WALKER, C. H. (1994) Comparative toxicology. In HODGSON, E. and LEVI, P. (eds) *Introduction to Biochemical Toxicology*, pp. 193–218. Norwalk, CT: Appleton and Lange.

WALKER, C. H. and LIVINGSTONE, D. R. (1992) *Persistent Pollutants in Marine Ecosystems.* Oxford: Pergamon Press.

WALKER, C. H and NEWTON, I. (1999) Effects of cyclodiene insecticides on raptors in Britain – correction and updating of an earlier paper. *Ecotoxicology* 8, 185–189.

WALKER, C. H. and THOMPSON, H. M. (1991) Phylogenetic distribution of cholinesterases and related esterases. In MINEAU, P. (ed.) *Cholinesterase-*

inhibiting Insecticides: Impact on Wildlife and the Environment, pp. 1–17. Amsterdam: Elsevier.

WALKER, C. H., GREIG-SMITH, P. W., CROSSLAND, N. O. and BROWN, R. (1991a) Ecotoxicology. In BALLS, M., BRIDGES, J. and SOUTHEE, J. (eds) *Proceedings of FRAME Meeting; 'Animals and Alternatives in Toxicology – Present Status and Future Prospects'*, pp. 223–252. London: Macmillan.

WALKER, C. H., MACKNESS, M. I., BREARLEY, C. J. and JOHNSTON, G. O. (1991b) Toxicity of pesticides to birds; the enzymic factor. *Biochemical Society Transactions* 19, 741–5.

WALKER, C. H., JOHNSTON, G. O. and DAWSON, A. (1993) Enhancement of toxicity due to the interaction of pollutants at the toxicokinetic level in birds. *Science of the Total Environment* 134(s), 525–31.

WALKER, C. H., KAISER, K., KLEIN, W., LAGADIC, L., *et al.* (1998) Alternative testing methodologies for ecotoxicity. *Environmental Health Perspectives* 106 (supplement – Alternative Testing Methodologies), 441–51.

WALTON, K. C. (1986) Fluoride in moles, shrews and earthworms near an aluminium reduction plant. *Environmental Pollution* 42A, 361–71.

WARD, D. V., HOWES, B. L. and LUDWIG, D. F. (1976) Interactive effects of predation pressure and insecticide (Temefos) toxicity on populations of the marsh fiddler crab *Uca pugnax*. *Marine Biology* 5, 119–26.

WARREN, C. E. and DAVIS, G. L. (1967) Laboratory studies on the feeding of fishes. In GERKING, S. D. (ed.) *The Biological Basis of Freshwater Fish Production*, pp. 175–214. Oxford: Blackwell Scientific Publications.

WARWICK, R. M. and CLARKE, K. R. (1998) Taxonomic distinctness and environmental assessment. *Journal of Applied Ecology* 35, 532–43.

WATERMAN, M. R. and JOHNSTON, E. F. (eds) (1991) *Methods in Enzymology*, 206. San Diego: Academic Press.

WATT, W. D., SCOTT, C. D. and WHITE, W. J. (1983) Evidence of acidification of some Novia Scotia rivers and its impact on Atlantic salmon, *Salmo salar*. *Canadian Journal of Fisheries and Aquatic Science* 40, 462–73.

WAYNE, R. P. (1991) *Chemistry of Atmospheres*, 2nd edn. Oxford: Clarendon Press.

WEEKS, J. M. and SVENDSEN, C. (1996) Neutral red retention by lysosomes from earthworms (*Lumbricus rubellus*) coelomocytes: a simple biomarker of exposure to soil copper. *Environmental Toxicology and Chemistry* 15, 1801–5.

WESELOH, D. V., TEEPLE, S. M. and GILBERTSON, M. (1983) Double-crested cormorants of the Great Lakes: egg-laying parameters, reproductive failure and contaminant residues in eggs, Lake Huron 1972–1973. *Canadian Journal of Zoology* 61, 427–36.

WHITTIER, J. B. and MCBEE, K. (1999) Use of flow cytometry to detect genetic damage in mallard dosed with mutagens. *Environmental Toxicology and Chemistry* 18, 1557–63.

WIDDOWS, J. and DONKIN, P. (1991) Role of physiological energetics in ecotoxicology. *Comparative Biochemistry and Physiology* 100C, 69–75.

WIDDOWS, J. and DONKIN, P. (1992) Mussels and environmental contaminants: bioaccumulation and physiological aspects. In GOSLING, E. (ed.) *The Mussel Mytilus: Ecology, Physiology, Genetics and Culture*. Amsterdam: Elsevier.

WIDDOWS, J., BURNS, K. A., MENON, K. R., PAGE, D. S. and SORIA, S. (1990) Measurements of physiological energetics (scope for growth) and chemical contaminants in mussels (*Arca zebra*) transplanted along a contamination gradient in Bermuda. *Journal of Experimental Marine Biology and Ecology* **138**, 99–117.

WIGFIELD, D. C., WRIGHT, S. C., CHAKRABARTI, C. L. and KARWOWSKA, R. (1986) Evaluation of the relationship between chemical and biological monitoring of low level lead poisoning. *Journal of Applied Toxicology* 6, 231–5.

WILES, J. A. and FRAMPTON, G. K. (1996) A field bioassay approach to assess the toxicity of insecticide residues on soil to Collembola. *Pesticide Science* 47, 273–85.

WILKINSON, C. F. (1976) Insecticide interactions. In WILKINSON, C. F. (ed.) *Insecticide Biochemistry and Physiology*. New York: Plenum Press.

WILLIAMS, R. J. P. (1981) Natural selection of the chemical elements. *Proceedings of the Royal Society of London* **213B**, 361–97.

WOLFE, M. F., SCHWARZBACH, S. and SULAIMAN, R. A. (1998) Effects of mercury on wildlife: a comprehensive review. *Environmental Toxicology and Chemistry*, 17, 146–60.

WONG, S., FOURNEIR, M., CODERRE, D., BANSKA, M. and KRZYSTYNIAK, K. (1992) Environmental immunotoxicology. In PEAKALL, D. B. (ed.) *Animal Biomarkers as Pollution Indicators*, pp. 166–89. London: Chapman & Hall.

WOOD, M. (1995) *Environmental Soil Biology*, 2nd edn. Glasgow: Blackie.

WOOD, R. J. and BISHOP, J. A. (1981) Insecticide resistance: populations and evolution. In BISHOP, J. A. and COOK, L. M. (eds) *Genetic Consequences of Man-made Change*, pp. 97–127. London: Academic Press.

WOODFORD, J. E., KARASOV, W. H., MEYER, M. W. and CHAMBERS, L. (1998) Impact of 2,3,7,8-TCDD on survival, growth, and behavior of ospreys breeding in Wisconsin. *Environmental Toxicology and Chemistry* 17, 1323–31.

WRIGHT, J. F., FURSE, M. T. and ARMITAGE, P. D. (1993) RIVPACS – a technique for evaluating the biological quality of rivers in the UK. *European Water Pollution Control* 3, 15–25.

YOKOI, K., KIMURA, M. and IROKAWA, Y. (1990) Effect of dietary tin deficiency on growth and mineral status in rats. *Biological Trace Element Research* 24, 223–31.

Index